Framework MATHS 7 E

David Capewell Westfield School, Sheffield
Marguerite Comyns Queen Mary's High School, Walsall
Gillian Flinton All Saints Catholic High School, Sheffield
Geoff Fowler Maths Strategy Manager, Birmingham
Kam Grewal-Joy Mathematics Consultant
Derek Huby Mathematics Consultant
Peter Johnson Wellfield High School, Leyland, Lancashire
Penny Jones Mathematics Consultant, Birmingham
Jayne Kranat Langley Park School for Girls, Bromley
Ian Molyneux St. Bedes RC High School, Ormskirk
Peter Mullarkey School Improvement Officer, Manchester
Nina Patel Ifield Community College, West Sussex

OXFORD
UNIVERSITY PRESS

OXFORD
UNIVERSITY PRESS

Great Clarendon Street, Oxford OX2 6DP

Oxford University Press is a department of the University of Oxford.
It furthers the University's objective of excellence in research,
scholarship, and education by publishing worldwide in

Oxford New York

Auckland Bangkok Buenos Aires Cape Town Chennai
Dar es Salaam Delhi Hong Kong Istanbul Karachi Kolkata
Kuala Lumpur Madrid Melbourne Mexico City Mumbai Nairobi
São Paulo Shanghai Singapore Taipei Tokyo Toronto

with an associated company in Berlin

First published 2002

British Library Cataloguing in Publication Data

Data available

ISBN 0 19 914847 3

10 9 8 7 6 5 4 3 2 1

Typeset by Mathematical Composition Setters Ltd.

Printed and bound by G. Canale & C. S.p.A.-Turin.

Acknowledgements

The photograph on the cover is reproduced courtesy of Pictor International (UK).

The publishers and authors would like to thank the following for permission to use photographs and other copyright material: Corbis UK, pages 1, 2, 168, 220, and 235, Empics, page 155, London Aerial Photo Library, page 79, Science Photo Library, page 117.

Figurative artwork by Jeff Anderson.

The authors would like to thank

Sarah Caton, Karen Greenway, Lyn Lynam, David Shiers and Karl Warsi for their help in compiling this book.

About this book

This book has been written specifically for students who have gained Level 5 or above at the end of KS2. The content is based on the Year 8 teaching objectives from the Framework for Teaching Mathematics.

The authors are experienced teachers and maths consultants who have been incorporating the Framework approaches into their teaching for many years and so are well qualified to help you successfully meet the Framework objectives.

The book is made up of units which follow the Extension tier of the medium term plans that complement the Framework document, thus maintaining the required pitch, pace and progression.

The units are:

Each unit comprises double page spreads that should take a lesson to teach. These are shown on the full contents list.

Problem solving is integrated throughout the material as suggested in the Framework.

How to use this book

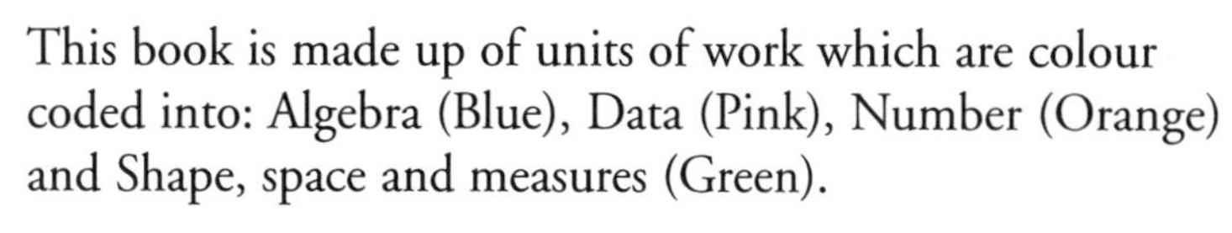

This book is made up of units of work which are colour coded into: Algebra (Blue), Data (Pink), Number (Orange) and Shape, space and measures (Green).

Each unit of work starts with an overview of the content of the unit, as specified in the Framework document, so that you know exactly what you are expected to learn.

This unit will show you how to:

- Understand a proof that the sum of the angles of a triangle is 180° and of a quadrilateral is 360°.
- Identify alternate angles and corresponding angles.
- Classify quadrilaterals by their geometric properties.
- Solve problems and investigate in the context of shape.
- Identify the necessary information to solve a problem.

The first page of a unit also highlights the skills and facts you should already know and provides Check in questions to help you revise before you start so that you are ready to apply the knowledge later in the unit:

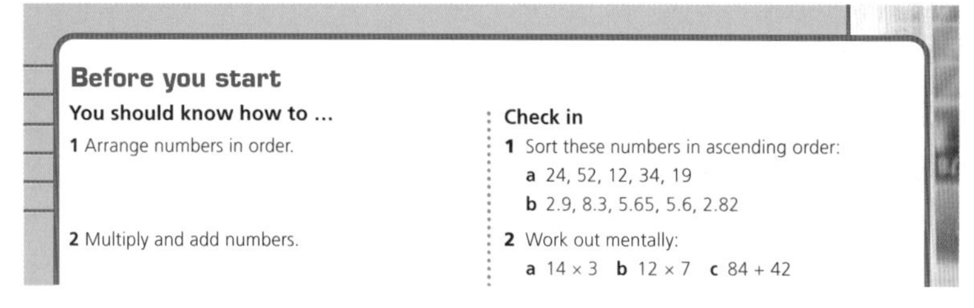

Before you start

You should know how to ...	Check in
1 Arrange numbers in order.	1 Sort these numbers in ascending order: a 24, 52, 12, 34, 19 b 2.9, 8.3, 5.65, 5.6, 2.82
2 Multiply and add numbers.	2 Work out mentally: a 14×3 b 12×7 c $84 + 42$

Inside each unit, the content develops in double page spreads which all follow the same structure.

The spreads start with a list of the learning outcomes and a summary of the keywords:

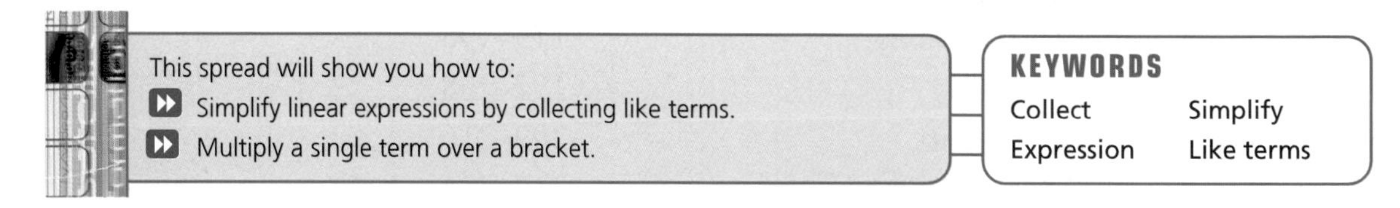

This spread will show you how to:

- Simplify linear expressions by collecting like terms.
- Multiply a single term over a bracket.

KEYWORDS

Collect Simplify
Expression Like terms

The keywords are summarised and defined in a Glossary at the end of the book so you can always check what they mean.

Key information is highlighted in the text so you can see the facts you need to learn.

▶ **An angle bisector divides an angle into two equal parts.**

Examples showing the key skills and techniques you need to develop are shown in boxes. Also hint boxes show tips and reminders you may find useful:

example

Expand $p(p+5)$

$p(p+5) = p \times p + p \times 5$
$= p^2 + 5p$

Remember: p^2 is called 'p squared'.

Each exercise is carefully graded, set at three levels of difficulty:

- The first few questions provide lead-in questions, revising previous learning.
- The questions in the middle of the exercise provide the main focus of the material.
- The last few questions are challenging questions that provide a link to the Year 9 learning objectives.

At the end of each unit is a summary page so that you can revise the learning of the unit before moving on.

Check out questions are provided to help you check your understanding of the key concepts covered and your ability to apply the key techniques.

N5 Summary

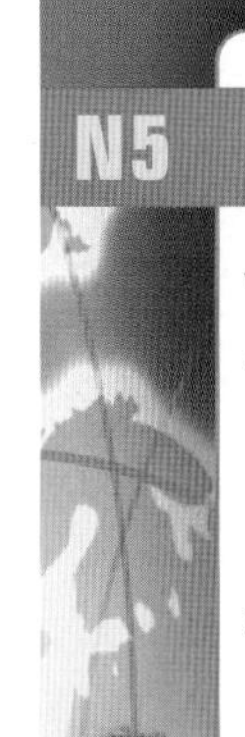

You should know how to ...

1 Calculate percentages and find the outcome of a given increase or decrease.

2 Use standard column procedures for multiplication and division of integers and decimals.

Check out

1 a Add 20% to £19.50
b Decrease £35 by 20%
c Find the cost of a TV at £790 + VAT when VAT is 17.5%

2 Show all your working to calculate:
a 97 × 86 b 125 × 31
c 8.62 × 3.4 d 16.7 × 8.36
e 513 ÷ 19 f 667 ÷ 23

The answers to the Check in and Check out questions are produced at the end of the book so that you can check your own progress and identify any areas that need work.

Contents

A1 Sequences

This unit will show you how to:

- Generate and describe integer sequences.
- Generate terms of a linear sequence using term-to-term and position-to-term definitions.
- Begin to use linear expressions to describe the *n*th term of an arithmetic sequence.
- Express simple functions in symbols.
- Represent mappings expressed algebraically.
- Solve word problems and investigate in the context of algebra.
- Suggest extensions to problems, conjecture and generalise.
- Identify exceptional cases or counter-examples.

The black and white keys on a piano form a sequence.

Before you start

You should know how to ...	Check in
1 Count on or back in steps.	**1 a** Count on from 5 in steps of 7: 5, _, _, _, _, _, _ **b** Count back from 19 in steps of 4: 19, _, _, _, _, _, _ **c** Count back from 12 in steps of 0.3: 12, _, _, _, _, _, _, _
2 Recall multiplication facts to 12 × 12.	**2** Write down the missing numbers: **a** 9 × 6 = _ **b** 12 × 7 = _ **c** 11 × _ = 121 **d** _ × 8 = 56 **e** 5 × _ = 45 **f** _ × 7 = 28
3 Double numbers.	**3** Double each number four times: **a** 4 **b** 10 **c** 2.5 **d** 6.1

A1.1 Introducing sequences

This spread will show you how to:

- Generate and describe integer sequences.
- Generate terms of a sequence given a rule for finding each term from the previous term.

KEYWORDS

Sequence   Rule

Term

Sequences are all around you ...

... door numbers follow a sequence

... the number of posts to make a fence form a sequence

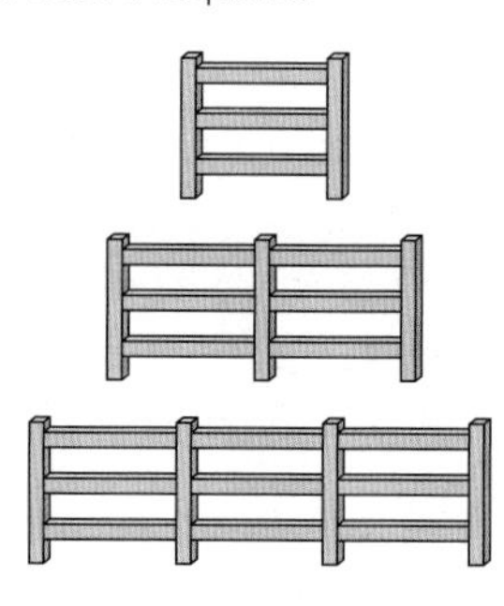

... the time at which the sun rises each day forms a sequence

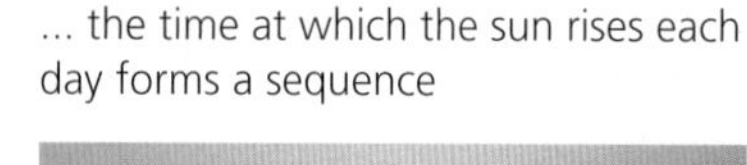

- A **sequence** is a set of numbers or objects that form a pattern.
- Each number in a sequence is called a **term**: 3, 9, 15, 21, 27, ...

The first term is 3 and the third term is 15.
The dots ... show the sequence goes on forever.

The terms are related by a **rule**.
Here are two examples:

3, 9, 15, 21, 27, ...   start at 3 and add 6.

3, 6, 9, 12, 15, ...   start at 3 and add 3,
or multiply the **position** by 3

- **You can generate terms of a sequence when you know the rule.**

Sometimes, finding the difference between terms can be useful to find the rule:

3, 14, 25, 36, 47, ...   +11 to get the next term.
(+11, +11, +11, +11)

2, 5, 10, 17, 26, ...   add consecutive odd numbers to get the next term.
(+3, +5, +7, +9)

example

Find the next 3 terms in the sequence: 1, 5, 9, 13, ...

The sequence goes up in 4s so the rule is 'add 4' to the previous term.

The next terms are: $13 + 4 = 17$

$17 + 4 = 21$

$21 + 4 = 25$

Exercise A1.1

1 Describe the rule for these sequences in words:

a 25, 21, 17, 13, ...
b 3, 10, 17, 24, ...
c 16, 19, 22, 25, ...
d 1.2, 1.8, 2.4, 3, ...
e 5, 10, 20, 40, ...

2 Find the value of each unknown letter:

a 4, 9, 14, 19, a, b, ...
b 60, 54, 48, c, 36, d, ...
c 1, 4, 9, 16, 25, g, h, ...
d 1, 1, 2, 3, 5, 8, j, k, ...
e 4, 8, 16, 32, m, n, ...
f 560, 56, 5.6, p, q, r, ...
g 5, 11, t, 26, v, 45, ...
h 3, w, x, y, 39, ...

3 Write down the first five terms of these sequences:

a The first term is 4, each term is 9 bigger than the one before.
b The first term is 11, each term is 3 smaller than the one before.
c The first term is 5, each term is double the previous term.
d The first term is 1, the second term is 3. Each term is the sum of the 2 previous terms.
e Each term is the term number, multiplied by 4, plus 1. (Hint: 5, 9, ...)
f Each term is the term number, multiplied by itself three times.

4 **a** Write three sequences in which both the 1st and 2nd terms are 5.
b Write three sequences with 3rd term 6 and 4th term 10.

5 In this sequence machine, the input has been multiplied by 3:

1, 2, 3, 4, ... → ×3 → 3, 6, 9, 12, ...

What has happened to the input in each of these machines?

a 1, 2, 3, 4, ... → 12, 24, 36, 48, ...
b 1, 2, 3, 4, ... → $^{-}1$, 0, 1, 2, ...
c 1, 2, 3, 4, ... → 3, 5, 7, 9, ...
d 1, 2, 3, 4, ... → 1, 4, 9, 16, ...
e 1, 2, 3, 4, ... → 1, 8, 27, 64, ...

6 Some sequences have special names. Can you write the first five terms of:

a The multiples of seven.
b The square numbers.
c The triangular numbers.
d The Fibonnaci numbers.
e The prime numbers.

7 These are 'fun' sequences. Can you work out what comes next?

a M, T, W, T, F, _, _ **b** O, T, T, F, F, _, _ **c** 31, 28, 31, 30, _, _

8 **Challenge:** Write these sequences to five terms:

a Each item is the term number, multipled by 4, subtract 1.
b Each item is the term number, multipled by 5, add 2.

Do these sequences ever have terms in common? Invetigate. Investigate further for your own sequences.

Sequences and rules

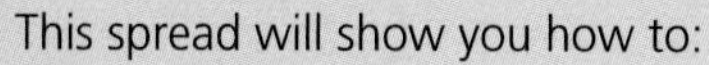

This spread will show you how to:

- Generate a sequence given a rule for finding each term from its position in a sequence.

KEYWORDS

Sequence Position
Term Rule

This sequence starts at 4 and grows by 3 each time:

4, 7, 10, 13, 16, ...

It is easy to find the next term from the previous term: you just add 3.
'Add 3' is a **term-to-term rule**.

- A term-to-term rule tells you how to find the next term from the previous term.

If you want the 123rd term it is easier to find a rule linking the term to the **position**.

Position	1	2	3	4	5
Term	4	7	10	13	16

You already know a sequence that grows by 3 each time: the multiples of 3.
If you add 1, you get the rule:

Position	1	2	3	4	5
Term	$1 \times 3 + 1$ 4	$2 \times 3 + 1$ 7	$3 \times 3 + 1$ 10	$4 \times 3 + 1$ 13	$5 \times 3 + 1$ 16

The **position-to-term** rule is: term = position × 3 + 1.

- A position-to-term rule tells you how to find a term from its position in the sequence.

In the sequence, the 8th term would be $8 \times 3 + 1 = 25$.

The 123rd term would be $123 \times 3 + 1 = 369 + 1 = 370$.

Check by adding on 3:
6th term is $16 + 3 = 19$
7th term is $19 + 3 = 22$
8th term is $22 + 3 = 25$

Exercise A1.2

1 Generate the first five terms of the sequences described by these term-to-term rules:

1st term(s)	**Term-to-term rule**
5	add 3
4	subtract 2
3	double
50 000	divide by 10
1, 1	add the two previous terms

2 Generate the first five terms of the sequences described by these position-to-term rules:

a add 3
b subtract 2
c double
d multiply by 3, then add 1
e multiply by 5, then subtract 4
f divide by 2, then add 3

3 **Investigation**

Here is a term-to-term rule:

1st term □, add □

Choose numbers to go in the boxes that will make all the terms:

a even
b odd
c multiples of 5
d all numbers ending with the digit 7

4 Describe each of these sequences using a position-to-term rule.

a 1, 5, 9, 13, 17, ...
b 2, 7, 12, 17, 22, ...
c 5, 8, 11, 14, 17, ...
d 8, 11, 14, 17, 20, ...
e 19, 16, 13, 10, 7, ...

Hint: Look at the **differences** between consecutive terms.

5 Match each sequence with its term-to-term and its position-to-term rule. An example is given.

Term-to-term	Sequence	Position-to-term
Add three	8, 14, 20, 26, 32, ...	Multiply by 6 and add 2
Add nine	23, 19, 15, 11, 7, ...	Multiply by 5 and subtract 4
Add six	9, 10, 11, 12, 13, ...	Multiply by 9 and subtract 6
Add one	1, 6, 11, 16, 21, ...	Multiply by 3 and add 4
Add five	3, 12, 21, 30, 39, ...	Add 8
Subtract four	7, 10, 13, 16, 19, 22, ...	Multiply by 4 and subtract from 27

6 **Challenge:**

Are there any sequences which have the same term-to-term and position-to-term rules? Investigate.

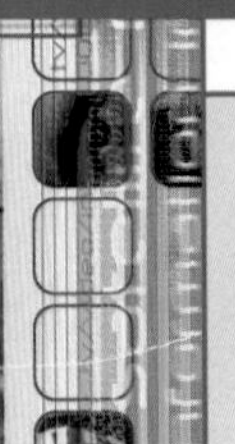

A1.3 Sequences in diagrams

This spread will show you how to:

- Generate sequences from practical contexts.
- Generate a sequence given a rule for finding each term from its position in a sequence.

KEYWORDS

Sequence, Rule, Position, Generate, Term

Sequences often occur in diagrams.

example

Here is a pattern made with matches:

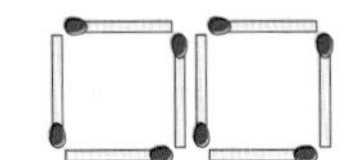
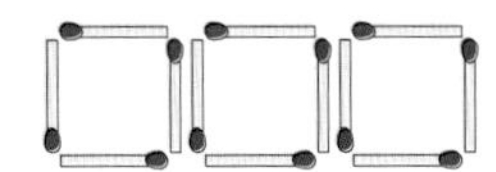

How many matches would you need for the pattern with 100 squares?

First you need to understand the pattern. It is useful to draw up a table:

Position (no of squares)	1	2	3	4
Term (no of matches)	4	8	12	16

+4 +4 +4

The term, or number of matches, goes up in fours.

The term-to-term rule is 'add 4'.

To find the 100 th term you could add 4 each time but that would take ages.
The sequence is 4, 8, 12, 16, or the 4 times table, so × the position by 4:

Position (no of squares)	1 ×4	2 ×4	3 ×4	4 ×4
Term (no of matches)	4	8	12	16

The position-to-term rule is: term = position × 4.

You would have 100 squares in the 100 th pattern.
For 100 squares you would need 100 × 4 matches = 400 matches.

▶ **You can generate any term of a sequence if you know the position-to-term rule.**

Exercise A1.3

1 Here is a sequence of matchstick patterns:

a Draw the next two patterns in the sequence.
b How many matches are there in each pattern?
c How many matches are there in the 10th pattern?
d Discuss why the 100th pattern in the sequence needs 102 matches.

2 Here is a sequence of tile patterns:

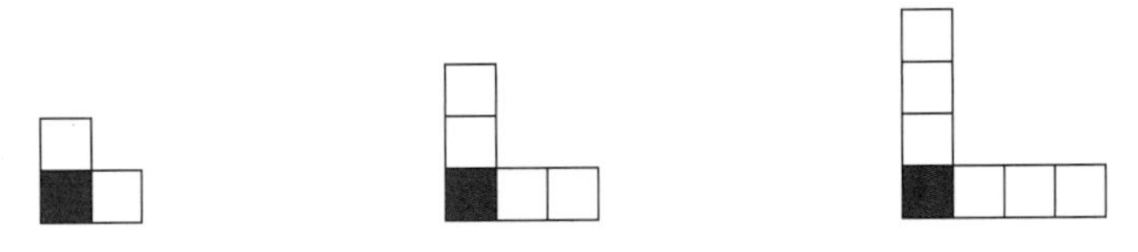

a How many white tiles are needed in each of the first 5 patterns?
b How many white tiles are needed in pattern number 10?
c What is the total number of tiles in pattern 100?

3 Here is a sequence of tile patterns:

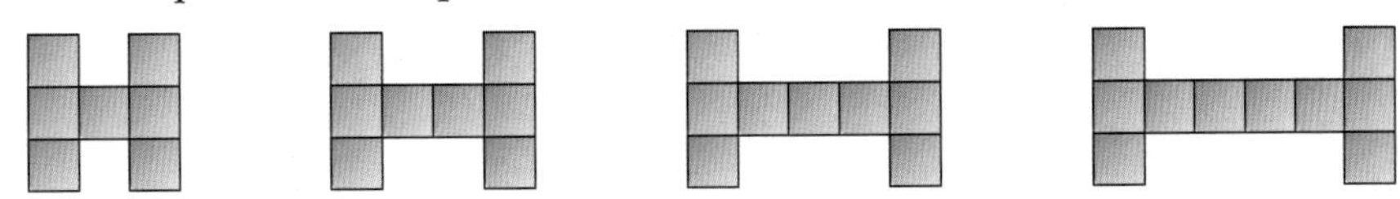

Without making the pattern, how do you know that the 100th pattern in this sequence will need 106 tiles?

4 Explain how you can know the number of matches in the 100th diagram in each of the following sequences:

a 100th diagram = 400 matches

b 100th diagram = 301 matches

c 100th diagram = 202 matches

5 Decorations need silver beads Ⓢ, gold beads Ⓖ and wire segments —.
For instance:

A huge decoration needs 100 silver beads. How many gold beads will it have? Explain.
How many wire segments will it need? Explain.

A1.4 Function machines

This spread will show you how to:

- Express simple functions at first in words and then using symbols.

KEYWORDS
Function
Output
Input

This sequence has a position-to-term rule: 'multiply by 4'

Position	1	2	3	4
Term	4	8	12	16

You can use a **function machine** to describe this rule:

Input | Function | Output

Position → [× 4] → Term

- A function machine shows a rule.
- In a function machine:
  - The **input** value is the value you put in to the machine.
  - The machine performs the **function.**
  - The **output** value is the result that the machine puts out.

You can show more than one input and output like this:

1, 2, 3, 4 → [× 4] → 4, 8, 12, 16

Function machines can have more than one operation:

example

This is the function machine for a sequence:

Position → [× 3] → [+ 1] → Term

Find **a** the 7th, and **b** the 23rd terms of the sequence.

a The 7th term has position 7:

$7 \times 3 = 21$
$21 + 1 = 22$

b The 23rd term has position 23:

$23 \times 3 = 69$
$69 + 1 = 70$

Exercise A1.4

1 For these function machines, find the outputs for each of the inputs 1, 2 and 3.

a 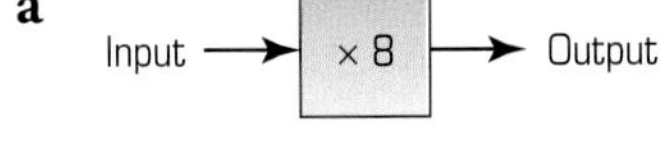

b 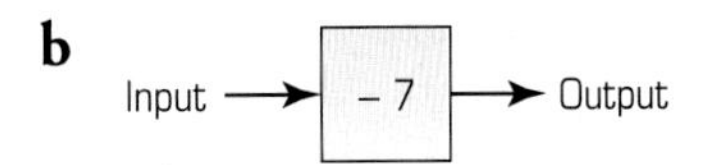

c 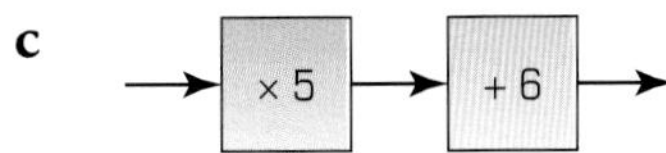

d 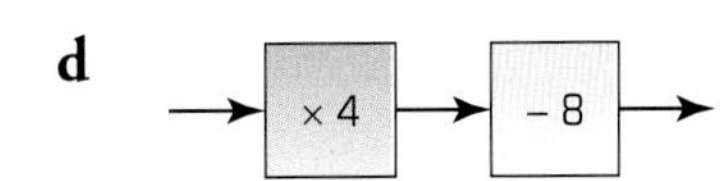

2 For these function machines, find the outputs for the inputs given.

a 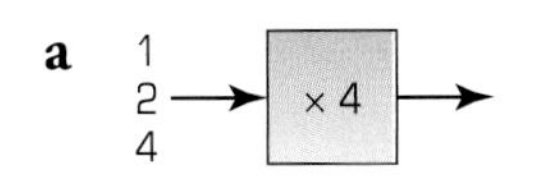

b ⁻3, 5, 9 → + 3 → × 4 →

c 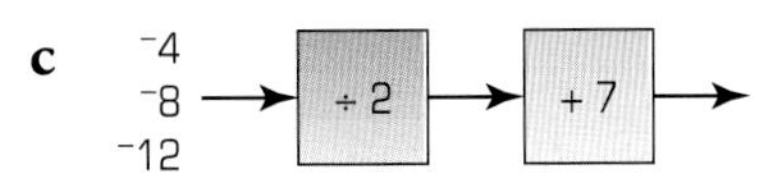

d 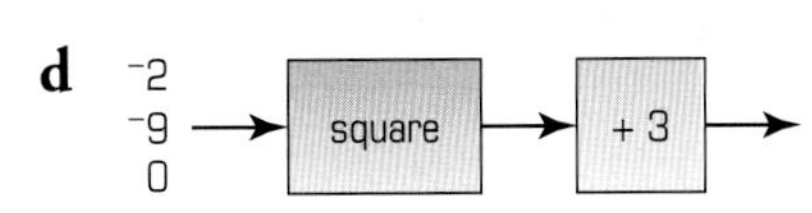

3 Find the missing inputs and outputs in the following.

a 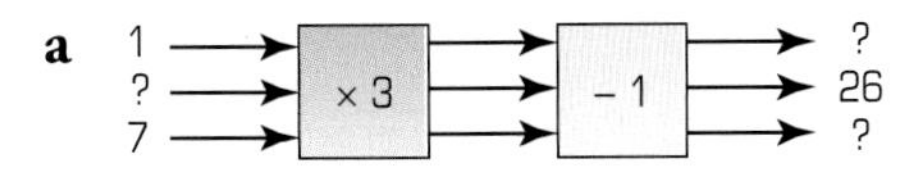

b 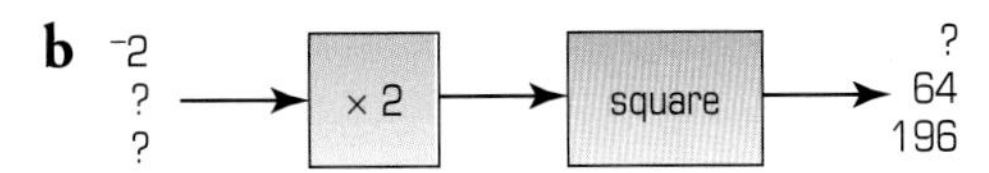

c 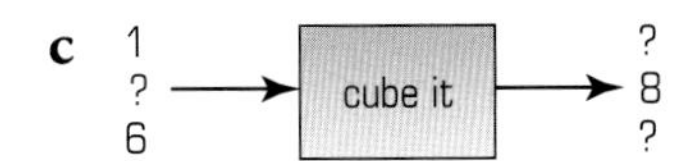

d 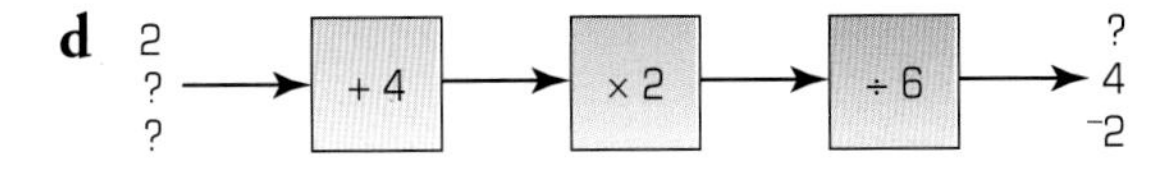

4 True or false: these function machines have two possible inputs that lead to the same output.

a 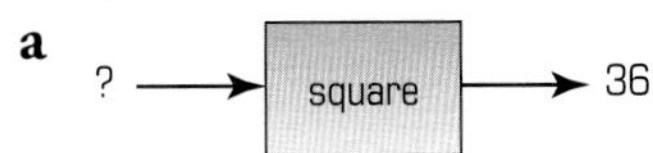

b 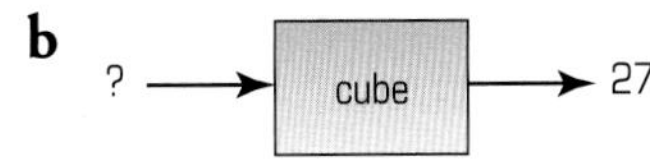

5 **Investigation**

a Work out the outputs for this function machine.

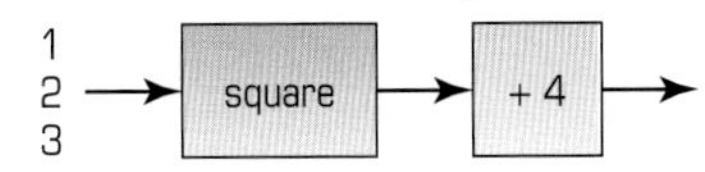

b Reverse the order of the functions (add 4 first).
Work out the outputs.
Are they the same as in **a**?

c *If you change the order of the functions you get a different answer.*
Investigate this statement for different functions.

A1.5 Finding the function

This spread will show you how to:
- Express simple functions at first in words and then using symbols.
- Given inputs and outputs, find the function.

KEYWORDS
Function
Rule
Mapping diagram

You can use function machines to solve problems in sequences.

example

What input gives an output of 10 for this machine?

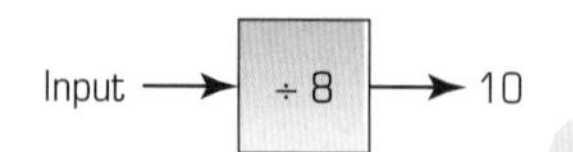

× 8 is the inverse operation to ÷ 8

Input ÷ 8 = 10   Input = 10 × 8 = 80

In this machine you know some input and output values but not the function:

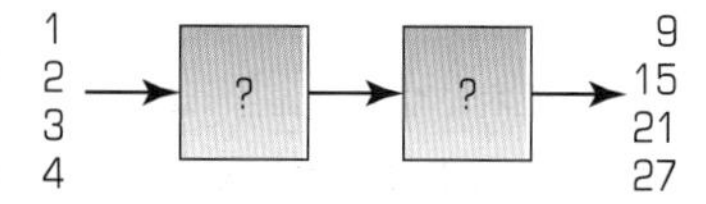

In this case it is '+4'.

Finding a function is similar to finding the position-to-term rule of a sequence.

It can be harder to spot the function when there is more than one operation. You should put the values into a table.

example

Find the operations for this function machine:

1, 2, 3, 4 → ? → ? → 9, 15, 21, 27

The output numbers go up in 6s so compare to the 6 times table:

Input	1	2	3	4
6 times table:	6	12	18	24
output numbers:	9	15	21	27

The output numbers are 3 more than the 6 times table.
The function is ×6 then +3.

Always check that your functions work!
1 × 6 + 3 = 9
2 × 6 + 3 = 15

▸ You can show more than one input and output using a **mapping diagram**.

For example, this function machine uses the inputs 1, 2, 3 and 4.

Redraw the inputs and outputs as number lines:

This is a mapping diagram for the function 'multiply by 5'.

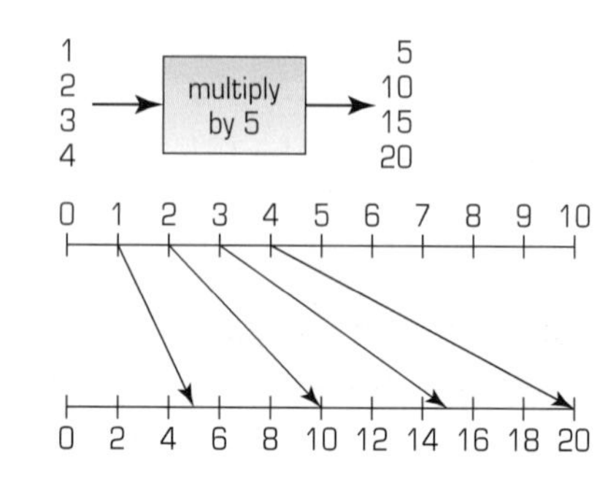

Exercise A1.5

1 For each of these function machines, work out the inputs for the given output.

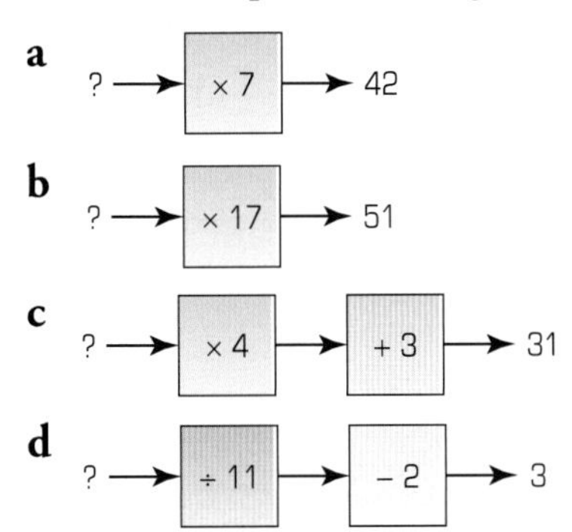

2 **Function Guessing Game**

- Player 1 thinks of a function.
- Player 2 writes down an input number, for example: 1
- Player 1 works out the output, and writes it down like this: $1 \rightarrow 5$
- Player 2 thinks of another number, for example: 3
- Play continues, with inputs and outputs being built up like this:

$1 \rightarrow 5$
$3 \rightarrow 7$
$6 \rightarrow 10$

... until Player 2 guesses the function correctly.

- The roles are then swapped and Player 2 has to think of a function.
- The winner is the player who has guessed the most functions correctly.

3 In these function machines, you are given three pairs of inputs and outputs.
Find the function.

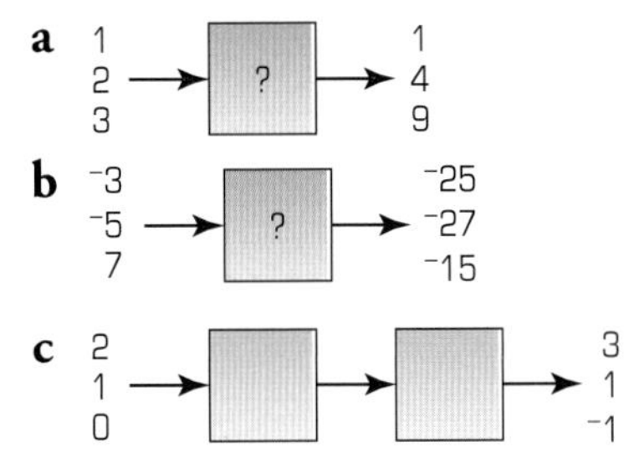

4 For the mapping diagram shown, draw a function machine, showing clearly **a** the inputs, **b** the function, and **c** the outputs.

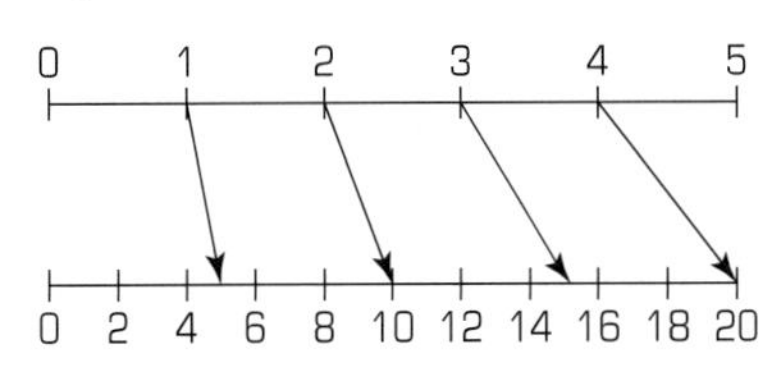

5 For each of the function machines in question 1, draw a mapping diagram.

6 **Investigation**

a Write down the outputs for this function machine.

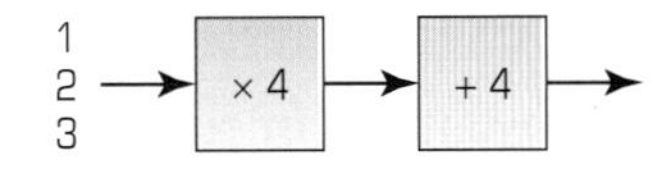

b Write down the outputs for this function machine:

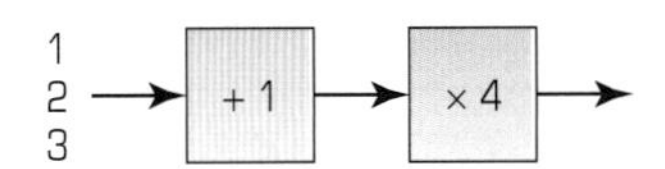

c What do you notice?

Challenge: find another pair of functions that give the same outputs.

7 Investigate whether each of these statements is true or false

a

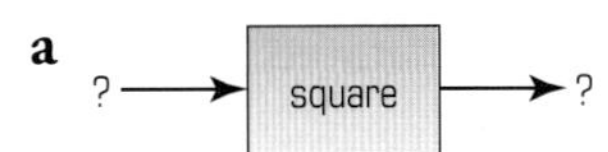

The input will always be smaller than the output.

b

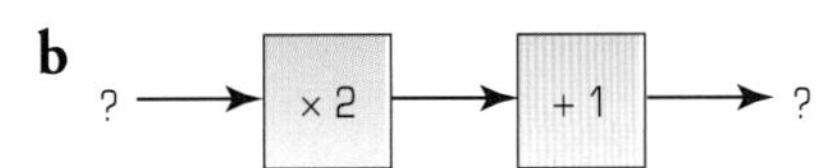

The output is never even.

c

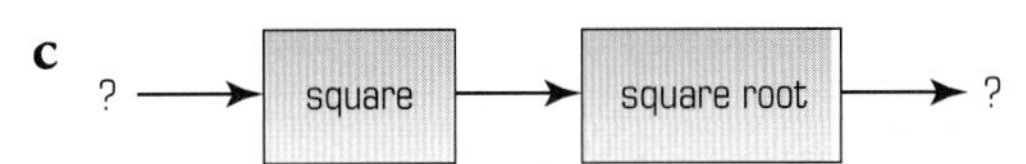

The input and output are always equal.

A1.6 Functions and algebra

This spread will show you how to:
- Use letter symbols.
- Begin to recognise algebraic conventions.
- Express simple functions in symbols.

KEYWORDS
Variable, Expression, Algebra, Term

The input to a function machine can vary – you can use a letter to represent all possible inputs:

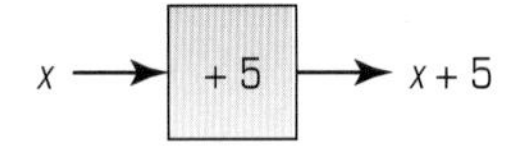

In this machine, x is a **variable**.

A **variable** is a value that can change.

Algebra is the branch of maths where you use letters to stand for unknown numbers or variables.

There are a few conventions or rules that you must always follow:

- Never use the multiplication '×' sign: write ab not $a \times b$
- When you multiply write numbers first then letters: $5r$ not $r5$.
- Never use the division sign: write $\frac{x}{10}$ not $x \div 10$

Hint: you say 5 rabbits (5*r*) not rabbits 5 (*r*5)!

example

Write the output for these function machines using algebra:

a $n \rightarrow$ [× 2] $\rightarrow$ output

b $n \rightarrow$ [× 2] $\rightarrow$ [+ 1] $\rightarrow$ output

a The output is $n \times 2$ – you write $2n$.

b the output is $2n + 1$

- A **term** in algebra is a letter, or a group of letters and numbers.
  ab, 5*r* and $\frac{x}{10}$ are terms.
- A collection of terms separated by + or – signs is called an **expression**.
  $xy + 2z$ is an expression.

- You can write the general term of a sequence using an algebraic expression.

For example, the position to term rule for a sequence is 'multiply by 3 then add 2'.

Write the inputs and outputs as mappings like this:

$1 \longrightarrow 3 \times 1 + 2 = 5$
$2 \longrightarrow 3 \times 2 + 2 = 8$
$3 \longrightarrow 3 \times 3 + 2 = 11$
$n \longrightarrow 3 \times n + 2$

The general rule of the sequence is $n \longrightarrow 3n + 2$

Exercise A1.6

1 For these function machines work out the outputs for the inputs 1, 2, 3 and x

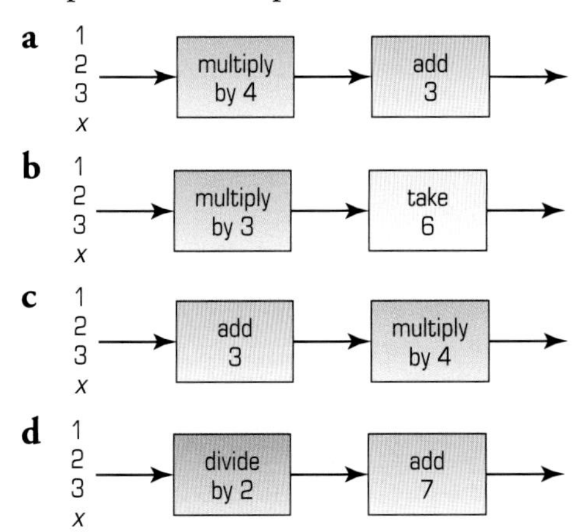

2 The numbers machines have different inputs. Write the output as an algebraic expression.

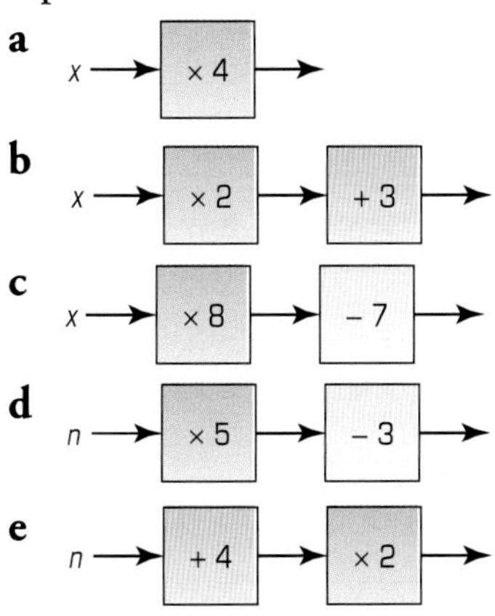

3 In these questions you are given the input and the output. Work out the operations in each function machine.

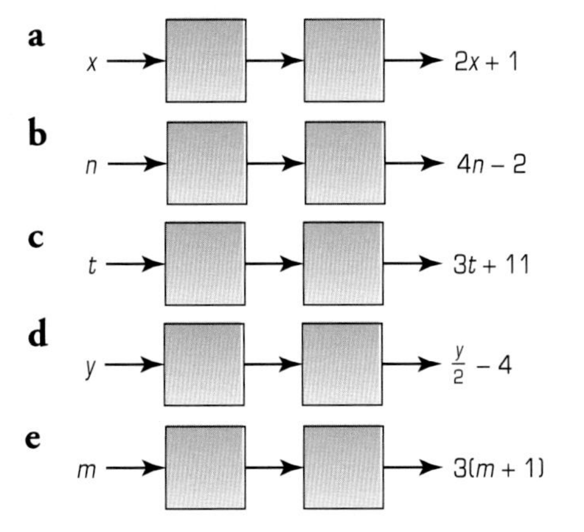

4 Find the rule in each of the function machines. Now write this as a mapping $n \longrightarrow ....$

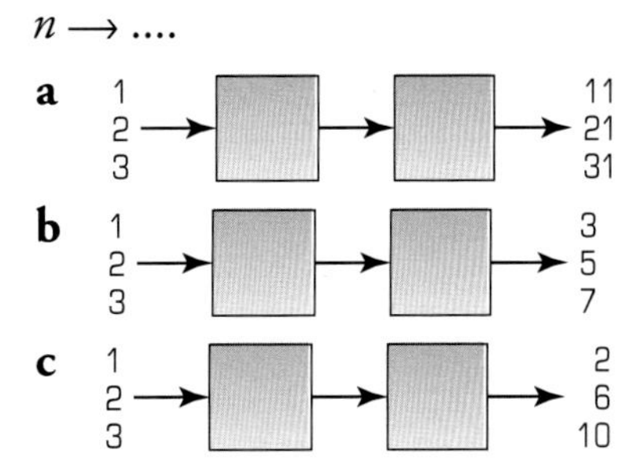

5 **a** Use these numbers and operations to complete the function machine:

1, 2, 3, +, −, ×.

You can only use each one once.
Write your solutions as mappings using algebra.

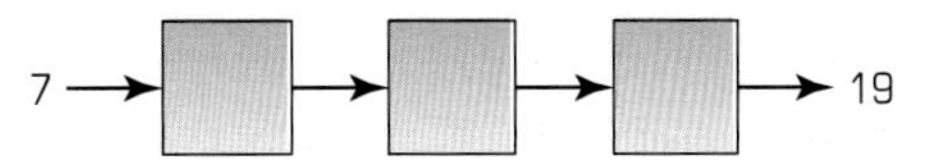

b Start with an input of 7 and the symbols 1, 2, 3, +, −, ×. What different outputs can you make?

6 Write these outputs as algebraic expressions:

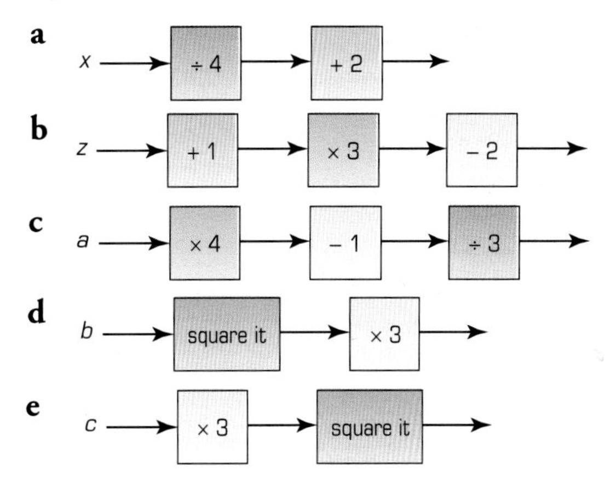

7 Write an algebraic expression and ask a partner to draw a function machine to represent it. Begin with one operation. With each new expression, add an extra operation.

A1 Summary

You should know how to ...

1 Generate and describe integer sequences.

2 Generate terms of a sequence using a term-to-term and position-to-term definition.

3 Solve word problems in the context of algebra.

4 Express simple functions in symbols.

Check out

1 Copy and continue each sequence for the next three terms:

a 3, 14, 25

b 6, 12, 24

c The square numbers: 1, 4

d 3.4, 4.1, 4.8

2 Generate the first five terms of each sequence:

a The first term 5 and each term is 13 more then the previous term.

b The first term is 7 and each term is 4 less than the previous term.

c Each term is the term number multiplied by 4 and subtract 2.

d Each term is the term number multipled by itself three times.

3 Explain why the 100th diagram in each sequence has the given number of matches:

a

100th term = 300 matches

b

100th term = 401 matches

4 Write the output for these function machines as mappings using algebra:

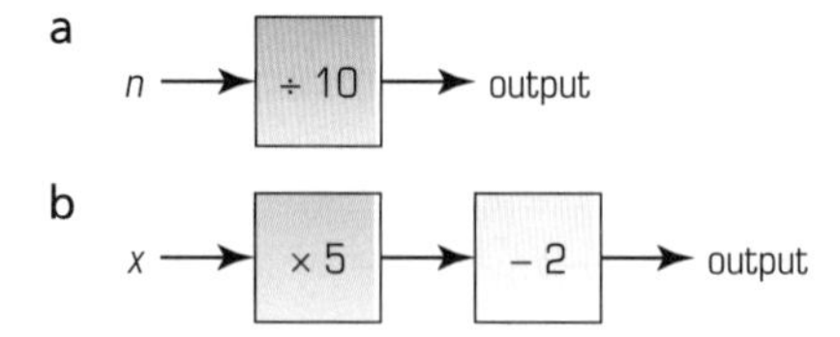

N1 Fractions, decimals and percentages

This unit will show you how to:

- Order decimals.
- Add subtract, multiply and divide integers.
- Consolidate and extend mental methods of calculation with decimals.
- Consolidate standard column procedures for addition and subtraction of integers and decimals with up to two places.
- Make and justify estimates and approximations of calculations.
- Carry out calculations effectively using the function key for sign change; use brackets and the memory.
- Enter numbers and interpret the display in different contexts.
- Use the order of operations, including brackets, with more complex calculations.
- Solve problems and investigate in the context of number.

Decimals describe parts smaller than a unit.

Before you start

You should know how to ...

1 Read and write whole numbers and decimals in figures.

2 Use decimal notation for tenths and hundredths.

3 Know how to multiply and divide by 10, 100 and 1000.

4 Know how to round a whole number to the nearest whole number and to 1 decimal place.

Check in

1 **a** Write 2004.6 in words
b Write eighty thousand and two in figures

2 Write as decimals **a** $\frac{1}{10}$ **b** $\frac{1}{100}$

3 **a** Multiply 2.3 kg by 10, 100 and 1000
b Divide £52 by 10, 100 and 1000

3 Round 4.567 to:
a the nearest whole number
b 1 decimal place

N1.1 Place value and ordering

This spread will show you how to:

- Identify decimal fractions on a number line.
- Order decimals.

KEYWORDS

Place value	Thousandth
Digit	Significant
Tenth	Less than
Hundredth	Greater than

- **Each digit in a number has a place value.**

For example: 2167.453 is

2000	100	60	7	0.4	0.05	0.003
thousands	hundreds	tens	units	tenths	hundredths	thousandths
1000 s	100 s	10 s	1 s	$\frac{1}{10}$ s	$\frac{1}{100}$ s	$\frac{1}{1000}$ s

You say 'two thousand, one hundred and sixty-seven point four five three'.

- **Using zeros in a number keeps the place values of the other digits:**

For example: 10.03 is 10 + 0.03.
If you forgot the zeros, it would become 1.3 which is 1 + 0.3!

- **Every number can be represented as a position on a number line.**

For example: 15.436 is between 15 and 16:
It is greater than 15.4 but less than 15.5:

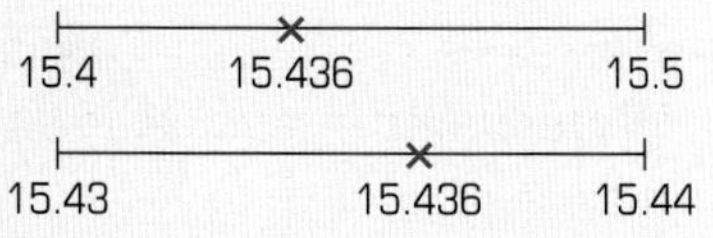

It is greater than 15.43 but less than 15.44:

15.43 — 15.436 — 15.44

You can imagine a number line to help you order decimals.

- **To compare two or more numbers first compare the place value of the most significant digit.**

For example: the most significant digit in 0.0213 is the 2 with value 0.02.

The first non-zero digit is the most significant.

example

Place these numbers in ascending order. 11.2 10.34 10.127 10.314 11.15

- The most significant digit is the first non-zero digit and that is the same in all five numbers. **1**1.2 **1**0.34 **1**0.127 **1**0.314 **1**1.15
- Arrange by comparing the second significant digit: 0 is smaller than 1. 1**0**.34 1**0**.127 1**0**.314 1**1**.2 1**1**.15
- Now arrange by the third significant digit: For the 10s: 1 is smaller than 3 and for the 11s: 1 is smaller than 2. 10.**1**27 10.**3**4 10.**3**14 11.**1**5 11.**2**
- Arrange the last two numbers by the fourth significant digit: 1 is smaller than 4. 10.127 10.3**1**4 10.3**4** 11.15 11.2

The numbers are now in size order.
You can check using a number line.

Exercise N1.1

1 Write these numbers in words:

a 1 312 945
b 20 505
c 3 009 007
d 109.65

Write these numbers in figures:

e three thousand, two hundred and six
f forty-two thousand and five
g one million, eight hundred thousand and thirty-two
h twenty-three and seventeen hundredths

2 **a** Write this list of numbers in order of size, smallest first:
6.3, 6.8, 7, 6.1, 6

b Draw a number line from 5 to 8 and label the numbers in part **a** upon it.

3 Place > or < between these numbers:
(**Remember**: < means 'less than' and > means 'greater than').

a	12 436	12 300
b	1.76	1.67
c	2.6	2.62

4 These are the results of the 200 metres race at Veryclose School:

Ali	29.1 s
Brad	29.15 s
Charlie	30.12 s
Doug	29.06 s
Clive	30.2 s

Put the runners in order from fastest to slowest.

5 Each player has a 0–9 dice or 0–9 digit cards.
Each player rolls their 0–9 dice and writes the digit in one of the spaces beside the first target number. This continues until all the spaces are filled.
The player whose number is closest to the target number is the winner of that round.
The players complete four more rounds.
The player who wins the most rounds is the winner.

Round	Target number	
a	Highest number	□□.□□□
b	Smallest number	□□.□□□
c	Nearest to 40	□□.□□□
d	Farthest away from 40	□□.□□□
e	Nearest to **any** square number	□□.□□□

6 Copy and complete these number patterns:

a

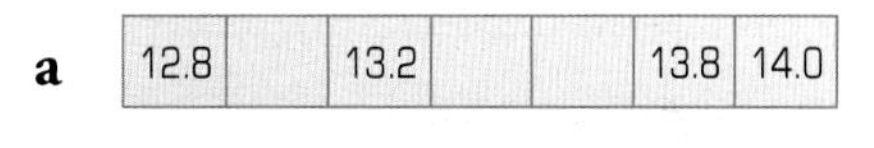

b

c

8.961		8.949			

d

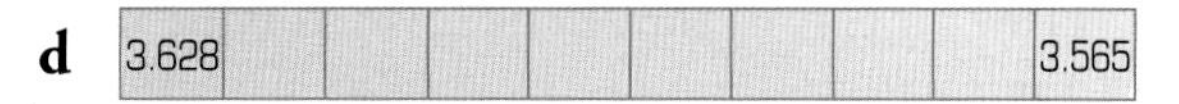

7 Given that $12.56 \leqslant y \leqslant 12.58$, give possible values for y:

a if y has two decimal places
b If y has three decimal places.

N1.2 Negative numbers

This spread will show you how to:

- Understand negative numbers as positions on a number line.
- Order, add and subtract integers in context.

KEYWORDS
Integer
Positive
Negative
Directed number
Add/Subtract

- **An integer** is a positive or negative whole number.

Integers below 0 are **negative** numbers. Integers above 0 are **positive** numbers.

$^{-}100$ $^{-}90$ $^{-}80$ $^{-}70$ $^{-}60$ $^{-}50$ $^{-}40$ $^{-}30$ $^{-}20$ $^{-}10$ 0 10 20 30 40 50 60 70 80 90 100

getting smaller getting bigger

- **You can use a number line to order integers.**
  For example: $^{-}70$ is smaller than $^{-}20$

You can use a number line to add and subtract integers.

Integers are often called **directed numbers** because the sign indicates the direction along the number line.

example

John uses £40 credit. He then spends a further £60.
How much does credit does he have altogether?

He uses £40 so he has $^{-}$£40 credit.
He spends another £60 so he adds $^{-}$£60 credit.
He now has $^{-}$£40 + $^{-}$£60 = $^{-}$£40 – £60 = $^{-}$£100 credit.

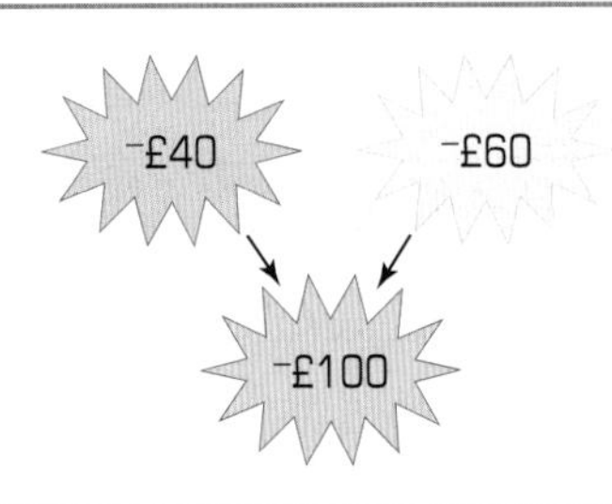

- **Adding a negative number is the same as subtracting a positive number.**
  For example: $^{-}2 + {}^{-}5 = {}^{-}2 - 5 = {}^{-}7$

example

Kim is looking over a cliff which is 70 metres above sea level.
How far is it to the sea bed which is 12 metres below sea level?

70 metres above sea level is +70 m

12 metres below sea level is $^{-}12$ m

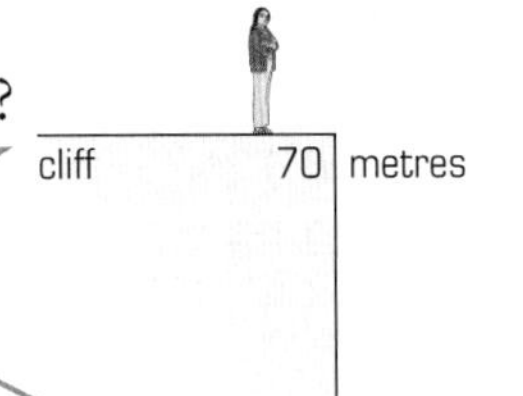

You want to know the difference between the two distances:
$70 - (^{-}12)$ metres.
From the diagram, this is the same as adding the two distances together:
$70 - (^{-}12) = 70 + 12 = 82$.
It is 82 metres from the top of the cliff to the sea bed.

- **Subtracting a negative number is the same as adding a positive number.**
  For example, $5 - (^{-}2) = 5 + 2 = 7$

Exercise N1.2

1 Copy and complete these number patterns.

a

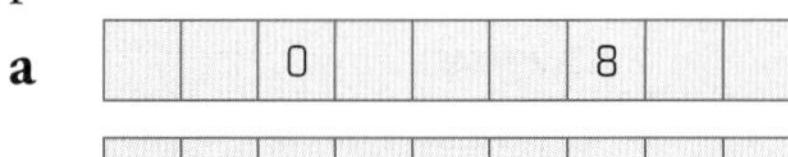

b

$^-15$			$^-3$				13	

c

	$^-72$				$^-36$			

2 Solve these problems:

a Mr. Cash has £53.75 in his bank account. He spends £48.21, deposits £26 in the account, and then spends a further £68.99. By how much is he overdrawn?

b A submarine is at a depth of 1926 m. If it rises by 278 m, what is its new depth?

3 These statistics were taken from a Geography book:

Location	mid-summer temperature °C	mid-winter temperature °C
Dawson	17	$^-30$
Verkhoyansk	15	$^-45$
Montreal	21	$^-9$
London	20	4
Rangoon	30	25
Buenos Aires	25	10
Moscow	20	$^-12$
Vienna	21	$^-2$
Marseilles	22	8
Helsinki	18	$^-8$
Bucharest	22	$^-3$

Find the range of temperatures in these cities. (The range is the difference between the mid-summer and mid-winter temperatures).

4 Put these sets of numbers in order from lowest to highest.

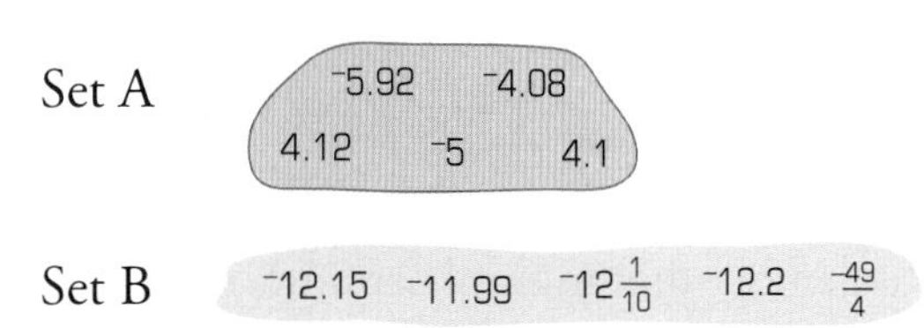

5 **Investigation**

a Extend these patterns:

i $8 + 4 = 12$ $\quad 8 + 3 = 11$
$8 + 2 = 10$ $\quad 8 + 1 = 9$
$8 + 0 = 8$ $\quad 8 + {}^-1 = 7$

ii $8 - 4 = 4$ $\quad 8 - 3 = 5$
$8 - 2 = 6$ $\quad 8 - 1 = 7$
$8 - 0 = 8$ $\quad 8 - {}^-1 = 9$

b Write down what you notice.

6 In 'Negative Countdown' you can add or subtract the numbers given, to make the target number. You must use all of the numbers.

a target = $^-5$ $\quad$ $^-3$ $\;$ 9 $\;$ 7

b target = 2 $\quad$ 8 $\;$ $^-4$ $\;$ 7 $\;$ 5

c target = $^-4$ $\quad$ $^-4$ $\;$ $^-2$ $\;$ 7 $\;$ $^-3$ $\;$ $^-6$

7 In a pyramid each brick is the sum of the two bricks beneath it. Copy and complete this pyramid.

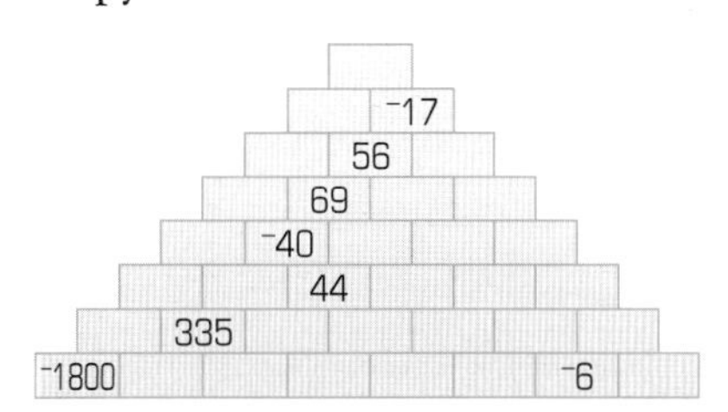

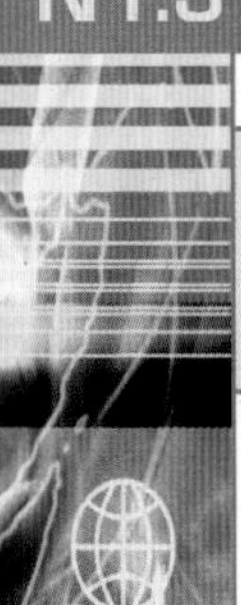

N1.3 Negative numbers: multiplication and division

This spread will show you how to:

- Multiply and divide positive and negative integers.

KEYWORDS

Inverse

Multiply/Divide

You can multiply and divide using negative numbers.
Remember that multiplication is repeated addition so:
$3 \times {}^-2$ means 3 lots of $^-2$, or $^-2 + {}^-2 + {}^-2 = {}^-6$
$^-6 \times 2$ is the same as $2 \times {}^-6$ which means $^-6 + {}^-6 = {}^-12$

- **When you multiply a negative number by a positive number you get a negative answer. For example, $^-4 \times 7 = {}^-28$.**

Notice that the size of the answer is the same as when you multiply positive numbers together:
If $4 \times 7 = 28$ then
$^-4 \times 7 = {}^-28$

example

Two whole numbers multiply to $^-20$. What are they?

This is the same as finding two numbers that multiply to 20, but one of them is negative.
These are all possibilities: $^-1 \times 20$ $^-2 \times 10$ $^-4 \times 5$ $^-5 \times 4$ $^-10 \times 2$ $^-20 \times 1$

Remember that division is the inverse of multiplication: $3 \times 4 = 12$ $12 \div 3 = 4$

- **When you divide a negative number by a positive number you get a negative answer. For example, $^-4 \times 7 = {}^-28$ so $^-28 \div 7 = {}^-4$**

  Notice that the inverse of $3 \times 4 = 12$ can be $12 \div 3 = 4$ or $12 \div 4 = 3$.
  Using negative numbers:
  The inverse of $^-4 \times 7 = {}^-28$ is $^-28 \div 4 = {}^-7$ or $^-28 \div {}^-7 = 4$

- **When you divide a negative number by another negative number you get a positive answer. For example, $^-28 \div {}^-7 = 4$**

Hint:
This is like saying 'how many $^-7$ s are in $^-28$?'

example

Mike loses £3 every time he gets a quiz question wrong.
He loses £27. How many questions did he get wrong?

He has $^-$£27 and has won $^-$£3 each time. That's $^-$£27 ÷ $^-$£3 = 9 questions altogether.

These tables summarise what happens when you multiply or divide two integers, *a* and *b*:

a	*b*	*a* ÷ *b*
24	4	6
$^-24$	4	$^-6$
24	$^-4$	$^-6$
$^-24$	$^-4$	6

a	*b*	*a* × *b*
3	4	12
$^-3$	4	$^-12$
3	$^-4$	$^-12$
$^-3$	$^-4$	12

Notice that it follows that
$^-3 \times {}^-4 = +12$

- **Multiplying or dividing positive and negative numbers is the same as multiplying the positive numbers together and then looking at the sign: $^- \times {}^- = +$.**

Exercise N1.3

1 Continue these patterns to three more multiplications:

$1 \times 5 = 5$ $\quad$ $1 \times {}^{-}5 = {}^{-}5$

$0 \times 5 = 0$ $\quad$ $0 \times {}^{-}5 = 0$

${}^{-}1 \times 5 = {}^{-}5$ $\quad$ ${}^{-}1 \times {}^{-}5 = 5$

${}^{-}2 \times 5 = {}^{-}10$ $\quad$ ${}^{-}2 \times {}^{-}5 = 10$

2 Complete these multiplications:

a ${}^{-}2 \times \square = {}^{-}6$ $\quad$ **b** $3 \times \square = {}^{-}15$

c ${}^{-}3 \times \square = {}^{-}15$ $\quad$ **d** ${}^{-}4 \times \square = 20$

e $\square \times {}^{-}6 = 3$ $\quad$ **f** ${}^{-}2 \times \square \times {}^{-}3 = {}^{-}18$

3 Complete this multiplication table:

×	$^{-}3$	$^{-}2$	$^{-}1$	0	1	2	3
3	$^{-}9$						
2				0			
1		$^{-}2$					
0							
$^{-}1$							
$^{-}2$						$^{-}4$	
$^{-}3$							

Shade positive numbers, negative numbers, and zero using different colours. Describe any patterns that you notice.

4 Complete these divisions:

a ${}^{-}6 \div \square = {}^{-}3$ $\quad$ **b** ${}^{-}6 \div \square = 3$

c $\square \div 9 = {}^{-}4$ $\quad$ **d** $10 \div \square = {}^{-}20$

5 **a** How many $^{-}3$s are there in $^{-}12$?

b How many 7s are there in $^{-}35$?

c How many negative 9s make negative 72?

d The answer to a question is $^{-}36$. What could be the question?

6 Use the +/– key on your calculator to work out:

a $49 \times {}^{-}51$ $\quad$ **b** ${}^{-}72 \times 4$

c $10.86 \div {}^{-}15$ $\quad$ **d** ${}^{-}9 \times {}^{-}54.8$

e ${}^{-}2772 \div {}^{-}18$ $\quad$ **f** ${}^{-}2.1 \div {}^{-}0.05$

7 Complete this multiplication grid:

×		4	$^{-}9$	
		$^{-}8$	18	
$^{-}3$		$^{-}12$		
	35			$^{-}14$
				12

Try to find another way to do it.

8 Work these out, remembering to do the brackets first:

a ${}^{-}2 \times (5 + 4)$

b ${}^{-}4 \times ({}^{-}6 + 3)$

c $8 \div (5 - 7)$

d $({}^{-}2 \times {}^{-}3) + ({}^{-}1 \times 5)$

e ${}^{-}12 \div (3 \times {}^{-}2)$

f $(16 \div {}^{-}4) - ({}^{-}6 \div {}^{-}2)$

9 **Investigation**

a Find the four factor pairs of 15.
Explain why there is an even number of factor pairs.
Can you find any numbers with an odd number of factor pairs?
What do these numbers have in common?

b Find as many answers as you can to $\sqrt{25}$.
Can you find $\sqrt{{}^{-}25}$?
Explain your answer.

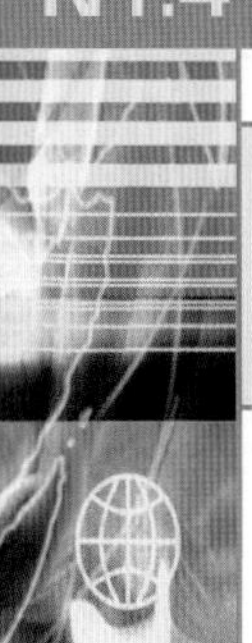

N1.4 Mental strategies

This spread will show you how to:

- Partition and deal with the most significant numbers first.
- Use compensation.

KEYWORDS
Compensation
Partitioning

There are many different ways of working out a calculation mentally.
Two of the most useful are partitioning and compensation.
Both methods involve making the numbers easier to work with in your head.

Partitioning

- **Partitioning involves breaking a number down into parts to help you calculate more efficiently.**

Hint
A number line can help you decide what size parts to use.

example

Use partitioning to calculate mentally:

a $781 + 146$　　　b $19.34 - 6.67$

a Break 146 down into its parts.
Add the most significant part first:

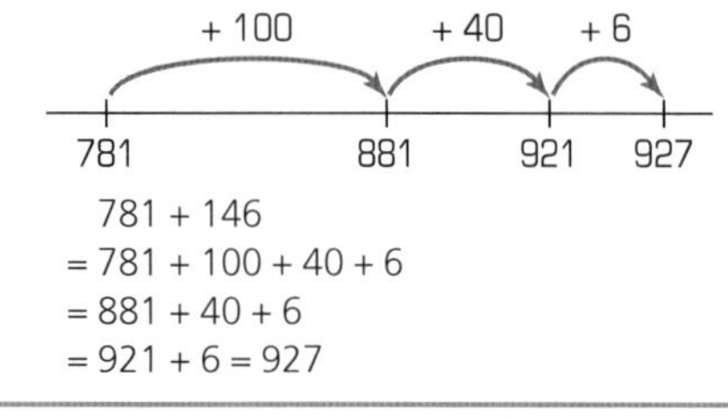

$781 + 146$
$= 781 + 100 + 40 + 6$
$= 881 + 40 + 6$
$= 921 + 6 = 927$

b Work with the whole numbers first:
then break down 0.67 into 0.34 and 0.33:

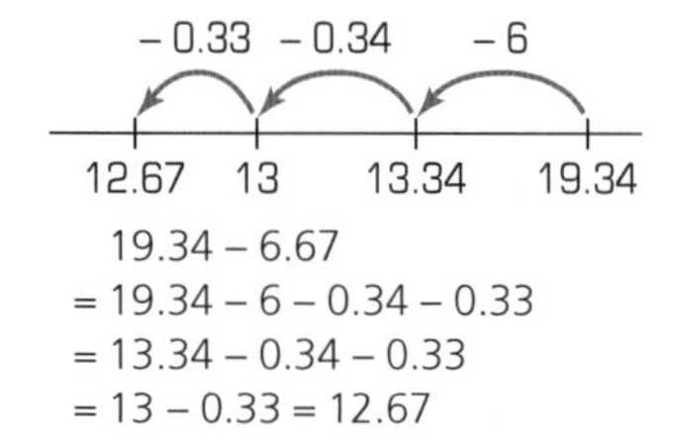

$19.34 - 6.67$
$= 19.34 - 6 - 0.34 - 0.33$
$= 13.34 - 0.34 - 0.33$
$= 13 - 0.33 = 12.67$

Compensation

- **Compensation involves rounding a number up or down and then compensating by adding or subtracting the extra amount.**

example

Use compensation to calculate mentally:

a $923 + 192$　　　b $^{-}7.3 - 2.8$

a round 192 up to 200 then compensate by subtracting the extra 8:

$923 + 192$
$= 923 + 200 - 8$
$= 1123 - 8$
$= 1115$

b round 2.8 up to 3.0 and then compensate by adding the extra 0.2:

$^{-}7.3 - 2.8$
$= {}^{-}7.3 - 3 + 0.2$
$= {}^{-}10.3 + 0.2$
$= {}^{-}10.1$

It can be complicated working with negative numbers.
It helps if you imagine a number line:

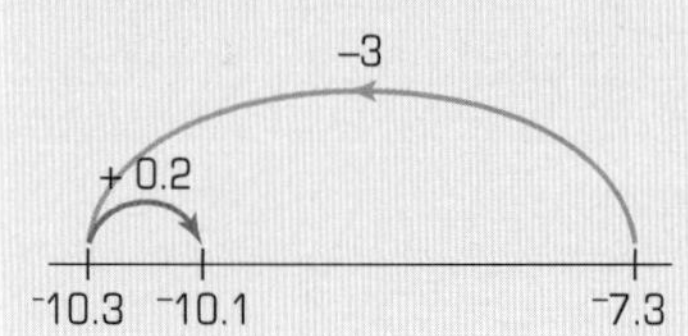

Exercise N1.4

1 Making Numbers Game:

Each player has a 0–9 dice. The players take it in turns to roll their dice and write their number in one of their boxes in Round 1.

After all the boxes have been filled the players calculate their answers.

The player with the highest answer wins the round.

The game is repeated for each of the other rounds.

The player who wins most rounds is the winner.

Round 1 □.□□ + □.□□ =

Round 2 □.□□ − □.□□ =

Round 3 □□.□□ + □□.□□ =

Round 4 □□.□□ − □□.□□ =

Round 5 □.□□□ + □.□□□ =

2 Investigation

Here are three cards. Each card has a number on both sides. These are the numbers on the front:

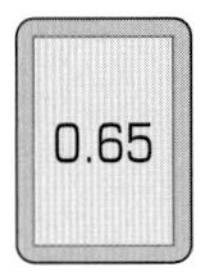

The total of the cards as shown is:

$0.65 + 0.75 + 0.85 = 2.25$

Here are all the possible totals of the cards.

0.6, 1.15, 1.7, 2.25

Work out what number is on the back of each of the three cards.

3 Two numbers have a sum of 4.85 and a difference of 2.31. What are the two numbers?

Explain how you worked out your answer.

4 In an arithmagon the number in a square must be the sum of the numbers on each side. Solve these arithmagons.

a

31

49 80

b

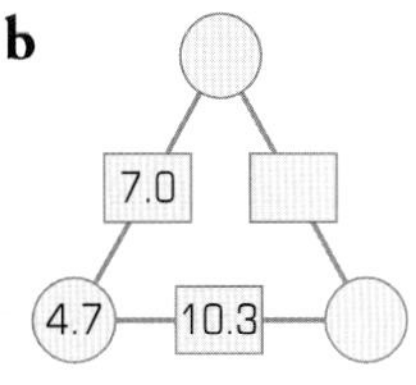

c

1.1

1.2 1.4

d

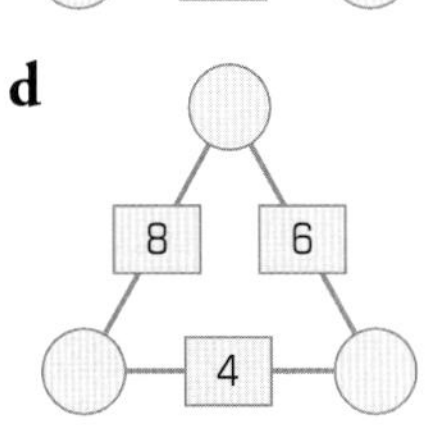

5 In a Pyramid the brick which sits directly above two bricks is the sum of the two bricks.

Copy and complete the following Pyramids.

a **b**

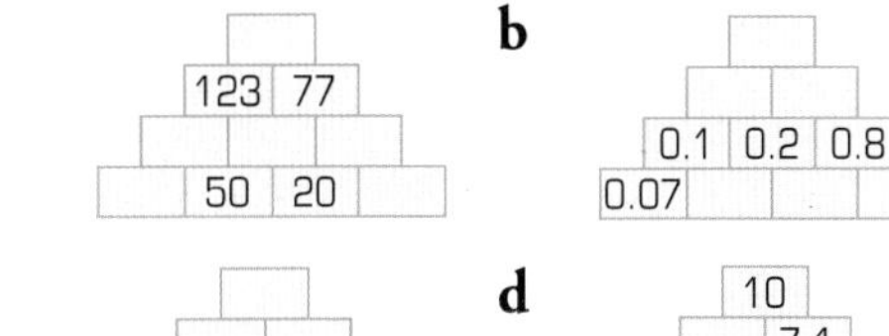

c **d**

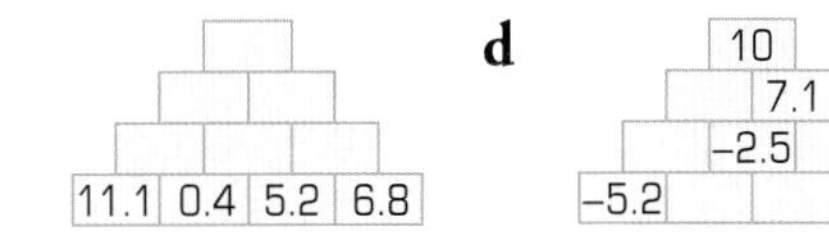

6 Here is a number pattern made by adding consecutive pairs.

5 4.32 9.32 13.64 22.96

$5 + 4.32 = 9.32$

$4.32 + 9.32 = 13.64$

$9.32 + 13.64 = 22.96$

Copy and complete these number patterns using the same rule.

a 2.71 6.53

b 4.34 12.15

c 3.79 14.33

Explain how you worked out your answer for part **c**.

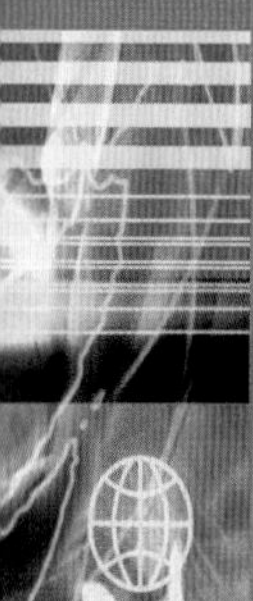

N1.5 Adding and subtracting decimals

This spread will show you how to:

- Use efficient column methods for addition and subtraction.
- Extend methods to include decimals with up to two decimal places.

KEYWORDS
Add
Estimate
Subtract

When numbers are too difficult to calculate in your head you need to use a written method.

- In column addition and subtraction you must always line up the units.
  This keeps all the digits in the correct place order.

Remember: subtracting a negative number is the same as adding a positive number. For example, $2 - (^{-}3) = 2 + 3$.

example

Calculate:

a $174.9 + 263.6$ **b** $401.2 - (^{-}26.85) + 0.71$
c $324.9 - 37.53$ **d** $14.24 + (^{-}8.79)$

a Estimate: $174.9 + 263.6 \approx 170 + 260 = 430$

$$\begin{array}{r} 174.9 \\ +263.6 \\ \hline 438.5 \\ \hline {}_{1\ 1} \end{array}$$

b $401.2 - (^{-}26.85) + 0.71$ is equivalent to $401.2 + 26.85 + 0.71$
Estimate: $401.2 - (^{-}26.85) + 0.71 \approx 400 + 30 + 1 = 431$

$$\begin{array}{r} 401.20 \\ 26.85 \\ +\ 0.71 \\ \hline 428.76 \\ \hline \end{array}$$

Fill in the missing digits with zeros.

c Estimate: $324.9 - 37.5 \approx 320 - 40 = 280$

$$\begin{array}{r} 324.90 \\ -37.53 \\ \hline 287.37 \\ \hline \end{array}$$

d $14.24 + (^{-}8.79)$ is equivalent to $14.24 - 8.79$
Estimate: $14.24 + (^{-}8.79) \approx 14 - 9 = 5$

$$\begin{array}{r} 14.24 \\ -8.79 \\ \hline 5.45 \\ \hline \end{array}$$

Exercise N1.5

1 **Investigation**

A palindromic number is one that has the same value when its digits are reversed. Examples are 161 or 23 932.

- ▶ Write down a 2 digit number, for example 91
- ▶ Reverse it 19
- ▶ Add the numbers together 110
- ▶ Reverse it 11
- ▶ Add the numbers together 121
- ▶ STOP when you reach a palindromic number

Investigate how many steps it takes for two-digit numbers to become palindromic.

Repeat for three-digit and four-digit numbers.

2 Calculate the following.

a 184 − 57 **b** 131 − 79

c 1608 + 2173 **d** 3619 − 2099

e 2002 − 189 **f** 3895 + 403

g 2661 − 1087 **h** 3002 − 2995

3 **a** The attendances at four football matches were:

35 003 28 271 12 043 and 66 487

What was the total attendance at these matches?

b A car is due for a service at 45 000 miles. It has currently travelled 42 643 miles. How many more miles is it until a service is due?

c A theatre has 1984 seats. On Monday 737 seats were taken. How many seats were empty?

d Calculate the perimeter of this garden.

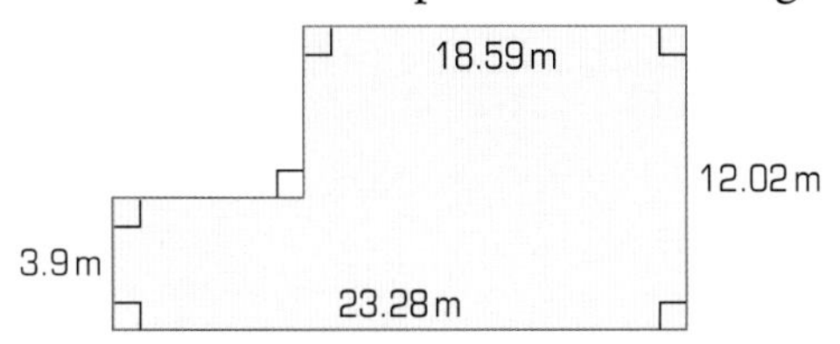

4 Calculate the following. (You should be able to use a mental method for some questions).

a 24.2 + 32.5

b 8.56 + 6.27

c 23.8 − 15.9

d 24.3 − 14.7

e 127.5 + 28.73

f 35.13 − 28.6

g 13.5 − 19.73 + 0.54

h 28.05 − (19.3 + 6.02)

i 100.02 − 70.025

5 This map shows 9 towns with the distance along major roads between them given in kilometres.

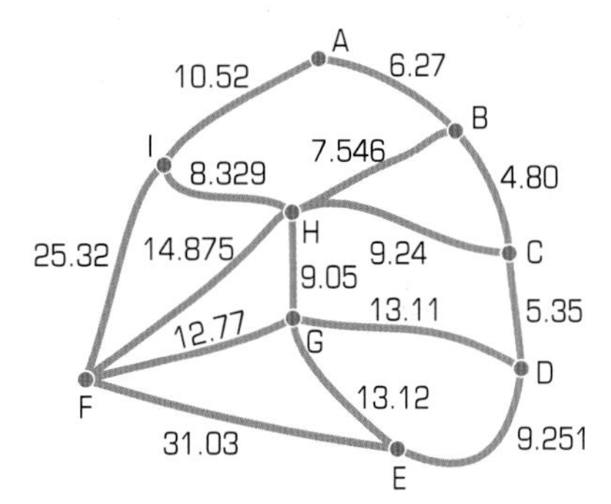

Calculate the shortest distance to get from:

a A to E

b I to D

c A to A visiting each town

d B to B visiting all the towns which are *not* vowels.

6 Calculate:

a $56.23 + {}^{-}2.57$

b $9.7 - 4.32$

c $8.14 - 17.67$

d $11.37 + {}^{-}23.54$

e $126.5 - {}^{-}39.57 + 19.24$

f $32.51 - {}^{-}1.625 - 9.72$

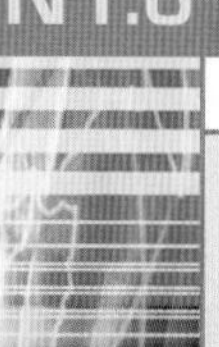

N1.6 Using a calculator

This spread will show you how to:

- Input a negative number.
- Use bracket keys.
- Interpret the display on a calculator in different contexts.

KEYWORDS
Calculate, Negative, Display, Memory, Brackets

Calculators are useful when the numbers are difficult.

example

Round 197.675 to:

a the nearest 10 **b** the nearest unit **c** 1 decimal place **d** 2 d.p.

Consider the number line.

a 197.675 is 200 to the nearest 10: 190 — 197.675 — 200

b 197.675 is 198 to the nearest unit: 197 — 197.675 — 198

c 197.675 is 197.7 to 1 decimal place: 197.6 — 197.675 — 197.7

d 197.675 is 197.68 to 2 d.p.: 197.67 — 197.675 — 197.68

It is important that you can use your calculator effectively.

- **You use the bracket keys or the memory function to solve problems with more than one step.**
- **You usually input negative numbers using +/–.**
  This shows that you are inputting a negative not subtracting!

Check the memory is empty before you use it.

Follow the examples to make sure you know how your calculator works.

example

Calculate:

a $29.3465 - (19.3 + 4.32 + 0.114 + 0.00354)$
b $^{-}17.532 + 6.9725 - (^{-}5.237)$

a Estimate first: $29 - (19 + 4 + 0 + 0) = 29 - 23 = 6$

Using brackets

Input: 2 9 . 3 4 6 5 – (1 9 . 3 + 4 . 3 2 + 0 . 1 1 4 + 0 . 0 0 3 5 4) =

Using the memory function

Input: 1 9 . 3 + 4 . 3 2 + 0 . 1 1 4 + 0 . 0 0 3 5 4 = Min AC 2 9 . 3 4 6 5 – MR =

Or input: 1 9 . 3 + 4 . 3 2 + 0 . 1 1 4 + 0 . 0 0 3 5 4 = +/– + 2 9 . 3 4 6 5 =

Display: 5 . 6 0 8 9 6

b Estimate first: $^{-}18 + 7 - (^{-}5) = {}^{-}18 + 7 + 5 = {}^{-}6$

Input: 1 7 . 5 3 2 +/– + 6 . 9 7 2 5 – 5 . 2 3 7 +/– =

Display: ⁻5 . 3 2 2 5

Exercise N1.6

1 Solve these problems using a mental or written method and a calculator where appropriate. Write your answer in the correct units.

a A shirt costs £14.99, a tie costs £13.79 and a book costs £5.99. Is £34.80 sufficient to pay for all these items? Explain your answer.

b In one month Mr. Big spends the following amounts on petrol:
£53.81, £48.52, £39.01, £62.98, £31.92
Calculate the total amount he spent on petrol during the month.

c These are the weights of ingredients in a recipe:

flour	850 g
sugar	220 g
2 tsp vanilla	1.24 g
2 eggs	49.0 g
2.5 litres milk	571.75 g

i What is the total weight of the ingredients?

ii How many grams less than 1 kg is this?

d Peter is rewiring his house. For upstairs he needs a length of 14.34 m, for downstairs he needs a length of 11.87 m. He has a length of 26.29 m. Will this be enough for the two pieces he requires?
Calculate how much short or how much extra cable there will be.

2 Explain the error made by somebody using a calculator.

a £12.38 + 86p = £98.38

b £4.87 + £5.23 = £10 and 1p

c $^-$£12.83 − £6.70 = $^-$£6.13

3 **Make me zero**

This is a game for 2 players.

- The first player enters a number into the calculator.
- The players take it in turns to subtract a number from this, but only one digit can be changed at a time.
- For example, 3.**6**25 − 0.6 = 3.**0**25 is allowed because only the 6 has changed.
- The display must never show a negative number.
- The winner is the person who makes the display on the calculator zero.

Here is an example of a game:

Player	Button	Display
1		83.625
2	−80	3.625
1	−0.6	3.025
2	−3	0.025
1	−0.01	0.015
2	−0.004	0.011
1	−0.01	0.001
2	−0.001	0

Player 2 wins.

a Play the game a few times with a partner.

b Explain how to modify the game if the display could show a negative number.

4 Calculate the following using a mental or written method, or where appropriate, using your calculator. Carry out a mental approximation to check the answer on the calculator.

a $4.564 + 2.37 + {}^-4.32$

b $17.6 - (3.82 + 4.7 - 6.34 + 4.9 + {}^-2.37)$

c $368.72 + {}^-28.392 - (4.1 - 2.6)$

d $328.1 - 199.5 + (33.7 - 16.8 - 64.1)$

N1 Summary

You should know how to ...

1 Order decimals.

2 Add, support, multiply and divide integers.

3 Understand addition and subtraction of integers.

4 Enter numbers and interpret the display in different contexts.

5 Solve word problems in the context of number.

Check out

1 Write in order, smallest first:

a 0.04 km, 200 cm, 0.5 m

b 0.03 m, 30 cm, 3 mm

c $^{-}2.3$, $^{-}2.005$, $^{-}2.06$

2 Calculate:

a $^{-}3 + {}^{-}4$ b $^{-}2 - {}^{-}1$

c $4 - {}^{-}6$ d $^{-}8 \div 2$

e $^{-}12 \div {}^{-}6$ f $6 \times {}^{-}3$

g $^{-}2 \times {}^{-}4$ h $^{-}1 \times {}^{-}1$

3 Use a standard written method to work out:

a $2.356 + 4.3$

b $4.006 - 0.32$

4 Use your calculator to work out £7 ÷ 50 and explain what your answer means.

5 a Find three consecutive numbers that total 231.

b Find three consecutive odd numbers that total 453.

c Find four consecutive multiples of 5 that total 350.

S1 Perimeter and area

This unit will show you how to:

- Deduce and use formulae for the area of a triangle, parallelogram and trapezium.
- Calculate areas of compound shapes made from rectangles and triangles.
- Use units of measurement to estimate, calculate and solve problems in everyday contexts involving length, area and volume.
- Know rough metric equivalents of imperial measures in daily use.
- Know and use the formula for the volume of a cuboid.
- Calculate volumes and surface areas of cuboids and shapes made from cuboids.
- Make simple scale drawings.
- Solve problems and investigate in the context of shape and space.
- Give solutions to an appropriate degree of accuracy in the context of the problem.

You need to understand length and area to decorate.

Before you start

You should know how to ...

1 Recognise measurements on a diagram.

2 Recognise basic shapes and solids.

3 Calculate with decimals.

Check in

1 a

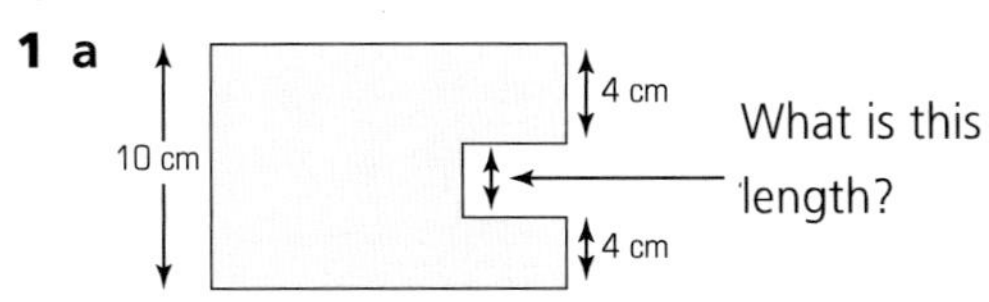

b Here is a 5 cm by 3 cm rectangle.

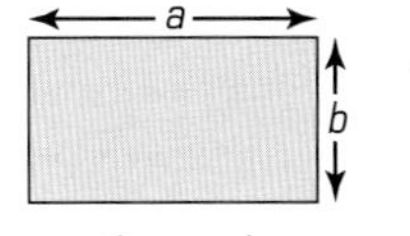

What are the lengths of *a* and *b*?

2 Name these shapes and solids.

a  **b**  **c** 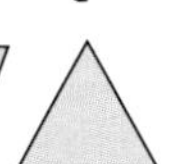 **d** 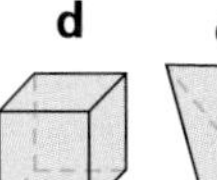 **e**

3 Multiply:

a 3.2 mm × 2.4 mm

b (5.9 m + 3.71 m) × 2

S1.1 Perimeter and area

This spread will show you how to:

- Deduce and use formulae for the perimeter and area of a rectangle.
- Calculate the perimeter and area of a shape made from rectangles.

KEYWORDS

Area, Dimension, Perimeter, Length, Rectangle, Width

- Area of a rectangle = length × width or $A = lw$
- The perimeter of a rectangle is the distance around the outside of it:
  Perimeter of a rectangle $= l + w + l + w = 2l + 2w = 2(l + w)$

Remember:
Lengths are measured in mm, cm, m etc.
Areas are measured in mm^2, cm^2, m^2, etc.

example

Use the formulae to find the area and perimeter of a 23 mm by 4 cm rectangle.

First you need to make the units match: 23 mm = 2.3 cm.
Area = lw = 2.3 cm × 4 cm = 9.2 cm^2
or 23 mm × 40 mm = 920 mm^2.
Perimeter = 2(2.3 cm + 4 cm) = 2 × 6.3 cm = 12.6 cm
or 126 mm

4 cm
23 mm

This shape is made up of more than one rectangle. It is a **compound shape**.

Split the shape into rectangles and work out their dimensions

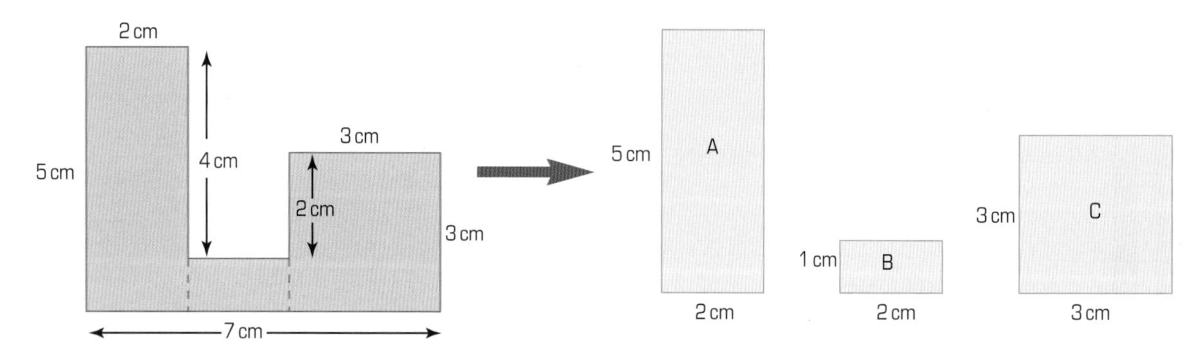

You find the area and perimeter of a compound shape like this:

Area

1 Find the area of each rectangle:
A = 5 cm × 2 cm = 10 cm^2
B = 1 cm × 2 cm = 2 cm^2
C = 3 cm × 3 cm = 9 cm^2

2 Add the areas together to find the total area of the compound shape:
10 cm^2 + 2 cm^2 + 9 cm^2 = 21 cm^2

Perimeter

1 Work out all the dimensions of the edges.

2 Add the lengths together.
Remember where you started!

2 + 4 + 2 + 2 + 3 + 3 + 7 + 5 = 28

Perimeter = 28 cm

Exercise S1.1

Find the areas and perimeters of the shapes in questions 1 to 3.
Remember the units.

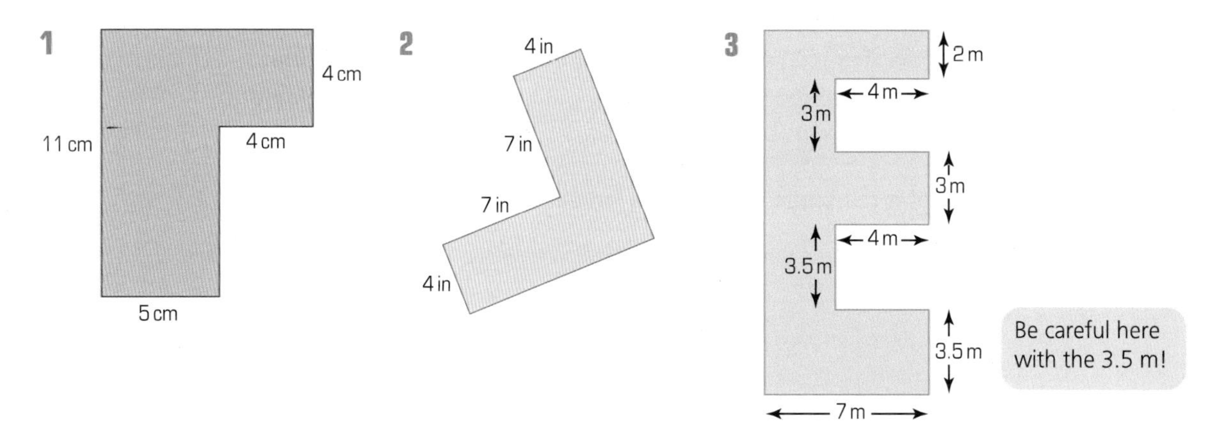

4 Find the area of the shaded region in each of these shapes. Do not use a calculator.

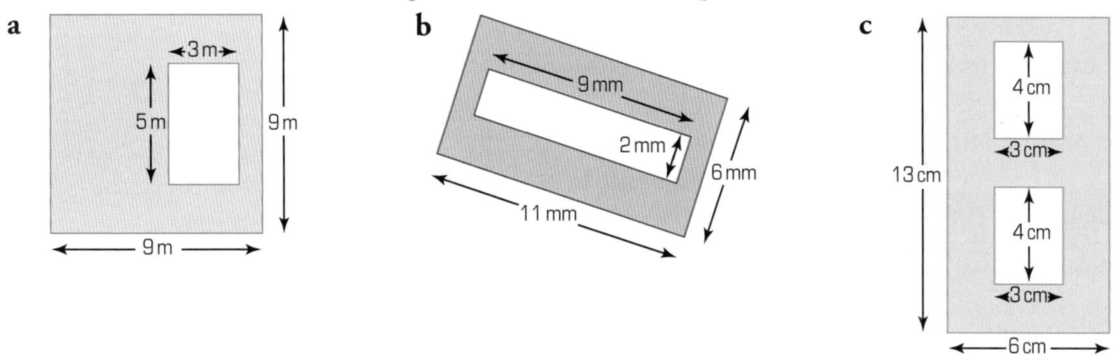

5 Find the areas and perimeters of these shapes. Give your answers in cm^2 and cm.

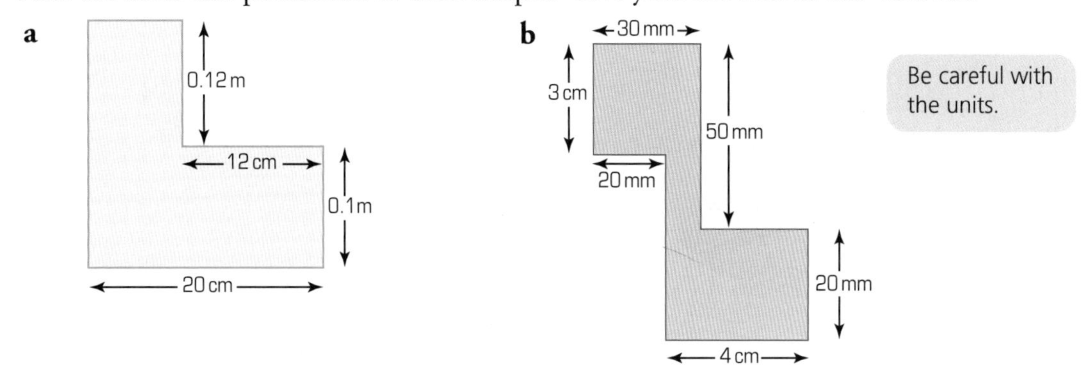

6 Draw as many rectangles as you can with an area of 20 cm^2.
What dimensions give the greatest perimeter?

7 What is the greatest area for a rectangle with perimeter 20 cm?

8 For any rectangle, if you halve the area does it halve the perimeter?

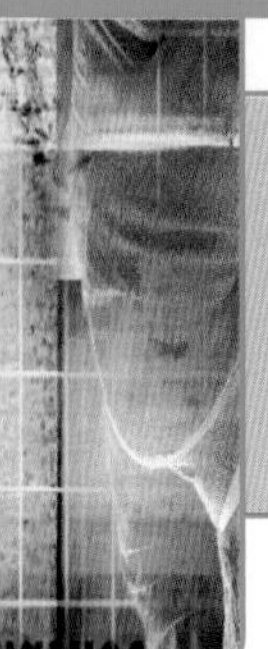

S1.2 More perimeter and area

This spread will show you how to:

- Derive and use a formula for the area of a right-angled triangle.
- Deduce formulae for the area of a parallelogram, triangle and trapezium.
- Calculate areas of triangles, parallelograms and trapezia.

KEYWORDS

Triangle, Perpendicular, Trapezium, Quadrilateral, Parallelogram

You can use the area of a rectangle to help you find the area of a triangle.

Draw a triangle on squared paper

Complete the surrounding rectangle

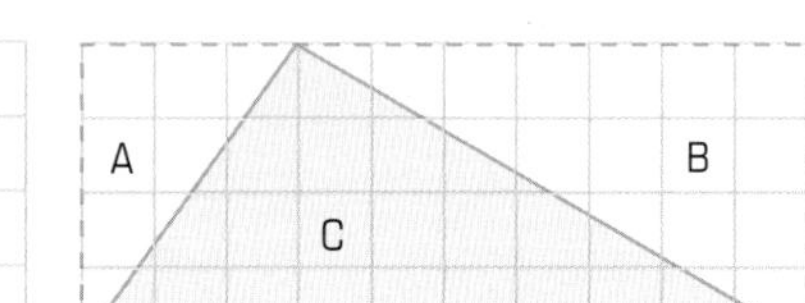

Fit triangles A and B into C.

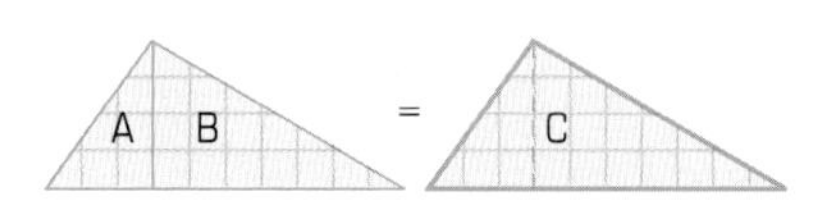

Area A + area B = area C

The area of a triangle is half the area of the surrounding rectangle.

- **Area of a triangle** $= \frac{1}{2}$ **(area of rectangle)**
  $= \frac{1}{2}$ **(base × perpendicular height)**

Note:
You must use the perpendicular height because the length and width are perpendicular.

You can use this formula to help you deduce the formulae for different quadrilaterals:

Parallelogram

A parallelogram has two pairs of parallel sides

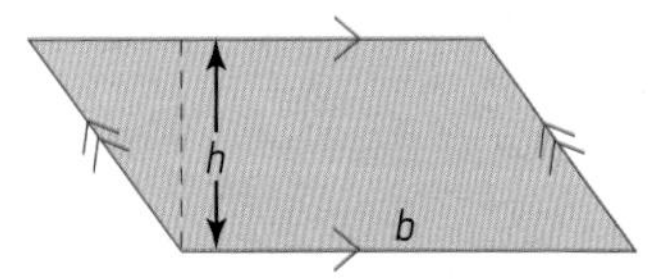

Take the triangle from one end and fit it to the other end:

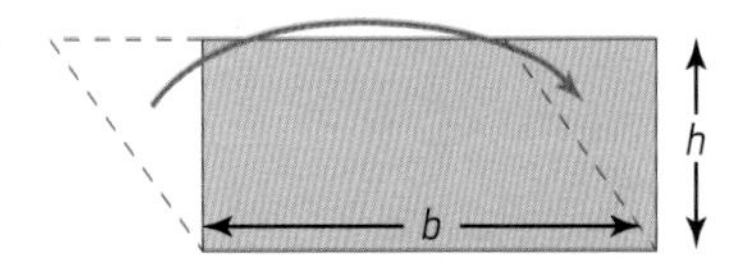

It makes a perfect rectangle.

- **Area of parallelogram**
  = area of rectangle
  = base × perpendicular height

Trapezium

A trapezium has one pair of parallel sides, a and b

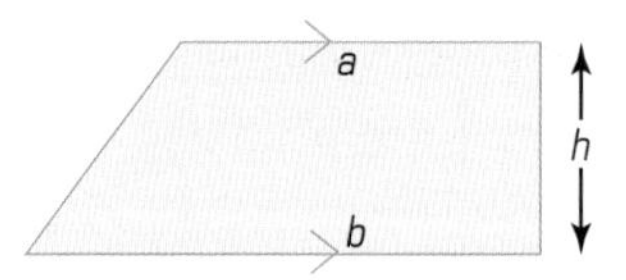

Split it into its component shapes:

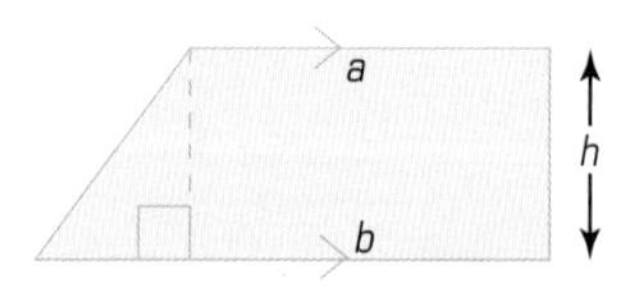

Find the area of each part and add them up:

$\frac{1}{2}(b - a)h + ah$
$= \frac{1}{2}bh - \frac{1}{2}ah + ah$
$= \frac{1}{2}bh + \frac{1}{2}ah$
$= \frac{1}{2}(a + b)h$

The work on page 72 will help you with this algebra.

- **Area of trapezium** $= \frac{1}{2}(a + b)h$

Exercise S1.2

1 Find the areas of these triangles.

a

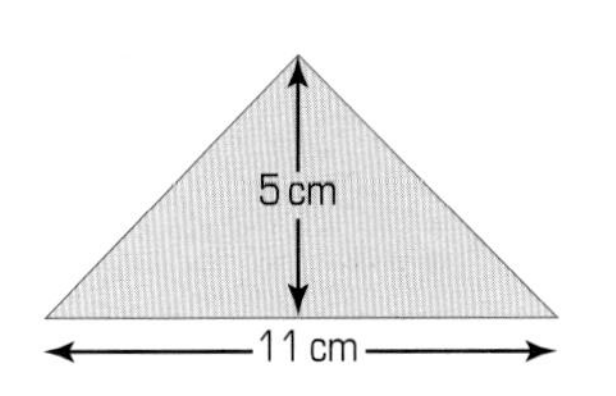

b

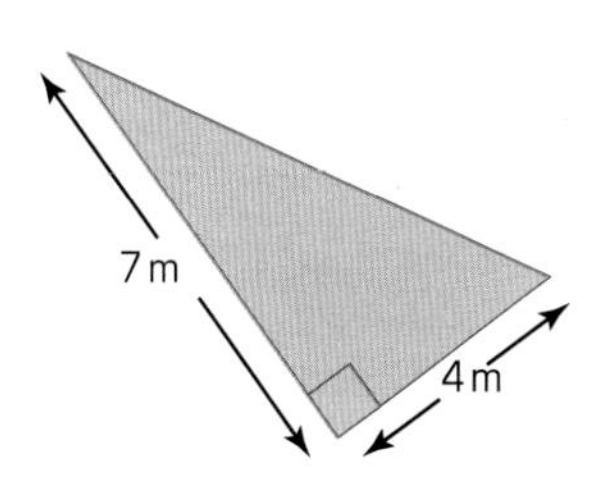

c

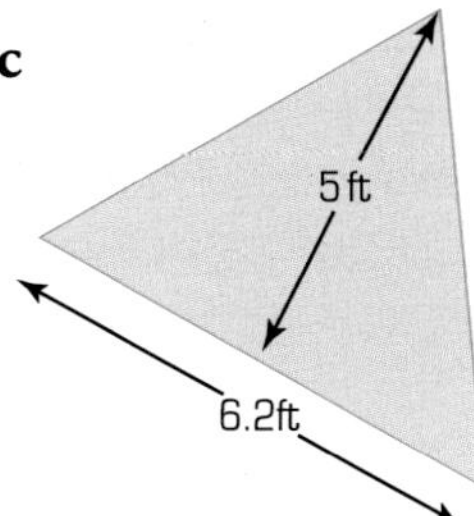

2 Find the areas of these quadrilaterals.

a

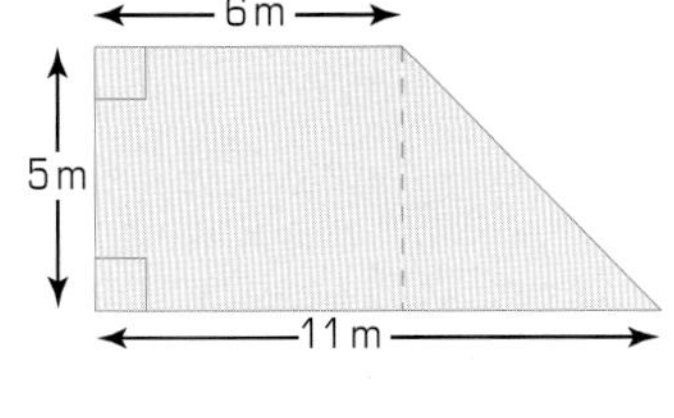

b

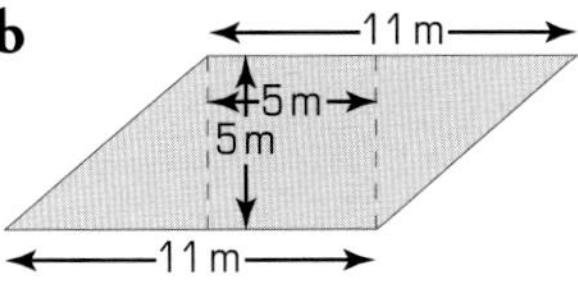

c

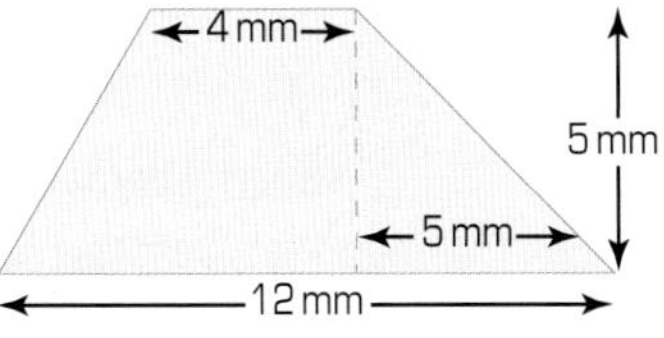

d

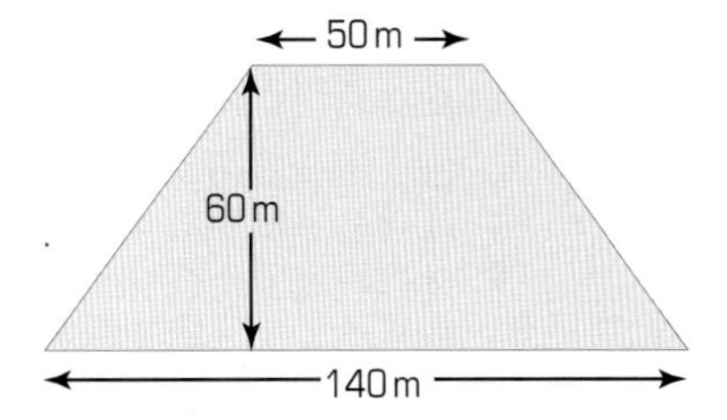

3 A rhombus is a parallelogram with sides of equal length:

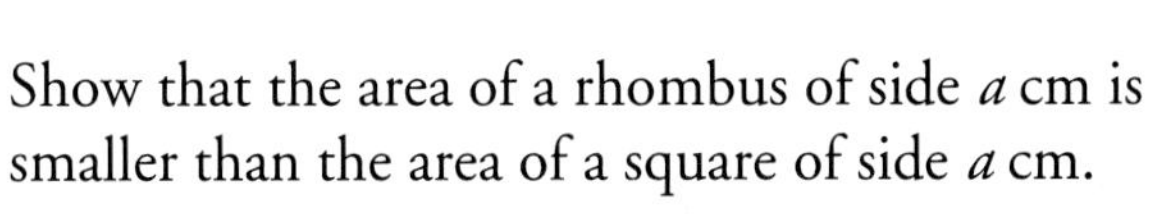

Show that the area of a rhombus of side a cm is smaller than the area of a square of side a cm.

4 Find the area of each of these kites by splitting them into triangles.

a

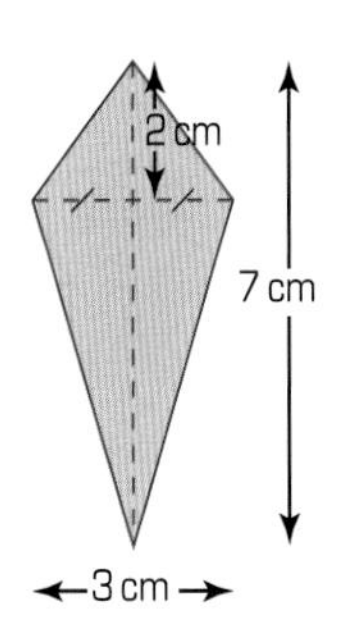

b

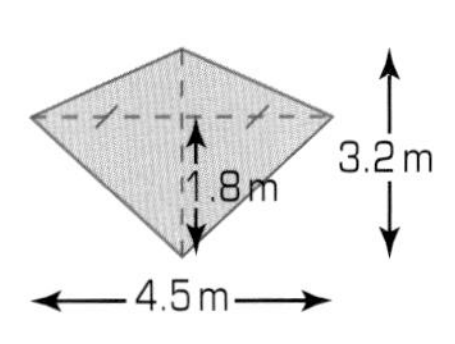

c

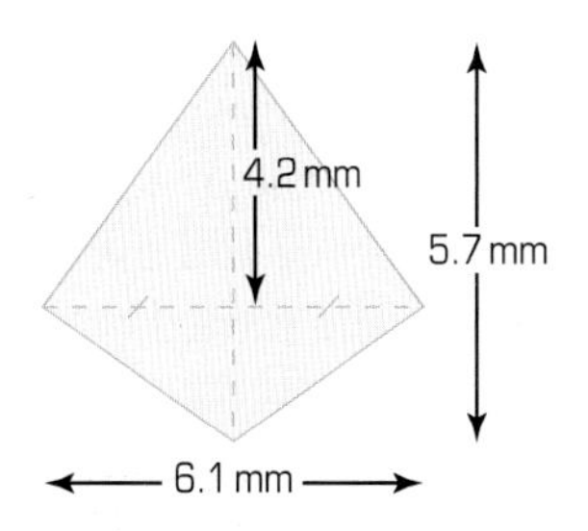

Can you work out a formula for the area of a kite?

Hint:
Work out 6.1 × 5.7 and compare with your answer to part **c**.

S1.3 Measurement and scales

This spread will show you how to:
- Use standard metric units.
- Suggest appropriate units and methods to estimate or measure length, area and volume.
- Use scales and simple scale drawings.

KEYWORDS
Millimetre, Kilometre, Centimetre, Volume, Metre, Scale

▶ You measure lengths and distances using metric units ...

millimetres (mm)
1 mm is about the width of a blade of grass

centimetres (cm)
1 cm is about the width of your little finger nail

metres (m)
1 m is about the width of a door

kilometres (km)
1 km takes about 15 minutes to walk

or imperial units ...

inches
1 inch is about 2.5 cm

feet
1 foot is 12 inches which is about 30 cm

yards
1 yard is 3 feet which is about 90 cm

miles
1 mile is about 1.6 km
It takes 25 mins to walk

▶ You measure areas using squares:

square millimetres (mm^2)
1 mm^2 is about the size of a sugar granule

square centimetres (cm^2)
1 cm^2 is smaller than a stamp

square metres (m^2)
1 m^2 is about half the area of a door

square kilometres (km^2)
1 km^2 is about the size of a large field

Volume is the amount of space taken up by a shape.

▶ You measure volume using cubes:

cubic millimetres (mm^3)
1 mm^3 is smaller than a grain of sand

cubic centimetres (cm^3)
1 cm^3 is the size of a sugar cube

cubic metres (m^3)
1 m^3 is about the size of a washing machine

the ^{2} shows area has two dimensions and the ^{3} shows volume has three dimensions.

When you are working with large units, you need to be able to use scale drawings.

example

Joanne sets out due North and walks for 2.5 km.
She then turns due East and walks for 1.6 km.
Using a scale of 1 km = 2 cm make a scale drawing of her journey and use it to find out how far she is from where she started.

For every 1 km you draw 2 cm
So for 2.5 km you draw 2 cm × 2.5 = 5 cm
And for 1.6 km you use 2 cm × 1.6 = 3.2 cm

The distance back is 6.0 cm which is 3.0 km

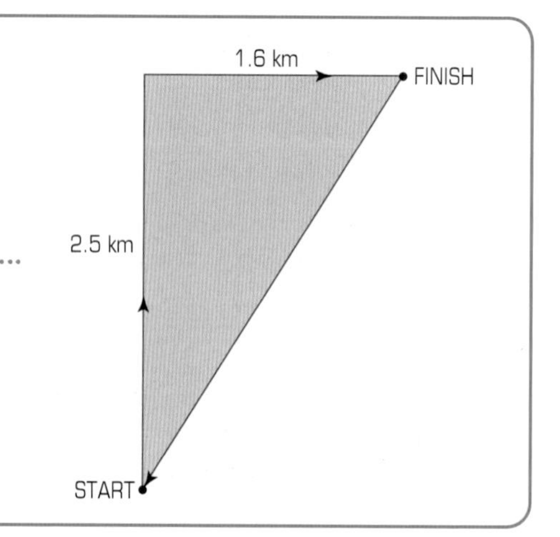

Exercise S1.3

1 **a** Find objects that satisfy each of these statements:

i 1.6 m < ______ < 2.6 m **ii** 3 cm < ______ < 3.7 cm **iii** 50 ft < ______ < 75 ft

b Copy and complete the range for each estimated height or length.

Use metric units.

i ______ < length of a double decker bus < ______ .

ii ______ < height of your classroom < ______

iii ______ < length of your pen < ______ .

Make your range as close to the actual value as possible.

2 Copy and complete this table by estimating lengths of the items.

- Choose your units from: mm, cm, m, km, inch, foot, mile
- Choose your equipment from: ruler, metre rule, tape measure, trundle wheel.
- Estimate then measure.

			Estimated			Actual		
Item	units	equipment	length	width	area	length	width	area
Ruler								
Eraser								
Your desktop								
Classroom door								
Calculator								
The front of this book								

3 What units and what equipment would you need to measure the length of a:

a bus **b** box of matches **c** swimming pool **d** ballpoint pen

4 Mary's bedroom floor measures 3 m by 5 m.

She wants to arrange her furniture to fit.

The base of each piece has these measurements:

- Bed – 1.2 m by 2 m
- Desk – 0.8 m by 1.3 m
- Wardrobe – 1.6 m by 1 m
- Chair – 80 cm square.
- Chest of drawers – 60 cm by 75 cm

Using a scale of 2 cm to represent 1 m, draw a scale diagram of Mary's bedroom and show her how to fit in her furniture making sure that all the drawers and doors can open.

5 Callum walks due West for 3.5 km.

He turns to face due South and walks for a further 2.8 km.

Using a scale of 1 km = 2 cm, make a scale drawing of Callum's journey and use it to find out how far he is from where he started.

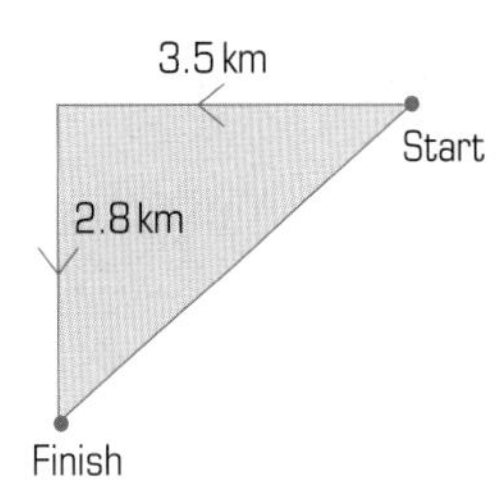

Three-dimensional shapes

This spread will show you how to:

- Use 2-D representations to visualise 3-D shapes and deduce some of their properties.
- Find the surface area of cuboids.
- Know the formula for the volume of a cuboid.

KEYWORDS

Dimensions, Cuboid, Prism, Surface area, Net, Vertex, Volume

Most everyday objects have three dimensions ... length, width and height. Some common 3-D shapes are:

Cubes

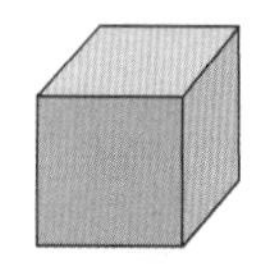

All faces square

Cuboids

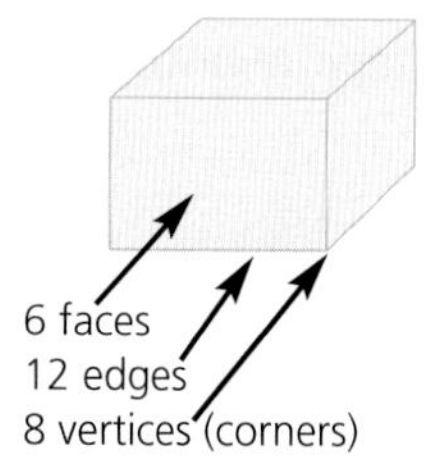

6 faces
12 edges
8 vertices (corners)

Prisms

Uniform cross-section (in this case triangular)

Pyramids

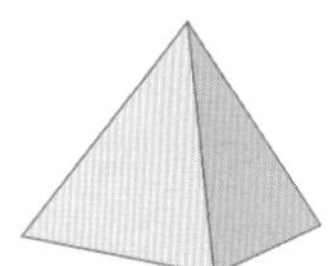

Tapers to a point

- A **net** is a flat shape that folds up into a 3-D shape.

You find the surface area of a 3-D shape by adding together the area of the faces.

example

Find the surface area of a cuboid measuring 2.7 mm by 5.3 mm by 3.8 mm.

Lay the shape out flat – draw a net:

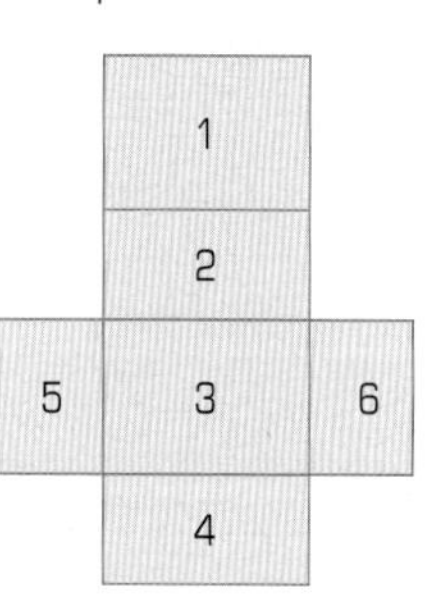

Find the area of each face:

→ ① and ③ = 5.3 × 3.8 = 20.14

→ ② and ④ = 5.3 × 2.7 = 14.31

→ ⑤ and ⑥ = 2.7 × 3.8 = 10.26

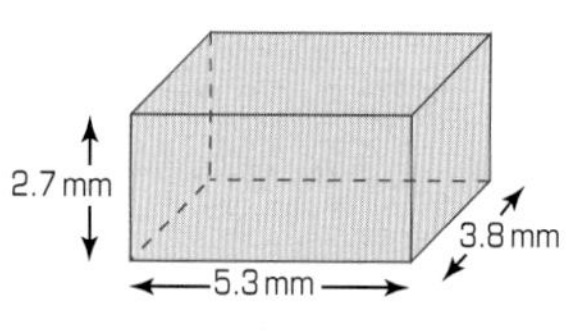

Add the areas up:

20.14 × 2 = 40.28
14.31 × 2 = 28.62
10.26 × 2 = 20.52 +
89.42

The surface area is 89.42 mm^2.

You can find the **volume** of a cuboid by counting cubes:

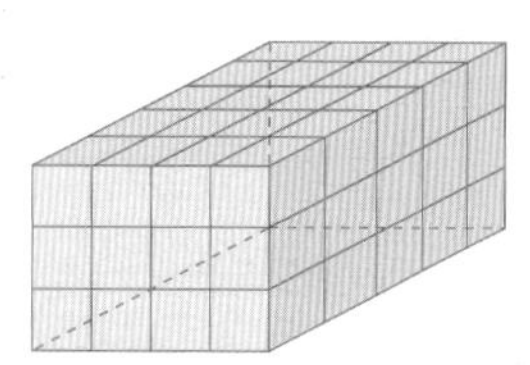

There are 4 × 5 cubes on the first layer.

There are 3 layers.

There are 4 × 5 × 3 = 60 cubes altogether.

Or you can use this formula:

- The volume of a cuboid = length × width × height

Exercise S1.4

1 Which of the following nets will make a cube?

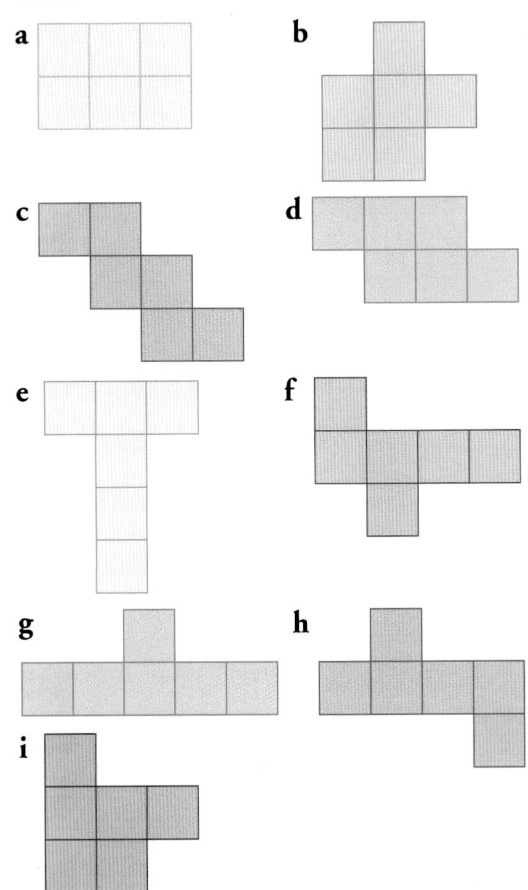

Six squares joined together like this are called **hexominoes**.
Draw all the hexominoes that will make cubes in your book.

2 Use four multilink cubes to make five different solids. Sketch your solids.

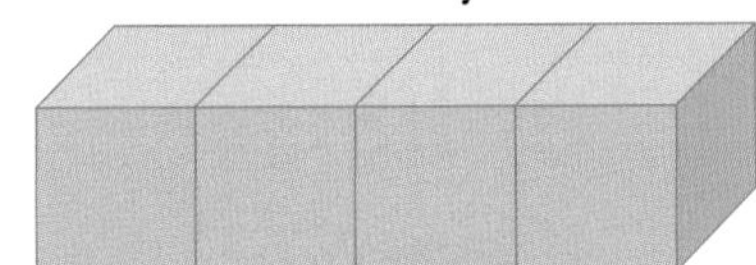

What is the surface area of each of your solids?

3 Construct the net of a cube with dimensions 5 cm by 5 cm by 5 cm.

a What is the surface area of this cube?

b What is the volume?

4 Construct the net of a cuboid with dimensions 2.8 cm by 3.8 cm by 4.5 cm.

a Find the surface area of this cuboid.

b What is its volume?

5 Work out the
- dimensions
- surface area
- volume of each cuboid.

Remember to state the units correctly.

a

3 in

2 in

4 in

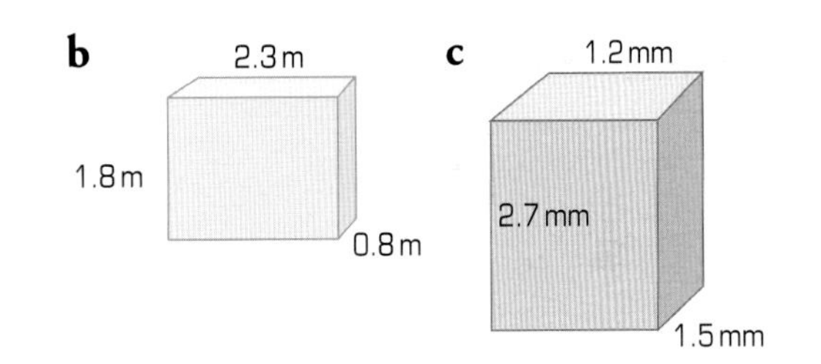

6 **a** Copy and complete this table.

Solid	Faces	Vertices	Edges
Cube			
Cuboid			
Square-based pyramid			
Triangular prism			

b Try to find a formula to link:
- the number of edges e
- the number of vertices v
- the number of faces f.

S1 Summary

You should know how to ...

1 Deduce and use formulae for the area of a triangle and parallelogram.

2 Find the area of compound shapes.

3 Know and use the formula for the volume of a cuboid.

4 Calculate volumes and surface areas of cuboids.

Check out

1 Find the area of each shape:

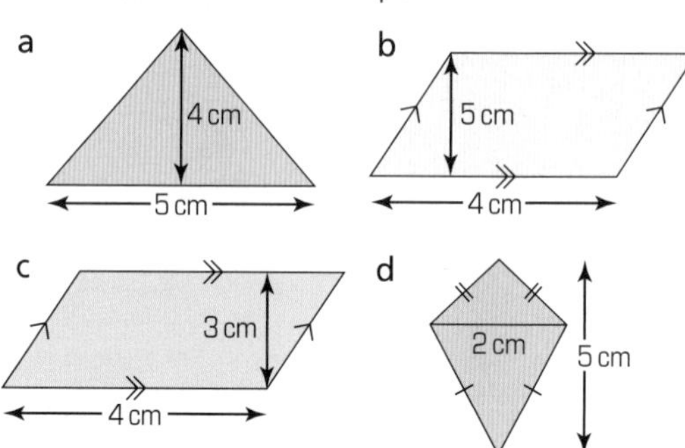

2 Find the area of:

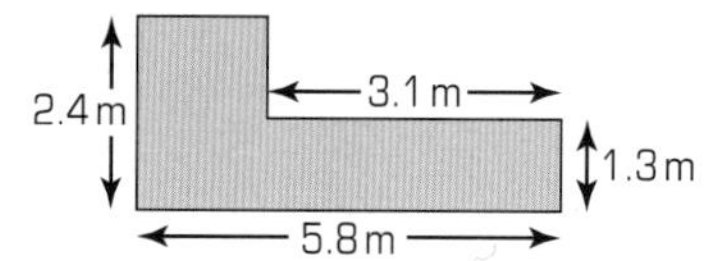

3 a What is the volume of a 1 m by 3 m by 4 m cuboid?

b What is the volume of a cube of side 4 cm?

4 What is the volume and surface area of this cuboid?

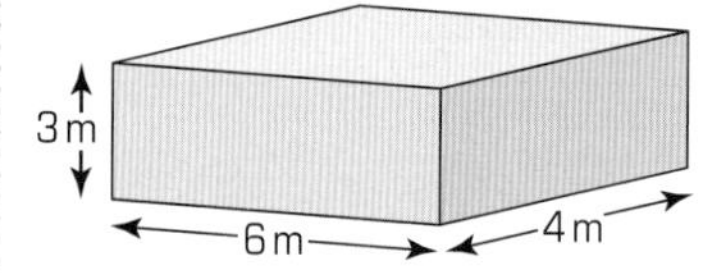

N2 Fractions, decimals and percentages

This unit will show you how to:

- Order fractions by writing them with a common denominator.
- Add and subtract fractions by writing them with a common denominator or by converting them to decimals.
- Use division to convert a fraction to a decimal.
- Calculate fractions of quantities.
- Multiply and divide an integer by a fraction.
- Understand addition and subtraction of fractions and integers.
- Interpret percentage as the operator 'so many hundredths of' and express one given number as a percentage of another.
- Calculate percentages and find the outcome of a given percentage increase or decrease.
- Use the law of arithmetic and inverse operations.
- Solve word problems and investigate in the context of number.
- Solve more complex problems by breaking them into smaller steps or tasks.

Knowing about fractions helps you to manage your money.

Before you start

You should know how to ...

1 Recognise when simple fractions, percentages and decimals are equivalent.

2 Cancel fractions to their lowest terms.

3 Know how to convert a fraction to a decimal using a calculator.

Check in

1 Write the equivalent fraction and percentage for:
a 0.24 **b** 0.3 **c** 0.02

2 Write these fractions in their simplest form:
a $\frac{2}{14}$ **b** $\frac{3}{12}$ **c** $\frac{4}{60}$

3 Convert $\frac{3}{7}$ to a decimal using a calculator. Write the display in full.

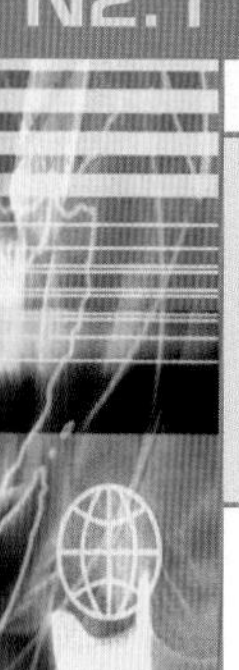

N2.1 Understanding fractions

This spread will show you how to:

- Use fraction notation to describe a proportion of a shape.
- Express a number as a fraction of another.
- Simplify fractions by cancelling.

KEYWORDS

Denominator Numerator
Fraction Simplify

This shape is made up of eight equal-sized cubes.

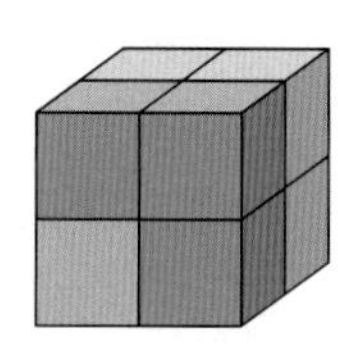

Three cubes are red.
$\frac{3}{8}$ of the shape is red.
$\frac{5}{8}$ of the shape is not red.

$\frac{3}{8}$ and $\frac{5}{8}$ are fractions.

- **A fraction describes part of a whole.**

The top number shows how many parts you have

The bottom number shows how many parts there are

$\frac{2}{3}$

The top number is the numerator

The bottom number is The denominator

Remember: the whole must be divided into equal-sized parts.

- **You can write a fraction in its simplest form by cancelling.**
  You divide the numerator and denominator by the highest common factor.
  If there is no common factor you say the fraction is in its lowest terms.

example

What fraction of:

a 120 kg is 90 kg? **b** £1 is 35p? **c** the months of a year contain the letter l?

a 90 kg as a fraction of 120 kg is $\frac{90}{120}$.
90 and 120 have a highest common factor of 30 so
$\frac{90}{120} = \frac{3}{4}$
90 kg is $\frac{3}{4}$ of 120 kg.

b 35p as a fraction of £1 is $\frac{35}{100}$.
35 and 100 have a highest common factor of 5 so
$\frac{35}{100} = \frac{7}{20}$
35p is $\frac{7}{20}$ of £1.

c The months of the year with an l in them are: April and July so there are 2 of them.
2 as a fraction of 12 is $\frac{2}{12} = \frac{1}{6}$
$\frac{1}{6}$ of the months of the year have an l in them.

Notice that the units have to be the same before you can compare them using fractions.

Exercise N2.1

1 **a** What fraction of this rectangle is shaded?

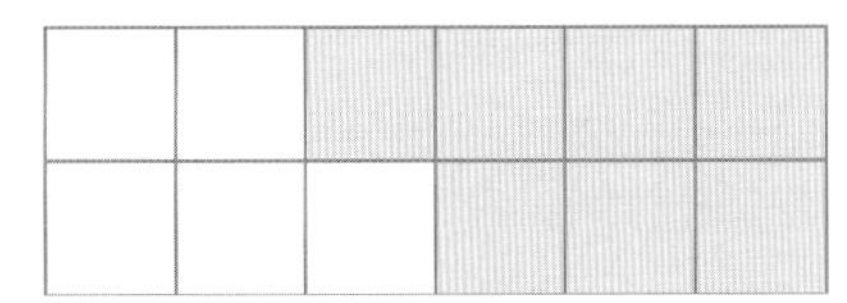

b What fraction of the rectangle is unshaded?

c What fraction of the shaded part is the unshaded part?

2 Write down the fraction of each shape that is shaded in its simplest form.

a

b

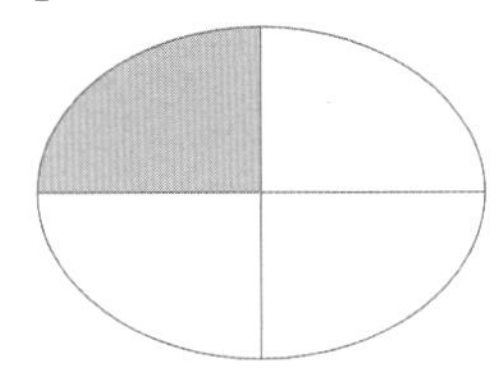

c

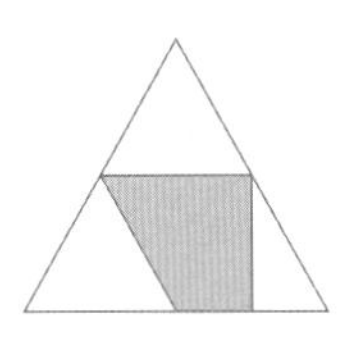

d

e

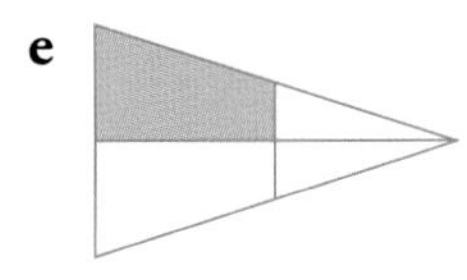

3 Copy these shapes and shade the fraction indicated.

a

Shade $\frac{3}{10}$

b

Shade $\frac{3}{32}$

4 A pair of trainers normally costs £80. During a sale they are reduced in price by £20. By what fraction has the price been reduced in the sale?

5 What fraction of a turn does a minute hand turn through between:

a 6:10 pm and 6:25 pm

b 6:25 pm and 7:15 pm

c 7:15 pm and 8:10 pm

d 9:12 pm and 10:08 pm

e 10:24 pm and 11:58 pm?

6 This pie chart shows the results of a vote to choose Head Girl in a school.

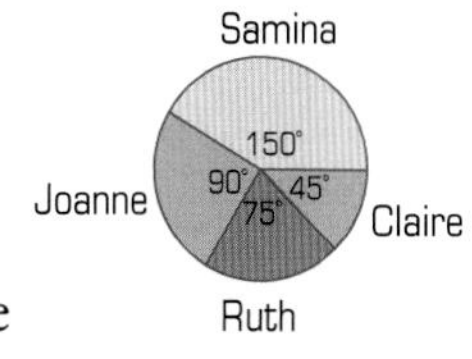

What proportion of the voters chose:

a Joanne

b Samina

c Claire

d Ruth

7 What fraction of:

a 1 year is March **b** 3 km is 900 cm

c £2.50 is 30p **d** 4 yards is 2 feet

e 8 kg is 1400 g **f** February is 1 week

g 1 day is 20 minutes?

Express your answers as fractions in their lowest terms.

8 What fraction of the whole square is each piece of this tangram?

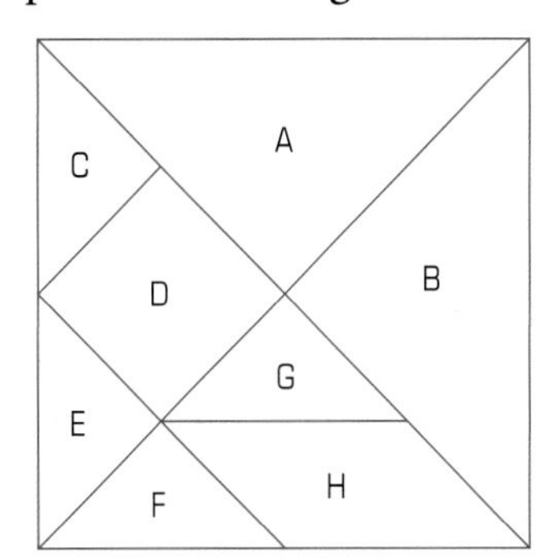

N2.2 Equivalent fractions

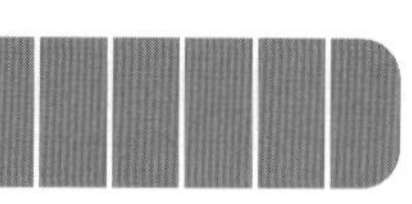

This spread will show you how to:

- Simplify fractions by cancellation and recognise equivalent fractions.
- Compare and order fractions.
- Add and subtract simple fractions.

KEYWORDS

Numerator
Denominator
Equivalent
Lowest terms
Simplify

$\frac{3}{8}$ of this rectangle is shaded

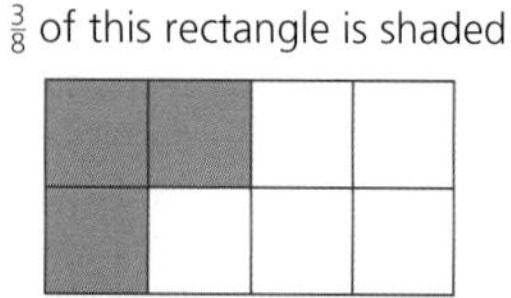

... that's the same as $\frac{6}{16}$

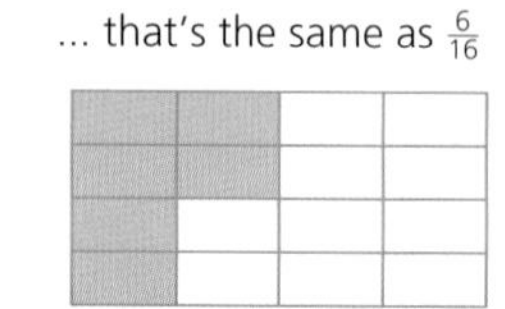

... and $\frac{12}{32}$

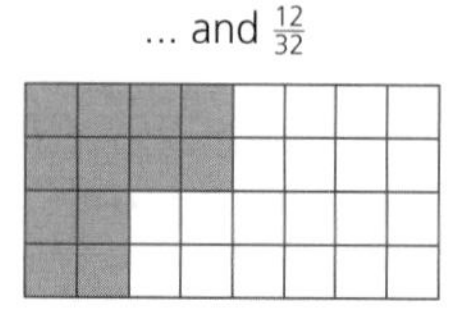

The fractions $\frac{3}{8}$, $\frac{6}{16}$ and $\frac{12}{32}$ are all equivalent. You can write: $\frac{3}{8} = \frac{6}{16} = \frac{12}{32}$

- **You can find equivalent fractions by multiplying the numerator and denominator by the same number.**

example

Find three equivalent fractions for $\frac{2}{3}$.

Multiply by 2:

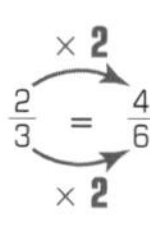

$\frac{2}{3} = \frac{4}{6}$ (× 2)

Multiply by 3:

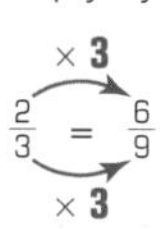

$\frac{2}{3} = \frac{6}{9}$ (× 3)

Multiply by 4:

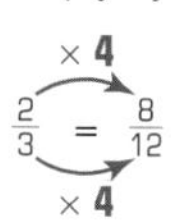

$\frac{2}{3} = \frac{8}{12}$ (× 4)

You can compare and order fractions with the same denominator.

example

Which is bigger and by what fraction?

a $\frac{1}{3}$ or $\frac{2}{5}$ **b** $\frac{3}{4}$ or $\frac{4}{5}$

a $\frac{1}{3} = \frac{5}{15}$ (× 5) and $\frac{2}{5} = \frac{6}{15}$ (× 3)

$\frac{2}{5}$ is bigger by $\frac{1}{15}$

b $\frac{3}{4} = \frac{15}{20}$ (× 5) and $\frac{4}{5} = \frac{16}{20}$ (× 4)

$\frac{4}{5}$ is bigger by $\frac{1}{20}$

When fractions have the same denominator you can add or subtract them.

example

Find in its simplest terms:

a $\frac{1}{6} + \frac{1}{6} + \frac{1}{6} + \frac{1}{6}$ **b** $\frac{3}{8} + \frac{1}{8}$ **c** $\frac{1}{2} + \frac{1}{4}$

a $\frac{1}{6} + \frac{1}{6} + \frac{1}{6} + \frac{1}{6}$
$= \frac{4}{6} = \frac{2}{3}$

b $\frac{3}{8} + \frac{1}{8}$
$= \frac{4}{8} = \frac{1}{2}$

c $\frac{1}{2} + \frac{1}{4}$
$= \frac{3}{4}$

Remember: You write a fraction in its lowest form or simplest terms by cancelling the highest common factor.

Exercise N2.2

1 Rearrange the letters into rows which contain equivalent fractions to find the names of 4 towns.

$\frac{5}{20}$ Y	$\frac{24}{39}$ R	$\frac{8}{18}$ A	$\frac{15}{35}$ E	$\frac{36}{81}$ T	$\frac{40}{90}$ S
$\frac{60}{135}$ N	$\frac{6}{14}$ L	$\frac{48}{108}$ M	$\frac{80}{130}$ C	$\frac{3}{12}$ O	$\frac{4}{9}$ E
$\frac{1}{4}$ R	$\frac{40}{65}$ E	$\frac{28}{63}$ E	$\frac{76}{171}$ H	$\frac{8}{13}$ W	$\frac{3}{7}$ S
$\frac{30}{70}$ D	$\frac{44}{99}$ C	$\frac{20}{80}$ K	$\frac{64}{104}$ E	$\frac{39}{91}$ E	$\frac{80}{180}$ R

2 Cancel the following fractions to their simplest form, expressing your answer as a mixed number where appropriate.

a $\frac{64}{320}$ **b** $\frac{2080}{1320}$ **c** $\frac{243}{720}$

3 For each pair of fractions insert a >, a < or an = sign.

a $\frac{2}{3}$ $\frac{3}{4}$

b $\frac{7}{12}$ $\frac{3}{4}$

c $\frac{8}{12}$ $\frac{2}{3}$

d $\frac{4}{5}$ $\frac{11}{15}$

e $3\frac{17}{24}$ $\frac{108}{29}$

f $\frac{35}{65}$ $\frac{53}{99}$

4 Use equivalent fractions to solve these problems:

a Find a fraction which is bigger than $\frac{2}{3}$ but less than $\frac{3}{4}$.

b Find a fraction which is halfway between $\frac{1}{6}$ and $\frac{1}{3}$.

c Say which is bigger: $\frac{7}{12}$ or $\frac{5}{9}$, and by how much.

d Find how many $\frac{1}{12}$s there are in $\frac{2}{3}$.

5 Find in its simplest terms:

a $\frac{1}{10} + \frac{3}{10}$

b $\frac{1}{12} + \frac{7}{12}$

c $\frac{1}{8} + \frac{3}{8} + \frac{5}{8} + \frac{7}{8}$

d $1\frac{1}{3} - \frac{2}{3}$

e $2\frac{1}{4} - 1\frac{1}{2}$

f $2\frac{1}{5} - \frac{4}{5}$

g $\frac{3}{5} + \frac{1}{10}$

h $\frac{2}{3} - \frac{1}{6}$

6 Rearrange these fractions in order starting with the smallest:

a $\frac{3}{5}$, $\frac{7}{20}$, $\frac{4}{15}$ and $\frac{17}{12}$

b $\frac{5}{8}$, $\frac{11}{15}$ and $\frac{15}{21}$

c $\frac{2}{7}$, $\frac{1}{3}$ and $\frac{4}{15}$

d $3\frac{2}{3}$, $\frac{36}{10}$, $3\frac{4}{7}$ and $\frac{28}{8}$

7 Find the fraction which is exactly halfway between $\frac{3}{11}$ and $\frac{5}{13}$.

8 What fraction is:

a 125 of 165

b 2 hours of 3 hours 15 mins

c 3 metres of 125 cm

d 130 eggs of 247 eggs

e 1 week of 483 840 seconds

f £2.75 of 425 pence?

Express your answers as fractions in their simplest form and as mixed numbers where appropriate.

9 **a** Find the largest number that will divide exactly into 216 and 424 (highest common factor).

b Use your answer to **a** to simplify the fraction $\frac{216}{424}$ to its lowest terms.

N2.3 Adding and subtracting fractions

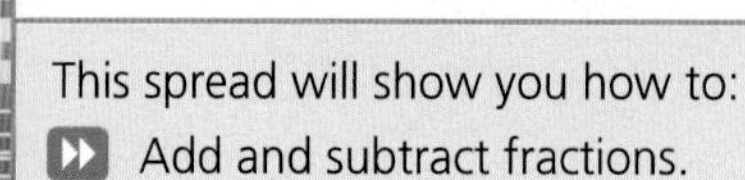

This spread will show you how to:

- Add and subtract fractions.

KEYWORDS

Convert, Equivalent, Denominator, Mixed number, Improper, Proper

You can use equivalent fractions to add fractions with different denominators together.
The diagram shows that:

$\frac{2}{3} + \frac{1}{4} = \frac{11}{12}$

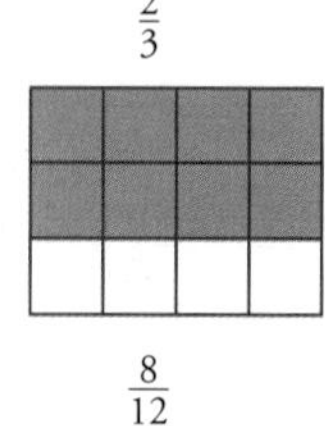

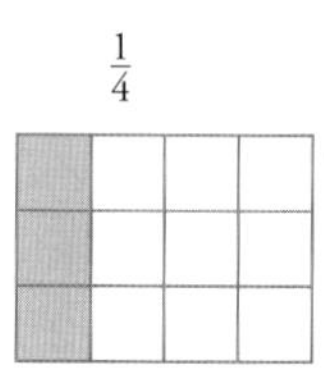

$\frac{8}{12} + \frac{3}{12} = \frac{11}{12}$

Notice that:
The multiples of 3 are 3, 6, 9, 12, 15, ...
The multiples of 4 are 4, 8, 12, 16, ...

12 is the lowest common multiple

example

Find:

a $\frac{2}{3} + \frac{1}{5}$

b $\frac{5}{6} - \frac{3}{4}$

a The multiples of 3 are 3, 6, 9, 12, 15, ...
The multiples of 5 are 5, 10, 15, ...
15 is the lowest common multiple.

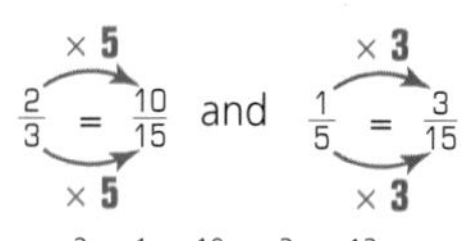

$\frac{2}{3} = \frac{10}{15}$ (× 5) and $\frac{1}{5} = \frac{3}{15}$ (× 3)

so $\frac{2}{3} + \frac{1}{5} = \frac{10}{15} + \frac{3}{15} = \frac{13}{15}$

b The multiples of 6 are: 6, 12, 18, 24, ...
The multiples of 4 are: 4, 8, 12, ...
12 is the lowest common multiple.

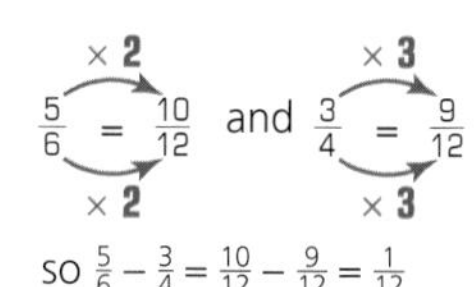

$\frac{5}{6} = \frac{10}{12}$ (× 2) and $\frac{3}{4} = \frac{9}{12}$ (× 3)

so $\frac{5}{6} - \frac{3}{4} = \frac{10}{12} - \frac{9}{12} = \frac{1}{12}$

- **To add or subtract fractions:**
  1 **Find equivalent fractions with a common denominator**
  2 **Add or subtract the numerators**

example

Find:

a $3\frac{1}{3} - 1\frac{1}{4}$

b $1\frac{1}{4} - \frac{2}{3}$

a $\frac{1}{3}$ is greater than $\frac{1}{4}$ so you can deal with the whole numbers and the fractions separately:
$3 - 1 = 2$ and $\frac{1}{3} - \frac{1}{4} = \frac{4}{12} - \frac{3}{12} = \frac{1}{12}$
so $3\frac{1}{3} - 1\frac{1}{4} = 2\frac{1}{12}$

b $\frac{1}{4}$ is smaller than $\frac{2}{3}$ so turn the mixed number into an improper fraction:
$1\frac{1}{4} - \frac{2}{3} = \frac{5}{4} - \frac{2}{3} = \frac{15}{12} - \frac{8}{12} = \frac{7}{12}$
so $1\frac{1}{4} - \frac{2}{3} = \frac{7}{12}$

Exercise N2.3

1 Work out the following leaving your answers as mixed numbers where appropriate.

a $\frac{3}{4}+\frac{3}{4}$ **b** $\frac{7}{3}-\frac{4}{3}$ **c** $2\frac{3}{8}-\frac{9}{8}$

d $1\frac{3}{7}+1\frac{5}{7}$ **e** $4\frac{11}{13}+3\frac{5}{13}$ **f** $5\frac{7}{23}-\frac{84}{23}$

g $12\frac{8}{17}-\frac{266}{17}$ **h** $23\frac{14}{21}+\frac{19}{3}$ **i** $3\frac{7}{8}-\frac{41}{8}$

j $\frac{77}{13}-(^{-}4\frac{2}{13})$

2 By converting to equivalent fractions identify which fraction in each pair is greater.

a $\frac{2}{3}$ or $\frac{5}{9}$ **b** $1\frac{3}{5}$ or $\frac{23}{15}$ **c** $\frac{4}{7}$ or $\frac{3}{5}$

d $2\frac{7}{12}$ or $\frac{30}{11}$ **e** $3\frac{9}{13}$ or $3\frac{11}{16}$ **f** $\frac{47}{12}$ or $\frac{15}{4}$

Which pairs of fractions were easier to compare? Explain why.

3 **Challenge** 'Mixed ferry crossings'.
Tom works for a ferry company. His job is to position the vehicles on the boat. His ferry can load 3 rows of vehicles.
Each row is a maximum of 10 m long.

Lane 1
Lane 2
Lane 3

← 10 m →

There are 11 vehicles waiting to use the ferry, and they have the following lengths in metres:

Car + Trailer	$5\frac{2}{3}$	$\frac{64}{15}$	$\frac{34}{9}$	$3\frac{3}{5}$
Car	$2\frac{4}{5}$	$2\frac{1}{3}$	$\frac{13}{6}$	$\frac{29}{10}$
Motorbike	$1\frac{1}{3}$	$\frac{27}{45}$	$\frac{25}{45}$	

Can Tom fit all the vehicles onto his ferry?

4 **Investigation** The fraction $\frac{5}{6}$ can be written as $\frac{1}{2}+\frac{1}{3}$.
The fraction $\frac{7}{10}$ can be written as $\frac{1}{3}+\frac{1}{5}+\frac{1}{6}$.

i Write these fractions as sums of different unit fractions (a unit fraction is a fraction with a numerator of 1):

a $\frac{5}{8}$ **b** $\frac{11}{12}$ **c** $\frac{9}{20}$

ii Investigate which other fractions can be written as the sum of unit fractions.

5 In each of these number patterns the empty boxes are filled by adding the numbers in the two preceding cells.

Example
2 2 ? ? ?
2 2 4 6 10

Copy and complete these number patterns.

a	2	$2\frac{1}{3}$		?	?	?
b	4	?	?	?	$17\frac{3}{4}$	
c	$\frac{2}{3}$	$\frac{3}{4}$	?	?	?	
d	$2\frac{3}{5}$		?	?	?	$1\frac{1}{5}$

6 Calculate the following:

a $2\frac{2}{3}+(^{-}4\frac{3}{5})$ **b** $(^{-}3\frac{4}{7})-4\frac{2}{3}$

c $\frac{11}{7}-(^{-}3\frac{4}{9})$ **d** $4\frac{2}{5}+3\frac{3}{7}-9\frac{2}{11}$

e $(^{-}5\frac{7}{8})-(^{-}3\frac{4}{5})+\frac{21}{17}$

f $\frac{-31}{12}-3\frac{3}{7}+(^{-}4\frac{12}{13})$

7 My bag weighs $1\frac{9}{20}$ kg. I put 3 books which each weigh $\frac{3}{5}$ kg into the bag. I put 2 folders which each weigh $\frac{3}{8}$ kg into the bag.

a What is the total weight of the three books?

b What is the total weight of the bag and all its contents?

c What fraction of the total weight of the bag are the books?

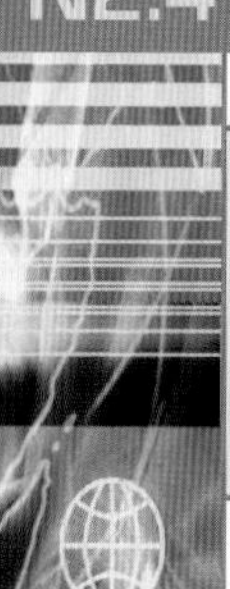

N2.4 Fractions and decimals

This spread will show you how to:

- Convert decimals to fractions.
- Convert fractions to decimals.
- Order fractions.

KEYWORDS

Decimal, Fraction, Terminating, Convert, Recurring

A fraction describes part of a whole. A decimal also describes part of a whole.
You can write a terminating decimal as a fraction like this:

A terminating decimal terminates or stops!

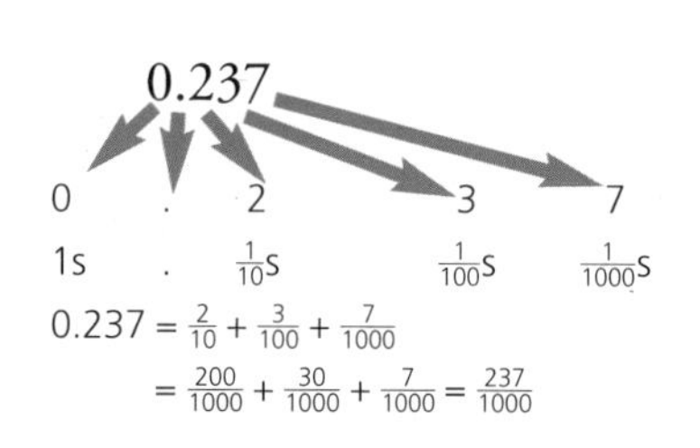

$0.237 = \frac{2}{10} + \frac{3}{100} + \frac{7}{1000}$
$= \frac{200}{1000} + \frac{30}{1000} + \frac{7}{1000} = \frac{237}{1000}$

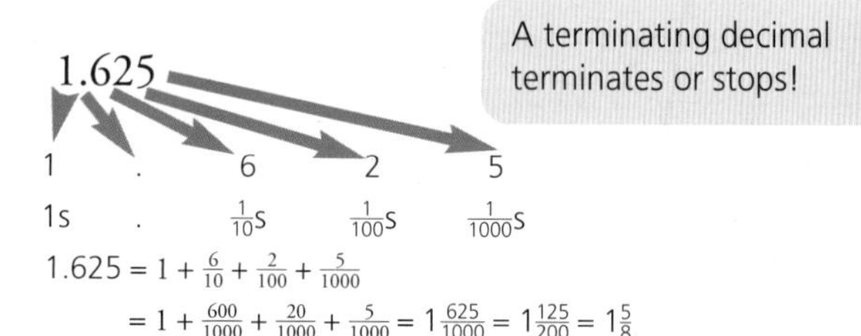

$1.625 = 1 + \frac{6}{10} + \frac{2}{100} + \frac{5}{1000}$
$= 1 + \frac{600}{1000} + \frac{20}{1000} + \frac{5}{1000} = 1\frac{625}{1000} = 1\frac{125}{200} = 1\frac{5}{8}$

▶ **You should know these common fractions and their decimal equivalents:**

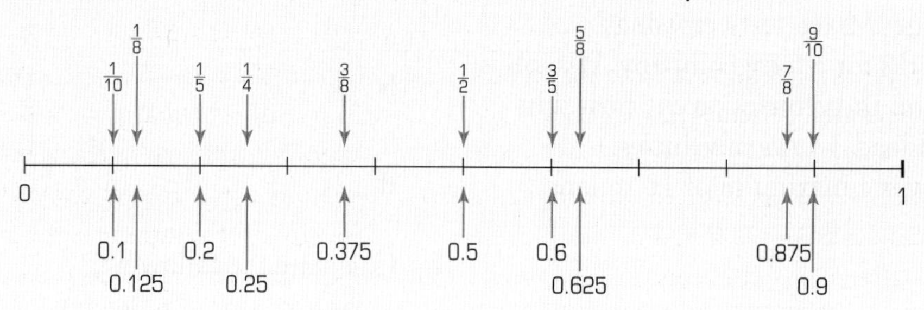

You can use these facts to convert other fractions to their decimal equivalents, for example: $\frac{2}{5} = 2 \times \frac{1}{5} = 2 \times 0.2 = 0.4$.

▶ **You can divide to convert a fraction to a decimal.**
For example $\frac{1}{2}$ means $1 \div 2 = 0.5$ and $\frac{3}{8}$ means $3 \div 8 = 0.375$

Remember that to find $\frac{1}{2}$ of a quantity you divide by 2.

It is often quicker to use decimals rather than fractions to compare fractional quantities.

example

Order these fractions, smallest first: $\frac{2}{3}$ $\frac{5}{6}$ $\frac{3}{4}$

Write them as decimals:
$\frac{2}{3} = 2 \div 3 = 0.66666666$
$\frac{5}{6} = 5 \div 6 = 0.83333333$
$\frac{3}{4} = 0.75$

The order is $\frac{2}{3}, \frac{3}{4}, \frac{5}{6}$

A decimal with a part that recurs is called a recurring decimal. All recurring decimals can be written as fractions.

Exercise N2.4

1 Convert these decimals to fractions in their simplest form, writing them as mixed numbers where appropriate.

a 0.625 **b** 0.125 **c** 0.345
d 0.732 **e** 1.215 **f** 3.7
g 11.42 **h** 7.848 **i** 5.725
j 2.432 **k** 8.75

2 Convert these fractions to decimals without using a calculator.

a $\frac{9}{12}$ **b** $\frac{11}{5}$ **c** $2\frac{6}{20}$ **d** $\frac{-72}{5}$
e $\frac{54}{15}$ **f** $\frac{176}{25}$ **g** $13\frac{18}{24}$ **h** $\frac{156}{40}$

3 Use short division to convert these fractions to their decimal equivalents.

a $\frac{3}{8}$ **b** $\frac{27}{8}$ **c** $4\frac{4}{5}$ **d** $\frac{17}{16}$

4 For each pair of fractions work out which one is greater. You will be able to do some of these mentally, others you might be able to do using a written method. Where the numbers are more difficult you may wish to use a calculator. Explain the method you have used for each question.

a $\frac{7}{8}$ or $\frac{4}{5}$ **b** $\frac{17}{8}$ or 2.2 **c** 1.4 or $\frac{23}{16}$
d $\frac{27}{19}$ or $\frac{53}{37}$ **e** $\frac{2}{3}$ or 0.6

5 These are the marks which Pella scored in each of her exams:
French $\frac{32}{50}$
German $\frac{13}{20}$
History $\frac{25}{40}$
Maths 66%
English $\frac{37}{57}$
Geography $\frac{42}{65}$

a In which subject did she do the best? Explain your answer.
b Put the marks for each subject in order, starting with the lowest first.

6 Put the following fractions in order of size from smallest to largest.

$\frac{1}{4}$, $\frac{6}{25}$, $\frac{2035}{8156}$, $\frac{3}{10}$, $\frac{26}{104}$, $\frac{147}{588}$

7 **a** Put these numbers in order from smallest to largest.

$\frac{23}{5}$, 4.62, $4\frac{11}{20}$, 4.7, $4\frac{5}{8}$

b Draw a number line and label each of these numbers as points on the number line.

8 **Investigation** (You will need a calculator)

a Convert the following unit fractions to decimals:
$\frac{1}{2}$, $\frac{1}{3}$, $\frac{1}{4}$, $\frac{1}{5}$, $\frac{1}{6}$, $\frac{1}{7}$, $\frac{1}{8}$, $\frac{1}{9}$, $\frac{1}{10}$
Write down what you notice.

b Try investigating what happens when you convert all the sevenths ($\frac{2}{7}$, $\frac{3}{7}$, $\frac{4}{7}$ and so on) from fractions to decimals
Write down what you notice.

c Choose some more fractions with a denominator greater than 10 (but say less than 20) and investigate what happens when you convert them to decimals. Try to write down what you have found out.

9 **Puzzle**

Janine wanted to convert a fraction to a decimal so she divided the numerator by the denominator.
Both numbers in her fraction were less than 40.
The answer came to 0.529411764 on her calculator. Can you find out the fraction with which she started?

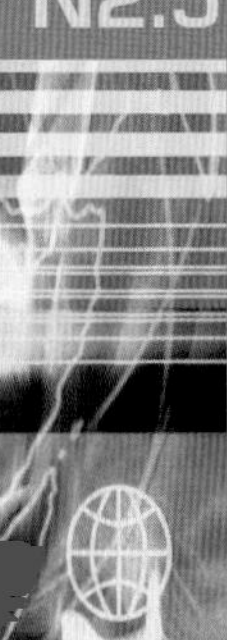

N2.5 Fractions of an amount

This spread will show you how to:

- Calculate fractions of numbers, quantities or measurements.
- Multiply a fraction by an integer or an integer by a fraction.
- Divide an integer by a fraction.

KEYWORDS

Denominator Numerator
Fraction Integer

This rectangle has an area of 20 cm^2:

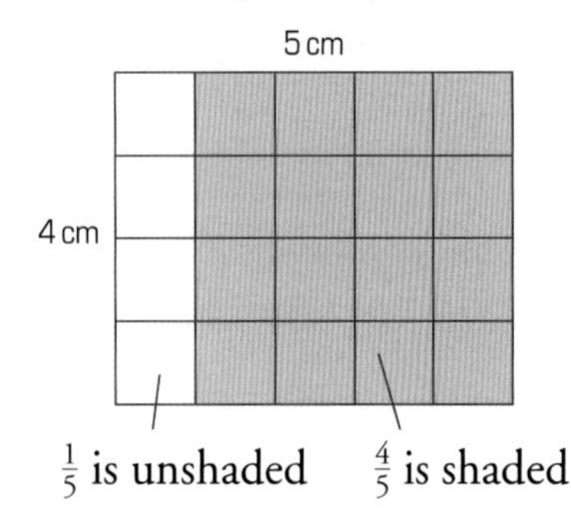

- **To find a fraction of an amount, multiply by the numerator and divide by the denominator.**
  For example, $\frac{4}{5}$ of $15 = 4 \div 5 \times 15 = 4 \times 15 \div 5 = 4 \times 3 = 12$

Sometimes it is better to multiply first then divide – it depends on the numbers.

example

Calculate

a $16 \times \frac{3}{5}$ **b** 4.4×30 **c** $\frac{3}{7} \times 15$

a $16 \times \frac{3}{5} = 16 \times 3 \div 5$
16 doesn't divide exactly by 5 so multiply first:
$48 \div 5 = 9\frac{3}{5}$ or 9.6

b $4.4 \times 30 = 4\frac{4}{10} \times 30$
$4\frac{4}{10} \times 30 = \frac{44}{10} \times 30$
$= 44 \div 10 \times 30$
$= 44 \times 30 \div 10$
$= 44 \times 3 = 132$

c $\frac{3}{7} \times 15 = 3 \div 7 \times 15$
3 doesn't divide exactly by 7 so multiply first:
$= 3 \times 15 \div 7$
$= 45 \div 7$
$= 6\frac{3}{7}$

You know that $3 \times 4 = 12$ can be expressed in many different ways:

	$3 \times 4 = 12$	or	$4 \times 3 = 12$	or	$12 \div 3 = 4$	or	$12 \div 4 = 3$.
Similarly,	$\frac{1}{4} \times 24 = 6$	so	$24 \times \frac{1}{4} = 6$	and	$6 \div \frac{1}{4} = 24$	and	$6 \div 24 = \frac{1}{4}$.

You can use these facts to divide an amount by a fraction.

example

Work out:

a $5 \div \frac{1}{3}$ **b** $24 \div \frac{1}{4}$

a Let $5 \div \frac{1}{3} = a$
Then $5 = a \times \frac{1}{3}$
$\frac{1}{3} \times 15 = 5$ so $5 \div \frac{1}{3} = 15$

b Let $24 \div \frac{1}{4} = b$
Then $24 = b \times \frac{1}{4}$
$\frac{1}{4} \times 96 = 24$ so $24 \div \frac{1}{4} = 96$.

Notice that when you divide by a fraction less than 1, the answer gets larger.

Exercise N2.5

1 Sajid earns £600 per month.
He spends his money as follows:

Clothes	$\frac{2}{8}$
Food	$\frac{3}{30}$
Rent	$\frac{1}{5}$
Social	$\frac{7}{60}$
Other spending	£200

a Calculate how much money Sajid spends each month on Clothes, Food, Rent and Social.
b What fraction of his earnings does Sajid spend on 'Other spending'?
c What fraction of his earnings does Sajid spend on Food and Rent?

2 Calculate $\frac{4}{7}$ of the area of this rectangle.

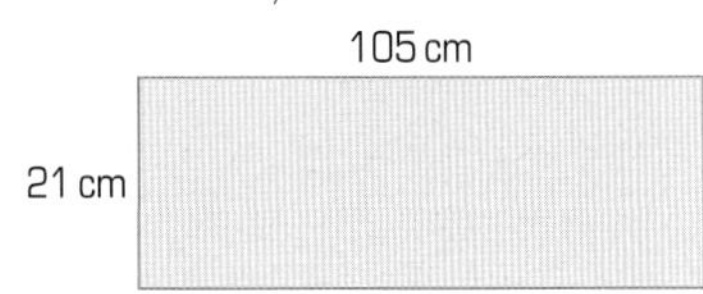

3 Calculate:
a $\frac{3}{5}$ of 17 m　**b** $\frac{3}{4}$ of £23
c $\frac{5}{8}$ of 122 cm　**d** $\frac{13}{8}$ of 140 km
e $4\frac{2}{5}$ of 32 hours　**f** $\frac{17}{25}$ of 108 kg
g 1.25 of 63 grammes　**h** 0.8 of 203°

4 Calculate these fractions of amounts.
a $\frac{7}{12}$ of 3 hours　**b** $\frac{7}{36}$ of 5 days

5 Calculate:
a $\frac{2}{3}$ of 140 cm　**b** $\frac{4}{7}$ of £80
(Give your answers to 2 decimal places)

6 Calculate the following:
a $18 \times \frac{5}{6}$　**b** $42 \times \frac{3}{7}$
c $\frac{4}{9} \times 72$　**d** $\frac{3}{8} \times 56$
e $7 \times \frac{8}{9}$　**f** $\frac{13}{5} \times 40$
g $\frac{5}{6} \times 42$　**h** $13 \times \frac{4}{5}$
i $\frac{3}{7} \times 23$　**j** $\frac{15}{8} \times 72$

7 Work out $2\frac{1}{4} \times 36$ cm.

8 Calculate, leaving your answers as mixed numbers where appropriate:
a $33 \times \frac{8}{11}$　**b** $48 \times 2\frac{1}{8}$
c $\frac{17}{19} \times 38$　**d** $\frac{3}{7} \times 15$
e $\frac{5}{12} \times 4$ hours　**f** $\frac{11}{36}$ of 2 days
g $\frac{4}{9} \times 1$ week　**h** $\frac{2}{3} \times \frac{3}{4} \times \frac{2}{5} \times 600$

9 Calculate the following:
a How many thirds are there in 8?
b Express 8 as an improper fraction with a denominator of 3.
c $8 \div \frac{1}{3}$
d $24 \div \frac{1}{4}$
e $3 \div \frac{1}{5}$
f $7 \div \frac{1}{10}$

10 Calculate the following:
a $8 \div \frac{1}{5}$　**b** $8 \div \frac{2}{5}$
c $8 \div \frac{3}{5}$　**d** $8 \div \frac{7}{5}$
Write down what you notice.

11 Calculate the following:
a $7 \div \frac{2}{3}$　**b** $13 \div \frac{3}{5}$
c $12 \div \frac{4}{7}$　**d** $23 \div \frac{5}{8}$
e $6 \div \frac{11}{5}$　**f** $9 \div \frac{7}{3}$

12 **Challenge:** 'Ben's weight'
Ben weighed 4 kg at birth.
On his fifth birthday Ben weighed $30\frac{3}{8}$ kg.
Ben's weight increased by the same fraction each year.
Find by what fraction Ben's weight increased each year.

13 **Investigation**
Consider the fraction $\frac{x}{y}$ where x and y are consecutive integers.
For example: $\frac{1}{2}, \frac{2}{3}, \frac{3}{4}, \ldots \frac{11}{12}, \ldots$
Convert each fraction to a decimal to investigate the pattern of numbers.

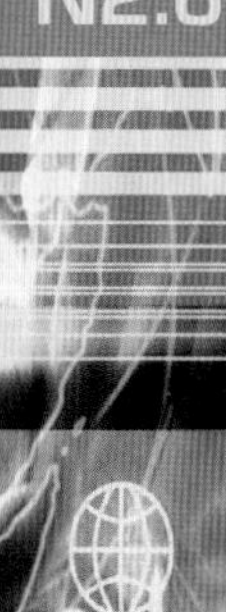

N2.6 Fractions, decimals and percentages

This spread will show you how to:

- Understand percentage as the operator 'so many hundredths of'.
- Recognise the equivalence of fractions, decimals and percentages.
- Calculate percentages of numbers, quantities and measurements.
- Find the outcome of a given percentage increase or decrease.

KEYWORDS

Decimal	Equivalent
Fraction	Increase
Percentage	Decrease

Another way of describing part of a whole is by using a percentage.
Percent means 'parts per hundred' and is written %.

- A percentage is a fraction written as a number of parts per 100.
  For example 60% = $\frac{60}{100}$ and 32% is $\frac{32}{100}$.

You convert a percentage to a fraction by writing it out of 100 then cancelling down.

example

Write as a fraction:

a 60% **b** 24% **c** 17.5%

a 60% means $\frac{60}{100}$
$\frac{60}{100} = \frac{6}{10} = \frac{3}{5}$

b 24% means $\frac{24}{100}$
$\frac{24}{100} = \frac{6}{25}$

c 17.5% means $\frac{17.5}{100}$
$\frac{17.5}{100} = \frac{175}{1000} = \frac{7}{40}$

You can use fractions to calculate simple percentages of amounts mentally.

example

a Calculate 10% of £270 **b** What is 75% of 240 sheep?

a 10% of an amount is the same as $\frac{1}{10}$ of it
10% of £270 = $\frac{1}{10}$ of £270 = £27

b 75% of an amount is the same as $\frac{3}{4}$ of it.
25% of 240 sheep = 60 sheep
75% of 240 sheep = 3 × 60 sheep = 180 sheep.

You increase an amount by a percentage by adding the extra amount on.
You decrease an amount by a percentage by taking the extra amount away.

example

a Increase £42 by 10% **b** Decrease £35 by 20%

a 10% is $\frac{1}{10}$ so 10% of £42 is £4.20
Increase means add on:
£42 + £4.20 = £46.20

b 10% of £35 is $\frac{1}{10}$ of £35 = £3.50
so 20% is £7
Decrease means take away:
£35 − £7 = £28

Exercise N2.6

1 Copy and complete this table:

Fraction		$\frac{19}{5}$				$\frac{17}{22}$
Decimal	0.35				0.333 ...	
Percentage			137%	37.5%		

2 Calculate the following amounts. You may find some of these questions easier if you convert between fractions and percentages as appropriate.

a 30% of £450 **b** 12.5% of 808 pigs
c 11% of 45.6 m **d** 250% of 346 mice
e 5% of 740 people

3 Investigation
Is 20% of 30 the same as 30% of 20?
Investigate for other percentage questions of your own.
Try to explain your results.

4 a What fraction of 3.2 m is 200 cm?
b What percentage of 3.2 m is 200 cm?
c Explain how you would check your answers to see if they were correct.

5 Calculate the new cost after the price increase or decrease:

a A pair of shoes costs £95. They are reduced in price by 20%. What is the new price?
b A worker in a factory earns £240 per week. His pay is increased by 5%. What is his new pay?
c A meal in a restaurant costs £27. The service charge is 11%. What is the total cost of the bill?
d My telephone bill is £68 plus VAT at $17\frac{1}{2}$%. How much is my total bill?

6 Game
This is a game for two players.

Box A	$12\frac{1}{2}$%	15%	$17\frac{1}{2}$%	20%	25%
Box B	£300	£350	£400	£450	£500

The first player selects a percentage from Box A. This can be a percentage increase or decrease. Then an amount is selected from Box B.

The player calculates a new amount based upon the % increase/decrease they have chosen.
For example, a 15% increase on £400 would give a new amount of £460.

The players take it in turns to select a percentage and an amount. Each player has 4 turns. The winner is the player who has the closest total to £2000 after 4 turns.

7 Challenge
Peter does a calculation in class to work out the new price of something after a 30% increase.
The answer is £286.
What was the original price?

8 Investigation
A shopkeeper has to add 15% to the price of an item in his shop.
He uses a single multiplication to work out the new price of the item.
Investigate what he has done.
Describe the single calculation he uses and why his calculation works.

N2 Summary

You should know how to ...	Check out
1 Add and subtract fractions.	**1** Work out: a $\frac{2}{3}+\frac{2}{3}+\frac{1}{3}$ b $\frac{2}{3}+\frac{5}{8}$ c $\frac{9}{10}-\frac{1}{5}-\frac{1}{2}$ d $1\frac{6}{7}-\frac{1}{3}$
2 Calculate fractions of quantities.	**2** Calculate: a $\frac{1}{5}$ of £93 b $\frac{2}{25}$ of 600 g c $\frac{5}{12}$ of an hour d $1\frac{2}{3}$ of 360°
3 Calculate percentages of quantities.	**3** Use the most efficient method to calculate: a 20% of £9.50 b 85% of £90 c 34% of 90 kg d 123% of £86 e 99% of £17 f 206% of 350 g
4 Find the outcome of a given percentage increase or decrease.	**4** a Increase £30 by 15% b Decrease 400 m by 32% c Increase £60 by 10%. Increase your answer by 10%. Is this equivalent to increasing £60 by 20%?
5 Solve word problems in the context of number.	**5** In a school of 1200 students, 540 are boys. What percentage of the school are girls?

11 Statistics and probability

This unit will show you how to:

- Calculate statistics, including with a calculator.
- Recognise when it is appropriate to use the range, mean, median, and mode and, for grouped data, the modal class.
- Interpret tables, graphs and diagrams for both discrete and continuous data, and draw inferences that relate to the problem being discussed.
- Use the vocabulary of probability when interpreting the results of an experiment.
- Appreciate that random processes are unpredictable.
- Know that if the probability of an event occurring is p, the probability of it not occurring is 1–p.
- Find and record all possible mutually exclusive outcomes for single events and two successive events in a systematic way, using diagrams and tables.
- Estimate probabilities from experimental data.
- Solve word problems and investigate in the contexts of handling data and probability.
- Compare and evaluate solutions

Averages level out any differences in a set of data.

Before you start

You should know how to …	Check in
1 Arrange numbers in order.	**1** Sort these numbers in ascending order: **a** 24, 52, 12, 34, 19 **b** 2.9, 8.3, 5.65, 5.6, 2.82
2 Multiply and add numbers.	**2** Work out mentally: **a** 14×3 **b** 12×7 **c** $84 + 42$
3 Use the language associated with probability.	**3** Describe the chance that: **a** it will rain tomorrow **b** from a bag of ten lemon and ten orange sweets I choose an orange sweet.

Finding the average

This spread will show you how to:

- Find the modal class of a set of data.
- Find the median of a large set of data.
- Find the range of a set of data.

KEYWORDS

Data	Modal
Average	Frequency
Typical	Median
Mode	Range

An **average** is a single value that is typical of a set of data.
There are three different measures of average. Here are two of them:

- **The mode, or modal value, is the value in the data that occurs most frequently.**
- **The median is the value in the middle of the data when it is arranged in order.**

The **range** is a measure of spread: it describes how the data is spread out.

- **The range = the highest value – the lowest value**

example

Find the mode, median and range of these data sets:

a Shoe sizes: 3, 4, 6, 3, 7, 5, 4, 5, 4

b 19, 21, 20, 17, 22, 18, 28, 27

a arrange in order:
3, 3, 4, 4, 4, 5, 5, 6, 7
The mode is 4 as it occurs most often.
The median is the middle value 4.
The range is 7 – 3 = 4

b arrange in order:
17, 18, 19, 20, 21, 22, 27, 28
There is no mode as each value occurs just once.
The median is the middle value, half-way between 20 and 21.
The median is 20.5
The range is 28 – 17 = 11

- **In general, when there are n pieces of data, the median is the $\frac{1}{2}(n + 1)$th value.**

example

Find the mode and median of this data:

Marks out of 10	Frequency
5	2
6	6
7	9
8	10
9	7

The mode is 8 because this class has the highest frequency.
The total frequency is 34.
The median is the $\frac{1}{2}(34 + 1)$th value = 17.5th value.
The 17th value is 7 and the 18th value is 8.
So the median mark is 7.5.

The mode is always a data value and is useful when you want to know how common something is, for example when ordering stock for a shop.
The median is the middle value and so is useful when describing data in terms of for example, 'more than half of the people prefer ... '.

Exercise D1.1

1 For each set of data:

- Find the median, the mode and the range.
- Decide which is the best average to use and give a reason why.

a number of sweets in a packet: 9, 12, 7, 13, 11, 8, 11, 14

b height of plants in a greenhouse: 20 cm, 24 cm, 21 cm, 25 cm, 22 cm, 19 cm, 23 cm, 22 cm, 24 cm, 21 cm

c time taken to run 100 metres: 10.2 s, 9.8 s, 10.0 s, 10.4 s, 9.7 s, 10.5 s, 10.3 s, 9.9 s, 10.4 s, 10.3 s

d weight of bags on a plane: 45 kg, 52 kg, 48 kg, 45 kg, 53 kg, 52 kg, 48 kg, 50 kg, 47 kg, 55 kg

e length of TV programmes: 30 mins, 55 mins, 30 mins, 1 hour, 45 mins, 1 hour, 1 hour 15 mins, 30 mins

2 Find the median, the mode and the range of these sets of data:

a

Shoe size	Frequency
4	3
5	4
6	5
7	2

b

Height (cm)	Frequency
156	21
157	34
158	42
159	37
160	19

c

Weight (kg)	Frequency
42	1
43	2
44	5
45	5
46	4
47	3

3 For each set of data:

- make a frequency table
- use it to find the modal value and the median
- write down the range
- give an example of what the data could be measuring.

Your tables should look like this:

Value	Frequency

a 2, 4, 5, 6, 4, 3, 6, 5, 5, 3, 4, 6, 4, 6, 5, 4, 4, 3, 2, 7, 4, 5, 4, 6, 5, 4, 5

b 20 cm, 24 cm, 21 cm, 24 cm, 22 cm, 24 cm, 25 cm, 22 cm, 24 cm, 23 cm, 22 cm, 24 cm, 21 cm, 22 cm, 24 cm, 20 cm

c 45 kg, 52 kg, 48 kg, 45 kg, 53 kg, 52 kg, 48 kg, 51 kg, 48 kg, 55 kg, 52 kg, 49 kg, 52 kg, 48 kg, 50 kg

d 100, 102, 103, 100, 105, 100, 103, 106, 100, 103, 106, 100, 102, 102, 101, 100, 103, 100, 100, 103, 103, 103, 101, 102, 104, 105, 100, 102, 102, 104, 101

e 6, 7, 9, 10, 12, 7, 9, 8, 12, 7, 9, 8, 10, 8, 12, 8, 9, 10, 12, 9, 8, 12

D1.2 The mean

This spread will show you how to:

- Calculate the mean for a set of discrete data.
- Calculate statistics.

KEYWORDS

Mean, Sum, Average, Range

The third average is called the **mean**. It is what most people mean when they say average!

- **The mean of a set of data is the sum of all the values divided by the number of values.**

Unlike the median and the mode, the mean uses every piece of data.
It gives you an idea of what would happen if there were equal shares.

example

Find the mean of each set of data:

a Temperatures in °C:
13, 13, 11, 14, 17, 19, 18

b

Number of goals scored	Frequency
0	13
1	21
2	11
3	8
4	6

a The sum of the values is 105
The number of values is 7
The mean is $105 \div 7 = 15$
So the mean temperature is 15°C.

Notice that the mean is not always one of the values.

b The sum of the values is:

13×0 goals $= 0$ goals
$+ 21 \times 1$ goals $= 21$ goals
$+ 11 \times 2$ goals $= 22$ goals
$+ 8 \times 3$ goals $= 24$ goals
$+ 6 \times 4$ goals $= 24$ goals

So the sum of the values is: $0 + 21 + 22 + 24 + 24 = 91$
The number of values is the total frequency:
$13 + 21 + 11 + 8 + 6 = 59$
The mean number of goals scored is $91 \div 59 = 1.54$ to 2d.p.

You can use the mean and also the range to find missing data values.

Remember:
range = highest value – lowest value

example

Jane has six number cards. Four of them show 10.
The mean of all six cards is 8 and they have a range of 6.
What could the other two cards show?

The six cards have a mean of 8 so the total must be $6 \times 8 = 48$.
Four of the cards total 40 so the other two must total 8.
So the highest value is 10. The range is 6 so the lowest value must be 4.
The two cards must both be 4.

Exercise D1.2

1 Find the mean average for these sets of data:

a the number of matches in a box: 30, 26, 32, 24, 28

b the number of letters delivered to a house one week: 4, 1, 0, 3, 1, 3

c the length of TV programmes shown one evening: 30 mins, 30 mins, 55 mins, 5 mins, 30 mins, 30 mins, 1 hour 35 mins, 1 hour 55 mins

d the number of smarties in a box: 15, 17, 14, 15, 13, 15, 16, 17, 14, 15

2 Find the mean of each set of data:

a shoe size of a class of students

Shoe size	Frequency
2	2
3	3
4	5
5	10
6	8
7	4
8	1

b temperature one April

Temperature in °C	Frequency
12	6
13	4
14	6
15	5
16	3
17	4
18	2

c age of quiz show contestants

Age	Frequency
19	8
20	12
21	20
22	15
23	25
24	30

3 Use your knowledge of the mean to answer these questions:

a Packets of seeds contain an average 14 seeds per packet.
You open three packets. One contains 16 seeds, another holds 13.
How many seeds would you expect the third packet to hold?

b Find four different numbers that have an average of 11.

c A chocolate manufacturer claims there are an average of 30 chocolates contained in a box.
Five boxes are tested – the first four contain 27, 31, 30 and 28 respectively.
How many would you expect the fifth box to hold?

d A runner's time over 200 m averaged 21 seconds in four races.
In his first three races his time was recorded as 19.9 s, 21.4 s and 22.0 s.
What was his time in the fourth race?

4 Smith scored an average of 3 goals per match during 5 matches.
Jones scored an average of 4 goals per match during 4 matches.

a Who scored the most goals?

b Smith scored 4, 1, 2 and 3 in 4 matches.
How many goals did he score in the 5th match?

5 6 friends have a mean of £4 in £1 coins in their pockets. 4 of them have £3, £5, £2 and £2. The range of amounts is £5.
How much do the other 2 friends have?

6 Jane has 6 number cards. Four of them show 10. The mean of all 6 cards is 10 and the range is 6.
What could the other cards be?

Interpreting diagrams

This spread will show you how to:

- Interpret diagrams, graphs and charts, and draw inferences based on the shape of graphs and simple statistics for a single distribution.

KEYWORDS

Interpret | Statistics
Pie chart | Line graph

You often find statistics presented in diagrams.
You need to be able to read and understand the diagrams.

- A pie chart uses a circle to display data.

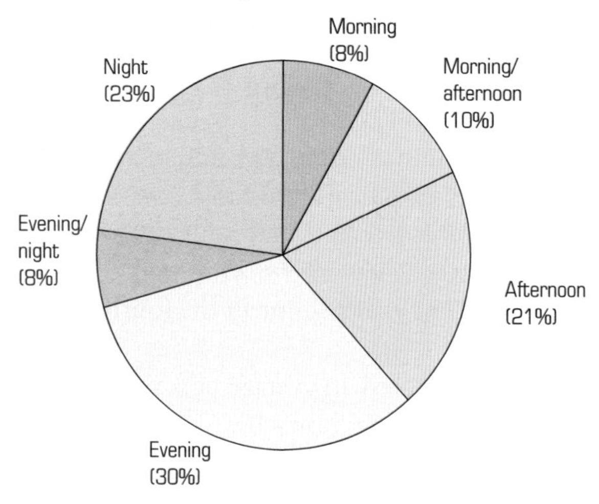

The pie chart shows that more burglaries were committed in the evening than at any other time of day.
If there were 200 notified burglaries, then 60 of these would have been committed in the evening.
If there were 24 morning burglaries, then 300 burglaries would have been notified in total.
8% represents 24 burglaries.
1% represents 3 burglaries,
So 100% is 300 burglaries.

- A line graph shows trends in data.

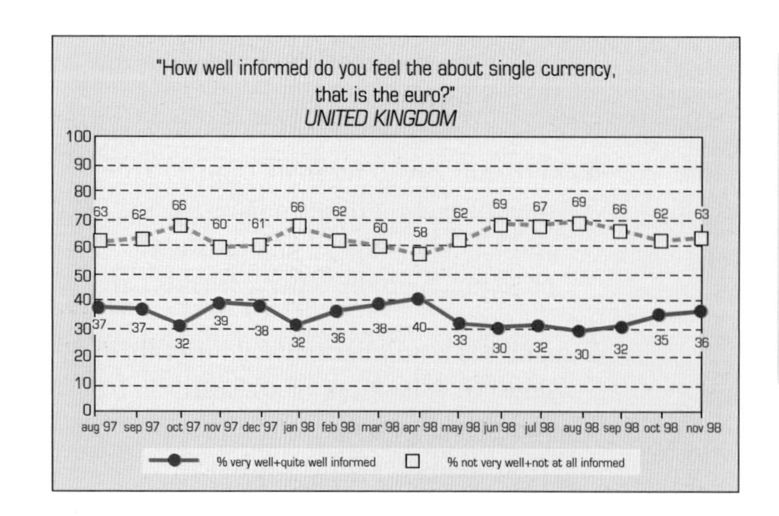

- The line graph shows that the time people felt most informed about the single currency was in April 1998, and least informed in June and August 1998.
- Notice that these dates correspond to the maximum and minimum percentages of people who did **not** feel well-informed. You can see this as the two lines almost mirror each other.

Exercise D1.3

1 The pie chart shows the results of a survey of schoolchildren to find out how they get to school.

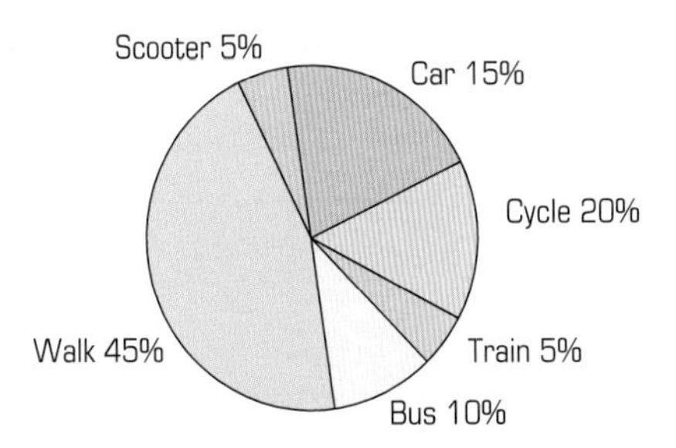

a If 135 children walked to school, how many children were involved in total?

b Find out the amounts of all the other categories, and draw a frequency table to show the results.

2 This bar chart shows the number of people using a cross-Channel ferry service during one particular week.

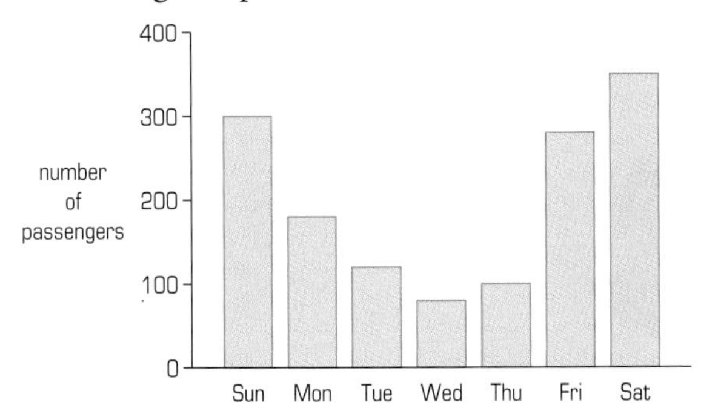

a On which day of the week was the service least used?

b On which day of the week was the service most used?

c Compare the total for Monday to Friday with the total for the weekend. Which total is the highest?

3 This bar chart shows a survey of the number of children per family.

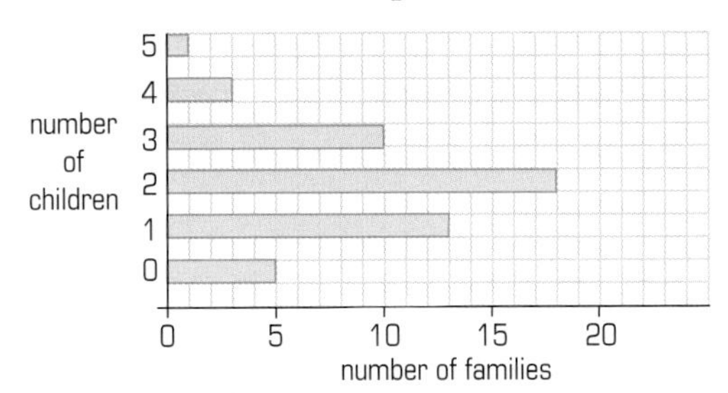

a How many families were involved in the survey?

b What is the modal number of children per family?

c Calculate how many children were involved in the survey.

4 This line graph shows how the population of a town has changed over a period of 50 years, recorded at 10-year intervals.

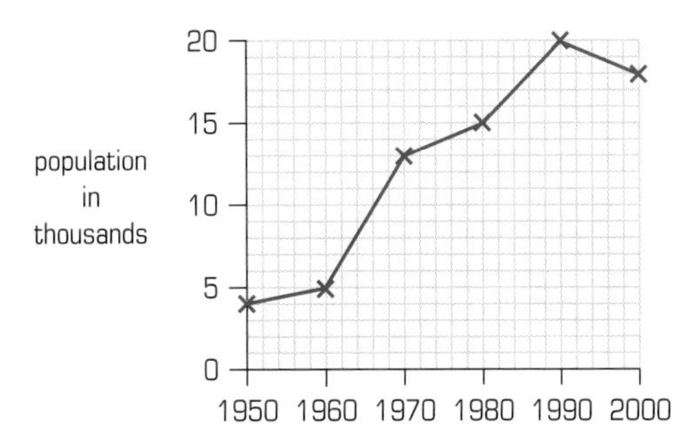

a How would you describe the change in population over the 50 year period? Try to include the word 'trend'.

b How much bigger was the population in 1990 than it was in 1950? Give your answer in thousands.

c Use the graph to estimate the population of the town in 1975.

d What might you expect the population to be in 2010? Why can this only be an estimate?

D1.4 Introducing probability

This spread will show you how to:

- Use vocabulary and ideas of probability.
- Understand and use the probability scale from 0 to 1.
- Know that if the probability of an event occurring is p then the probability of it not occurring is $1 - p$.

KEYWORDS

Probability, Impossible, Chance, Certain

The chance of an event happening is somewhere between certain and impossible:

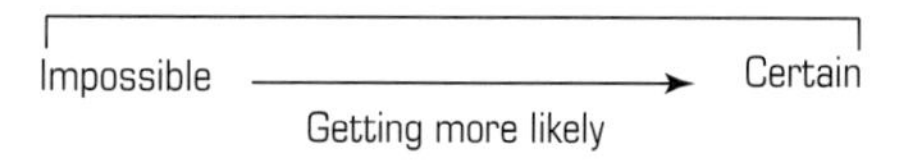

Note

The probability of an event can change in time. It is certain to be Friday tomorrow if it is Thursday today, but it is impossible if it isn't Thursday!

Probability is a measure of the chance of an event happening.

- **All probabilities are measured on a scale of 0 to 1.**

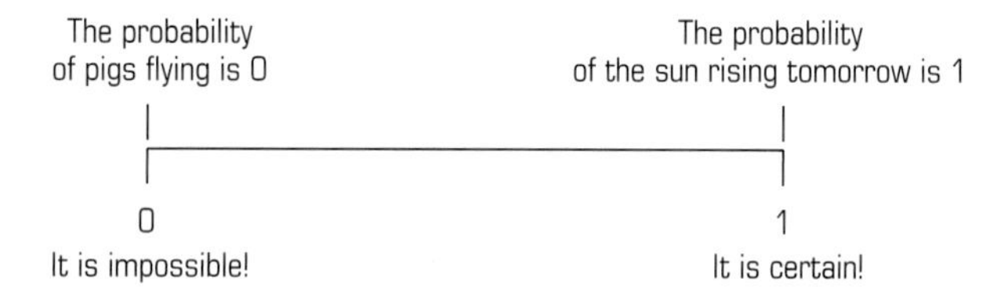

example

Six pupils enter a lucky draw at the school fete.

Their names are written on a blank dice and then the dice is rolled. The name at the top shows the winner.

Luke has a single chance of winning the first prize. The probability of Luke winning is $\frac{1}{6}$.

There are 5 other names on the dice. The probability of Luke **not** winning is $\frac{5}{6}$.

There are 2 girls' names in the hat. The probability of a girl winning is $\frac{2}{6}$ or $\frac{1}{3}$.

- **Probability of an event occurring = $\dfrac{\text{Number of ways the event can occur}}{\text{Total number of possible outcomes}}$**
- **If the probability of an event occurring is p, the probability of it not occurring is $1 - p$.**

Exercise D1.4

You can use any of these probability lines to show probabilities:

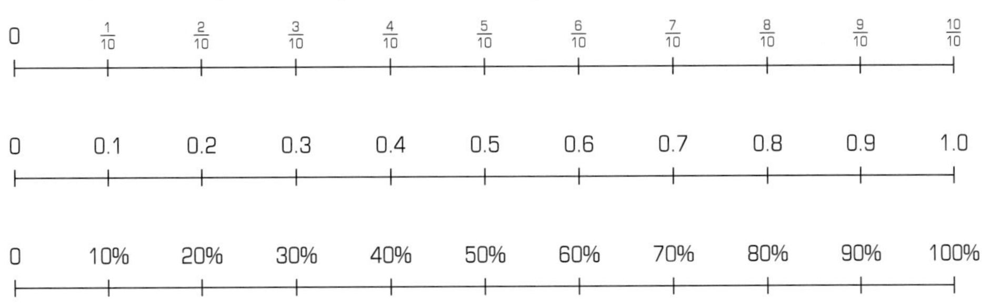

Choose the most appropriate probability line to show your answers to each of these questions.

1 The letters in the word ZEBRA are placed in a bag.
What is the probability of:
a picking an E
b picking an R
c picking a vowel
d not picking a vowel
e not picking an R?

2 A box of 10 chocolates contains:
3 caramel centres
1 nut centre
4 crème centres
1 caramel and nut centre
1 crème and nut centre
If you choose a chocolate at random, what is the probability that it will:
a have a crème centre
b have caramel in it
c not have a nut in it
d have just a nut centre
e have a nut or a caramel in it
f not have a nut or a caramel in it?

3 The letters of the word REINDEER are placed in a bag.
Draw an appropriate probability line and mark on it the probability of picking:
a a D **b** an R **c** an E **d** not an E **e** a vowel

4 A computer is programmed to produce a number from 1 to 40 at random.
What is the probability that the number produced is:
a greater than 10
b less than 20
c a multiple of 5
d not a multiple of 5
e a multiple of 2 and 7
f not a multiple of 7
g a prime number
h not a prime number?

D1.5 Calculating probabilities

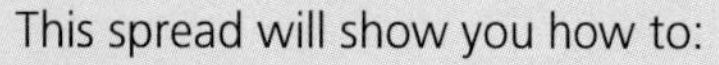
This spread will show you how to:

- Find and justify probabilities based on equally likely outcomes.
- Find and record all possible outcomes for single events and two successive events in a systematic way.

KEYWORDS

Probability Random
Outcome
Event

- Probability of an event occurring = $\frac{\text{Number of ways the event can occur}}{\text{Total number of possible outcomes}}$

example

Class 7S and class 7P each held a vote to elect a class captain.

In 7S five pupils, Edward, Hazel, Majid, Natalie and Steven, received equal votes. Their names were put into a hat and one chosen.

In 7P four pupils, Alfred, David, Josie and Ruby, received equal votes. Their names were put into a hat and one chosen.

Find the probability of choosing:

a Hazel **b** Two boys
c Natalie and Josie **d** Natalie or Josie.

You can show the possibilities in a table like this.

	Edward	Hazel	Majid	Natalie	Steven
Alfred	A & E	A & H	A & M	A & N	A & S
David	D & E	D & H	D & M	D & N	D & S
Josie	J & E	J & H	J & M	J & N	J & S
Ruby	R & E	R & H	R & M	R & N	R & S

If you list outcomes systematically, it is easier to find them all.

There are 20 possible outcomes altogether. Each outcomes is **equally likely**.
Each outcome has a $\frac{1}{20}$ change of occurring.

a There are **4** ways in which Hazel can be chosen. The probability of choosing Hazel is $\frac{4}{20}$.
b There are **6** ways in which two boys can be chosen. The probability of choosing two boys is $\frac{6}{20}$.
c There is only **1** way to elect Natalie and Josie. The probability of choosing Natalie and Josie is $\frac{1}{20}$.
d There are **8** ways to choose Natalie or Josie. The probability of choosing Natalie or Josie is $\frac{8}{20}$.

Exercise D1.5

1 A safe has a lock that uses the numbers 1 to 4 in different combinations. The owner remembers that the first digit is a 3 but can't remember any of the other numbers.

a How many different combinations are there?

b What is the probability that the owner will get it right first time?

2 At Happy Pizzas you can add two pizza toppings at no extra charge. The extra toppings are: Ham, Onion, Mushroom, Pepperoni and Olives.

a How many combinations of the two extra toppings could you choose?

b What is the probability of someone choosing Olives as an extra topping?

3 Mary wants to phone a friend but she can only remember the first three digits of the six-digit number: 362xxx.
She knows the other digits are 2, 9 and 7.

a List all the possible numbers it could be.

For example, it could be 362 279 or 362 297 or ...

Mary decides to try all the numbers until she gets it right.
The calls cost 35p each.

b How much will it cost if she has to try all the numbers?

What is the probability of her getting the number right:

c first time?

d after four tries?

e in three tries or less?

4 Geena has 5 coins in her purse:

She picks out two coins from her pocket at random.
List all the possible pairs or coins she can pick.

Hint:
When you have two coins the same they should be counted separately. You could call them A and B.

Use your list to find the probability of her picking:

a the 2p and a 20p

b a £1 coin

c a total of more than £1.

5 Simon and Anna are playing a game with two dice.
They take it in turns to roll the dice.

- Simon scores a point if the total score on the two dice is: 2, 3, 4, 10, 11 or 12
- Anna scores a point if the total score on the two dice is: 5, 6, 7, 8 or 9

The first to score 10 points wins the game.
Who is more likely to win the game?
Why?

D1.6 Experimental probability

This spread will show you how to:

- Collect data from an experiment and record in a frequency table.
- Estimate probabilities based on the data.

KEYWORDS

Probability, Experiment, Outcome, Biased, Estimate

To **calculate** the probability of an event you need to identify all the possible outcomes of the event and all the ways the event can occur:

- Probability of an event occurring = $\dfrac{\text{Number of ways the event can occur}}{\text{Total number of possible outcomes}}$

You can also **estimate** the probability of an event occurring using experimental data.

example

Peter and Lyn are carrying out an experiment to estimate the number of different coloured balls in a bag.

- They take it in turns to draw a ball from the bag.
- They note its colour and record it in a table.
- They then replace the ball in the bag.
- They repeat this 10 times.

Here are their results:

Colour	Tally	Frequency
Blue	///	3
Green	//	2
Red	////	4
Yellow	/	1

a Use the results to estimate the probability of drawing each colour ball from the bag.

The more times you repeat an experiment the more accurate your results should be.

The bag contains 20 balls in total.

b How many would you expect to be yellow?

a The total number of outcomes is 10.
The estimated probability of picking a:
blue is $\frac{3}{10}$
green is $\frac{2}{10}$
red is $\frac{4}{10}$
yellow is $\frac{1}{10}$

b If $\frac{1}{10}$ are estimated to be yellow, you would expect 2 balls to be yellow in a bag of 20.

Exercise D1.6

Experiment 1

- Put five different colour cubes in a bag then shake it.
- Predict what colour the first cube you draw out of the bag will be.
- Pick a cube from the bag at random (without looking). Look at the colour.
- Copy this table:

	Guesses			
	First	Second	Third	Fourth
1				
2				
3				
4				
5				
6				
7				
8				
9				
10				

- If your prediction was correct, put a tick in the first box. If not, put a cross.
- **Do not** put the cube back in the bag.
- Repeat until you have taken out four cubes.
- Repeat the experiment until you have filled the table.

a Draw a probability line from 0 to 1. Mark on it the probability of guessing the first cube correctly.

Write the probability as a fraction.

b Now mark the probability of guessing the second cube correctly.

c Mark the probability of guessing the third cube and the chance of guessing the fourth cube correctly.

d What is the chance of guessing the fifth cube correctly? Explain your answer.

Experiment 2 Test a dice to see whether it is fair or biased.

- Roll the dice 60 times. Note the score on the dice each time.
- Record your results in a frequency table.

Dice score	Frequency
1	
2	

a Which is the most likely score on your dice?

b According to your results, what is the probability of getting a 4 with your dice?

c What would you expect the frequency of each score to be if you had a fair dice?

d Do you think the dice is biased? Why?

e What could you do to check further?

f George rolls an ordinary dice three times and gets a six each time.

  i What is the probability that he will throw a six on his fourth throw?

  ii Do you think George's dice is fair? Give a reason for your answer.

Experiment 3

Devise experiments to test where:

A an outcome is certain

B an outcome is impossible

C an outcome has an even chance of occurring

D an outcome is unlikely but not impossible.

You will need to invent your own outcomes. Mark where the outcomes would go on a probability line.

Summary

You should know how to ...

1 Calculate statistics.

2 Represent problems and interpret solutions in graphical form.

3 use the vocabulary of probability.

4 Know that if the probability of an event occuring is p the the probability that the event does not occur is 1 – p.

5 Find and record all possible mutually exclusive outcomes in a systematic way.

Check out

1 Calculate the mean, median, mode and range of

a 6, 2, 3, 8, 9, 9, 5

b 20, 20, 21, 22, 69

2 Meteorologists measure cloudiness of the sky in eighths. An overcast sky is $\frac{8}{8}$ and a clear sky is $\frac{0}{8}$. The cloudiness of the sky was measured at noon each day for one month. The results are displayed on the graph.

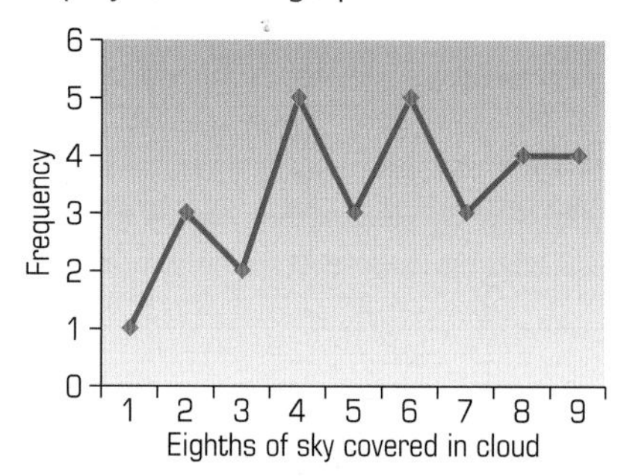

Which month do you think this graph represents? Give reasons for your answer.

3 Draw a probability scale

Impossible — Certain

and mark on it the events:

A: In ten years you will visit Mars.

B: It will rain tomorrow.

C: You will eat something today.

4 If the probability that you choose the winner of the grand national is 0.0135, what is the probability that you do not choose the winner?

5 'Choosy pizza' lets you choose your own pizza toppings from: Onion, Sweetcorn, Cheese, Mushroom, Peppercorn.
List all the possible combinations when you choose two toppings.

A2 Expressions and formulae

This unit will show you how to:

- Know that algebraic operations follow the same conventions and order as arithmetic operations.
- Simplify or transform linear expressions by collecting like terms.
- Use index notation for small positive integer powers.
- Use formulae from mathematics and other subjects.
- Substitute integers into simple formulae.
- Solve word problems and investigate in the context of algebra.
- Represent problems and interpret solutions in algebraic form.
- Use logical argument to establish the truth of a statement.
- Solve more complex problems by breaking them into smaller steps or tasks.

Las pesas que equilibran estas dos balanzas valen lo mismo.

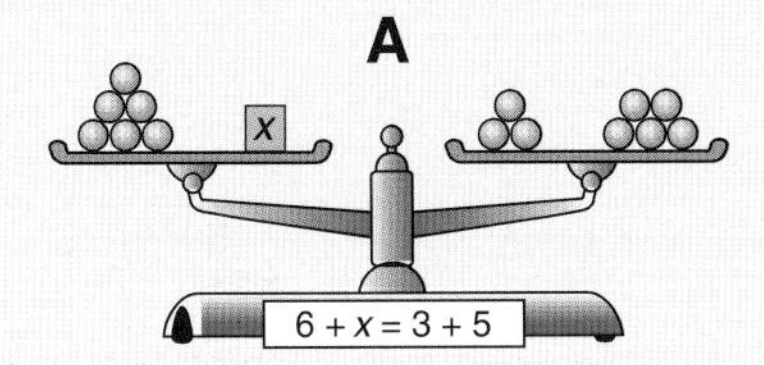

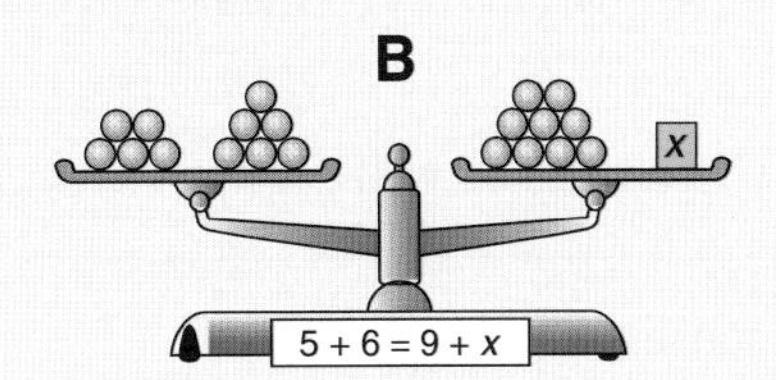

La solución de ambas ecuaciones es $x = 2$. En estos casos se dice que las ecuaciones son equivalentes.

Algebra is the same the world over.

Before you start

You should know how to ...

1 Use letter symbols to stand for numbers.

2 Calculate with negative numbers.

3 Square numbers.

Check in

1 Write the following statements using algebra.

a I think of a number x, multiply it by four and add 8.

b I think of a number y, subtract three and divide by nine.

c I think of a number z, and multiply it by itself.

2 Evaluate these in 1 minute:

a $^{-}3 + 5$ **b** $^{-}4 - ^{-}7$ **c** $^{-}5 \times ^{-}3$ **d** $^{-}15 \div 5$

e $5 + ^{-}4$ **f** $^{-}6 \times 7$ **g** $100 \div ^{-}4$ **h** $6 - 13$

3 Square each of these numbers:

a 4 **b** 12 **c** 15 **d** $\frac{1}{2}$ **e** 0.3 **f** $^{-}5$

A2.1 Using letter symbols

This spread will show you how to:

- Recognise the way multiplication and division are represented in algebraic expressions.
- Recognise that algebraic operations follow the same conventions and order as arithmetic operations.
- Use index notation for small positive integer powers.

KEYWORDS

Evaluate
Expression
Substitute
Symbol
Variable

Algebra is the language that mathematicians use to communicate with each other.

There are a few conventions or rules that you must always follow:

- **Never use the multiplication '×' sign:** **write *ab* not $a \times b$ so you don't confuse '×' with the letter *x*!**
- **When you multiply write numbers first then letters: 5*r* not *r*5.**
- **Never use the division sign: write $\frac{x}{10}$ not $x \div 10$**

Hint: you say 5 rabbits (5*r*) not rabbits 5 (*r*5)!

You can use these rules to write long sentences in a shorter way:

example

Write this sentence using algebra:
I think of a number, add 7, multiply by 5 then divide by 4.

Build up the expression in stages:

Think of a number, add 7: $n + 7$

Multiply by 5: $5(n + 7)$

Divide by 4: $\frac{5(n+7)}{4}$

The bracket means all multiplied by 5.
The line means all divided by 4.

Here are some more expressions and their sentences:
$5p - 4$ means *I think of a number, multiply it by 5, subtract 4.*
$15 - w$ means *I think of a number and subtract it from 15.*

Algebra is much easier if you read it. Always begin with 'I think of a number ...'

- **You can substitute a value into an expression to evaluate it.**

example

Evaluate $\frac{3x+7}{4}$ when $x = 3$.

The expression says *I think of a number, multiply it by 3 add 7 and then divide it all by 4.*
$3 \times 3 = 9 \quad 9 + 7 = 16 \quad 16 \div 4 = 4$
So the value of the expression when $x = 3$ is 4.

Exercise A2.1

1 **The Dice Run** – a game for 2 to 4 players. You need a dice.

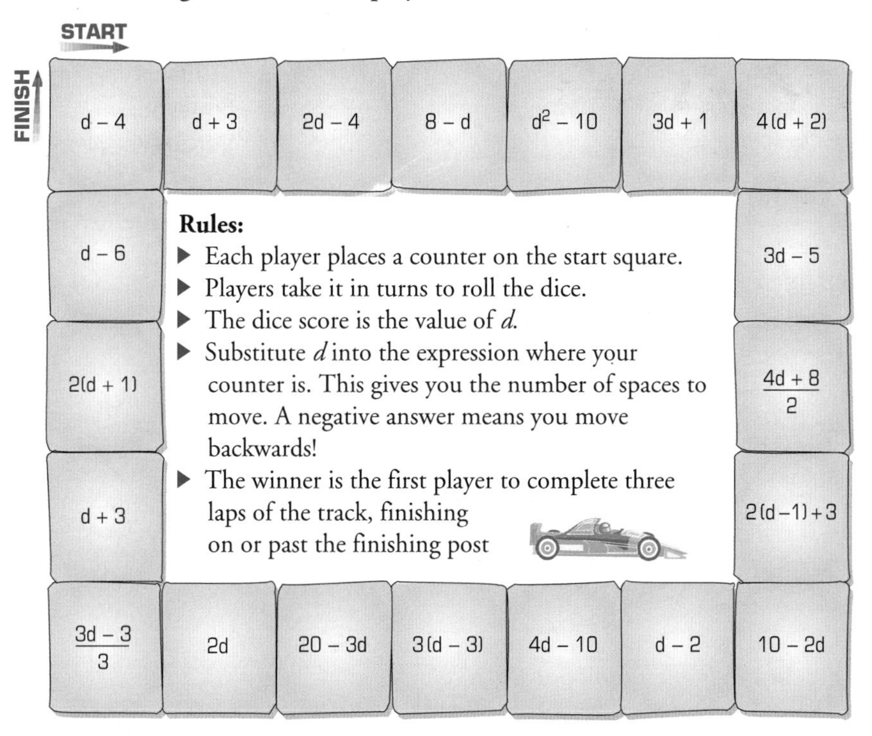

2 These cards show expressions based on the value of d.
d stands for the number showing on a dice.

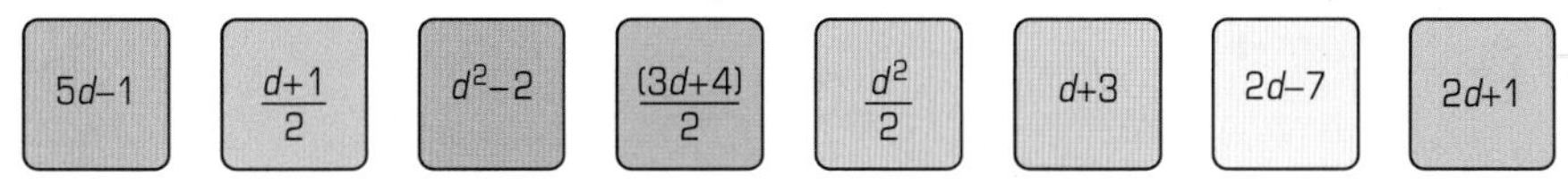

a Roll a dice to find a value for d. Use your value to find two cards from the set that give the same answer. Repeat for other values of d.

b Which card and which value of d give the smallest value? The largest value?

3 **Think of a number ...**

▶ Think of a number ... call it a. Think of a different number ... call it b.

▶ Which pairs of cards are always the same whatever values you choose for a and b?

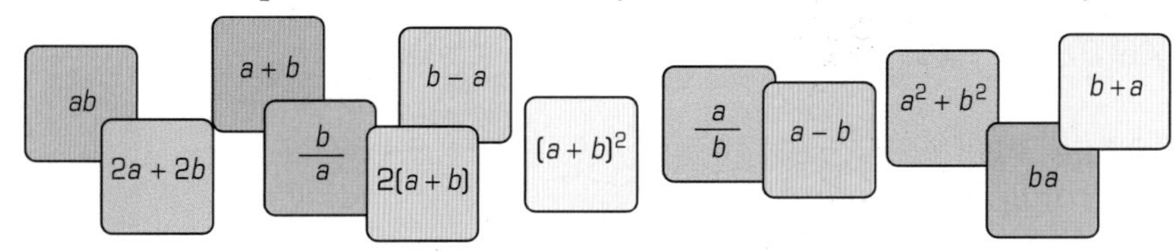

4 Find as many values for each of the unknowns as you can:

a $p^2 = 25$ **b** $3m = 12$ **c** $x + y = 8$

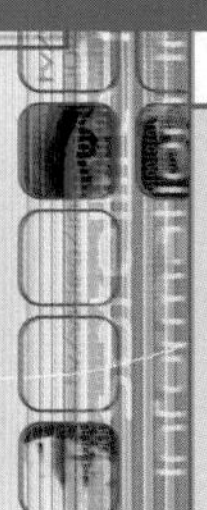

A2.2 Using the rules of algebra

This spread will show you how to:

- Know that an expression is formed from letter symbols and numbers.
- Use the equals sign appropriately and correctly.
- Explore general algebraic relationships.

KEYWORDS
Expression
Predict
Vertex/Vertices

You can use algebraic expressions to explain puzzles and solve problems.

This TOAN (think of a number) snake always takes you back to the number you started with.

If you use algebra you can explain why.

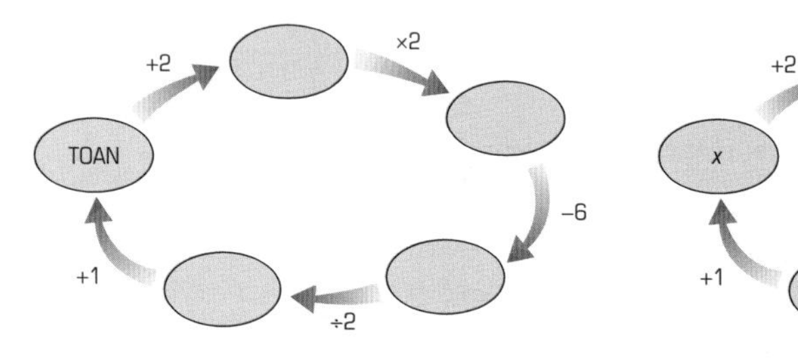

▶ **You can use algebra to make predictions.**

example

Take a triangle with any three numbers at the vertices: → Add the numbers together to make a side total → Add the side totals to make an overall total:

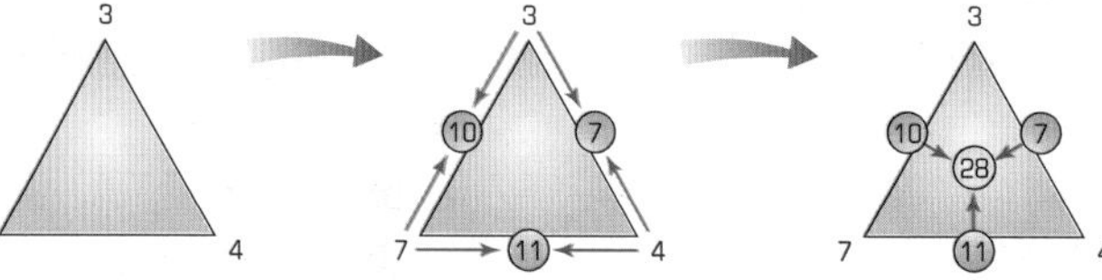

What do you notice about the total?
Can you predict what the total will be for any three starting numbers?

It is a good idea to try out a few examples to see if you can spot a pattern:

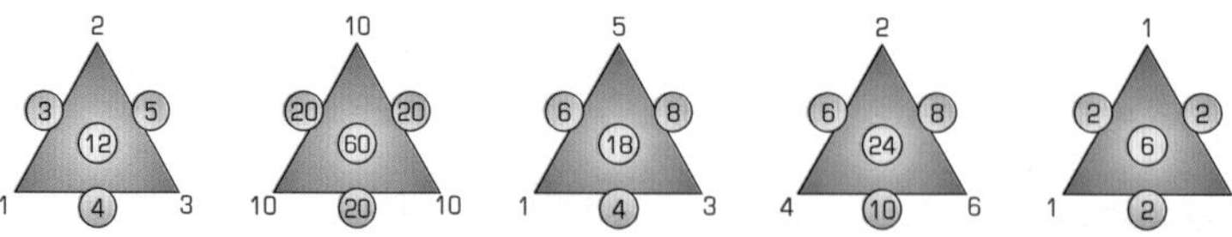

You may notice that the overall total is twice the sum of the starting numbers.
You can prove this using algebra. The overall total is:

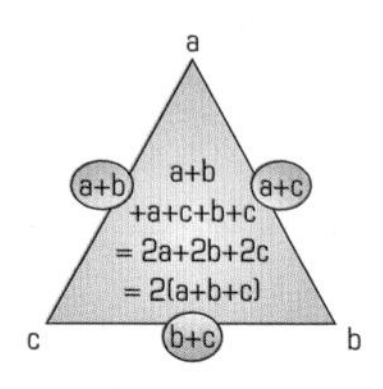

$$a + b + a + c + b + c$$
$$= 2a + 2b + 2c$$
$$= 2(a + b + c)$$
= twice the sum of the starting numbers!

Exercise A2.2

Solve these problems using algebra:

1 Pyramid totals

a This pyramid began with the numbers 2, 6 and 3 in the bottom row.
How was the pyramid total made?

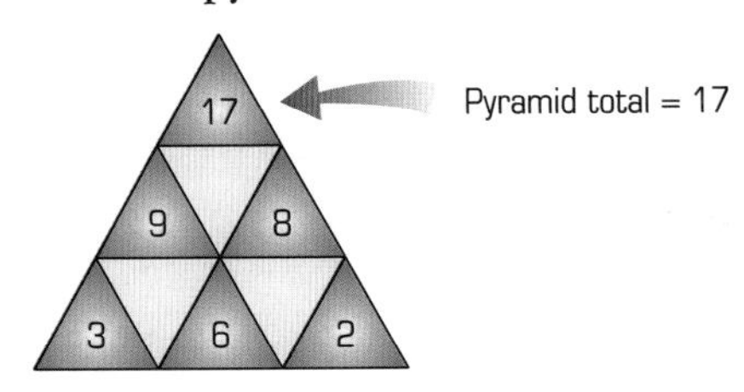

b Choose your own three starting numbers for the bottom row.
What is the biggest total you can make?
Does it matter which order you put the three starting numbers?

c What happens when you choose three consecutive starting numbers?
What happens if you choose three equal numbers?

d Try to predict the pyramid total without doing the additions.

e Use algebra to confirm any findings you have observed.

2 a Follow these instructions for different start numbers:

- Think of a number
- Double it
- Add ten
- Subtract eight
- Halve it
- Subtract your original answer.

b Explain your answers using a TOAN snake.

3 Truncated squares

A 10×10 square is taken ...

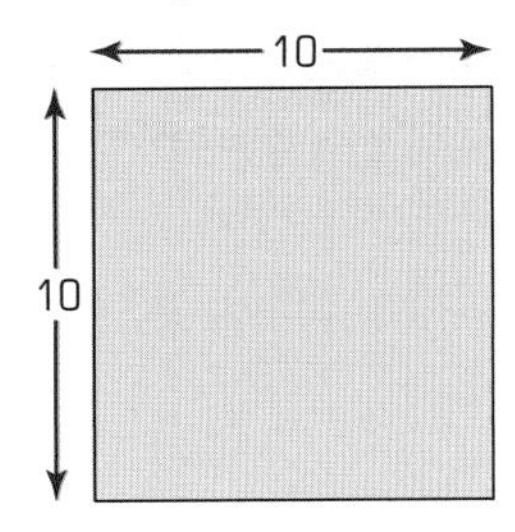

and a 3×3 square is cut from the corner:

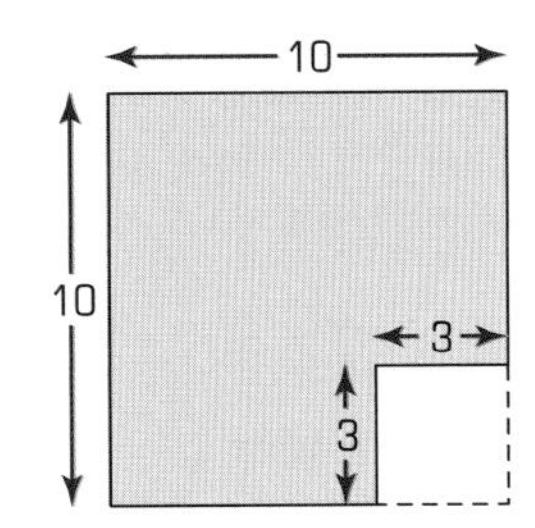

a Find the perimeter of the new shape.

b Repeat for different sized squares cut from the corner.
What do you notice about your solutions?
Why does this happen?

c Use algebra to confirm your findings.

d **Investigation**
Extend the questions into:

- different sized starting squares
- rectangles

4 Put four consecutive numbers in a 2×2 grid, for example:

13	14
15	16

a What do you notice about the diagonal sums?

b What do you notice about the vertical differences?

c **Challenge**
What do you notice about the diagonal products?

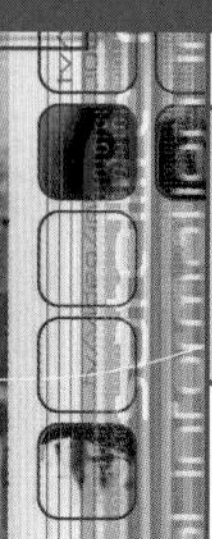

A2.3 Simplifying expressions

This spread will show you how to:
- Simplify linear expressions by collecting like terms.
- Multiply a single term over a bracket.

KEYWORDS
Collect Simplify
Expression Like terms

When you write algebraic expressions you should make them as short as possible.

The perimeter of the rectangle can be written using algebra:

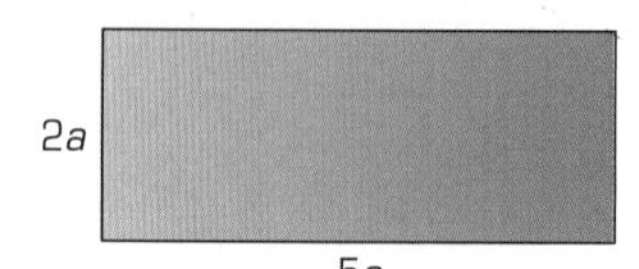

$P = 2a + 5a + 2a + 5a$
$P = 14a$

Each part of an algebraic expression is called a term:

$3x$	$+$	$4y$	$-$	z	$-$	3
term in x		term in y		term in z		number term

- You can collect together terms that use the same combination of letters. These are called **like terms**.

example

Write an algebraic expression for the perimeter of this rectangle:

b
6a

$P = 6a + b + 6a + b$

Collect like terms $P = 6a + 6a + b + b$

Simplify $P = 12a + 2b$

Note that $12a$ and $2b$ are not like terms and can't be collected.

Remember: the formula for the perimeter of a rectangle is 2(length + width).

You could say: $P = 2(6a + b) = 2 \times 6a + 2 \times b = 12a + 2b$.
You must be careful when the expression involves negative terms.

These expressions cannot be simplified:
$5y - 4 + 2z$
$6p - 6$
$b - b^2$
$2k + x + 3$

example

Write an expression for this perimeter as simply as possible:

8y − 2x
5x − 4y

$P = 8y - 2x + 5x - 4y + 8y - 2x + 5x - 4y$

$P = 8y - 4y + 8y - 4y - 2x + 5x - 2x + 5x$
$P = 4y \quad + 4y \quad + 3x \quad + 3x$
$P = 8y + 6x$

The negative signs stay with their terms.

Exercise A2.3

1 Copy the diagram. Find like pairs and add them together. Follow the given example. Be careful with negative terms.

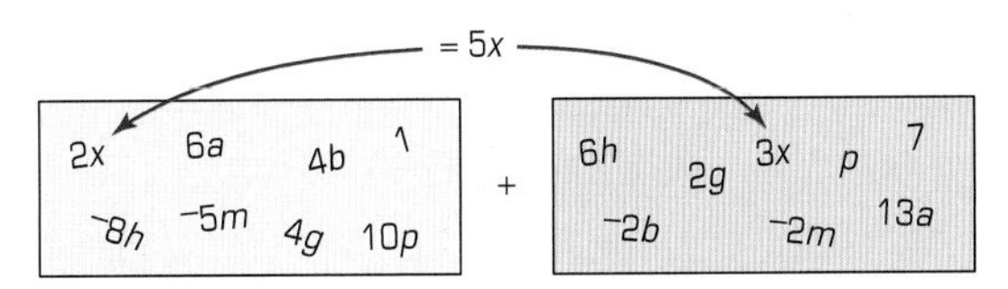

2 Simplify these expressions where possible.

a $y + y + y$ **b** $6z + 4z$

c $p + 3p + 2p$ **d** $9x - 9$

e $2h + 4h - 3h$ **f** $6b + 2m + 3b + 8m$

g $9x + 2y - 3x + y$ **h** $6a + 10 - 2a - 2$

i $9a + 2b - 8a - 7b$ **j** $4h - 2j + 8h + 5j$

k $5x + 11z - 2z + 9$ **l** $6q + 3y - 2q + 10w$

3 Two pieces of wood, A and B, have lengths as shown.

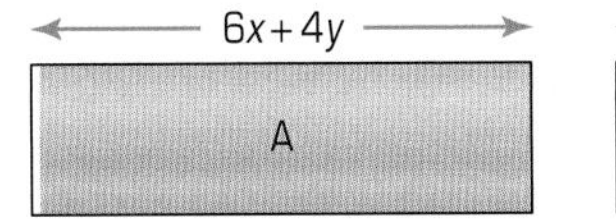

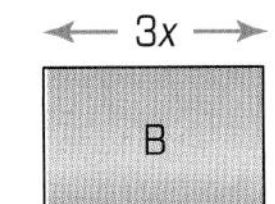

Give simplified lengths for each of the following:

a

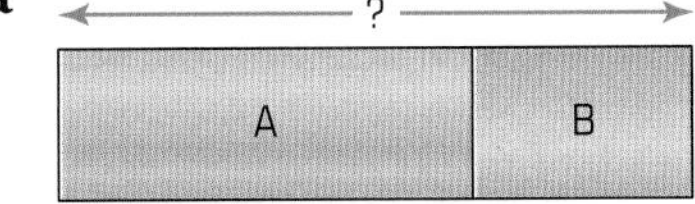

b

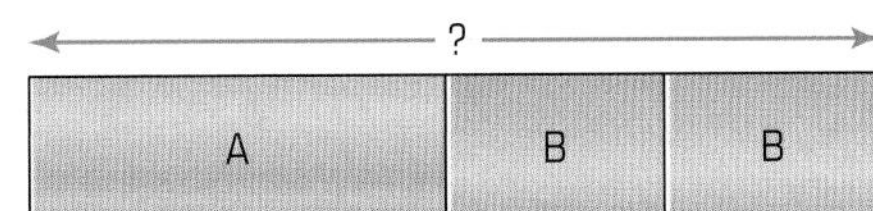

c

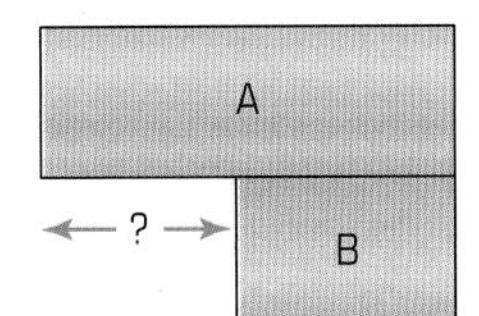

d

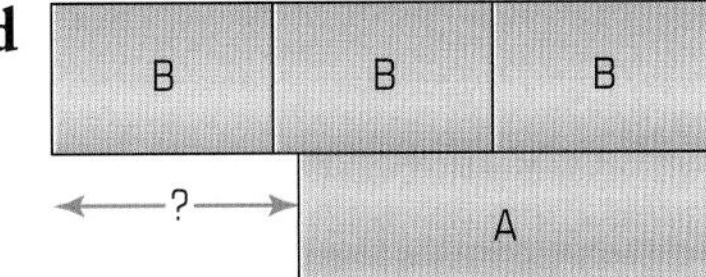

4 A simplified expression is $2a + 5b$. Write as many versions of the unsimplified expression as you can.

5 In this pyramid, the total in each triangle is made by adding the two triangles beneath it. Find the missing expressions, writing your answers as simply as possible.

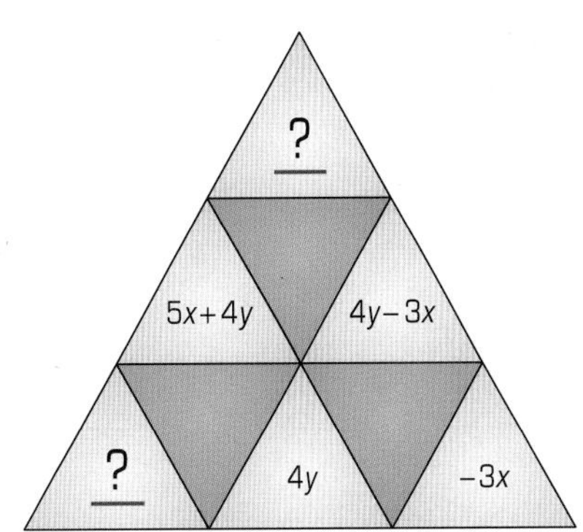

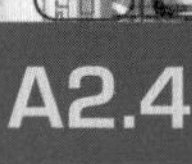

A2.4 Simplifying harder expressions

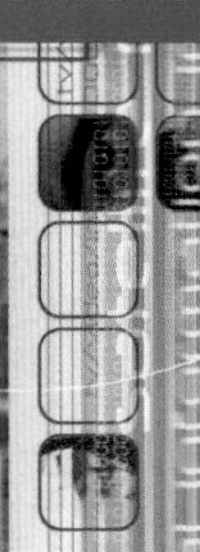

This spread will show you how to:

- Recognise that algebraic operations follow the same conventions and order as arithmetic operations.
- Use index notation for small positive integer powers.
- Represent problems and intepret solutions in algebraic form.

KEYWORDS

Expression Factor
Term
Simplify

Expressions such as $2x + 1$ cannot be simplified as the terms are different.

You can always simplify expressions that have been multiplied as you can at least remove the × sign.

The area of this rectangle can be written using algebra:

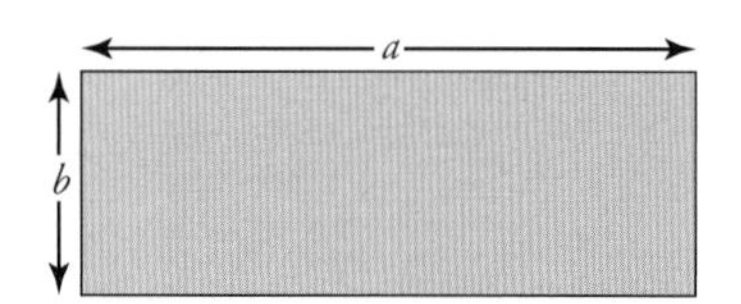

Area $= a \times b = ab$

With more difficult expressions it helps to deal with numbers first then the letters.

example

Simplify:

a $3p \times 4q \times 2r$ **b** $2p \times p \times 3$

a $3p \times 4q \times 2r = 3 \times 4 \times 2 \times p \times q \times r$
$= 24pqr$

b $2p \times p \times 3 = 2 \times 3 \times p \times p$
$= 6p^2$

You can often simplify expressions that involve divisions.

Remember: you write $p \div 3$ as $\frac{p}{3}$.

example

Simplify:

a $\dfrac{27ab}{3b}$ **b** $\dfrac{15b+5}{3}$

a $\dfrac{27ab}{3b} = \dfrac{27 \times a \times b}{3 \times b}$
$= 9a$

27*ab* and 3*b* have a common factor of 3*b* which will cancel.

b $\dfrac{15b+5}{3}$ cannot be simplified.

You can write:
$\dfrac{15b+5}{3} = \dfrac{15b}{3} + \dfrac{5}{3}$
$= 5b + \frac{5}{3}$
but this is not simpler.

Exercise A2.4

1 Decide if the statement on each card is true or false.
Any that you find to be false should be rewritten as correct statements:

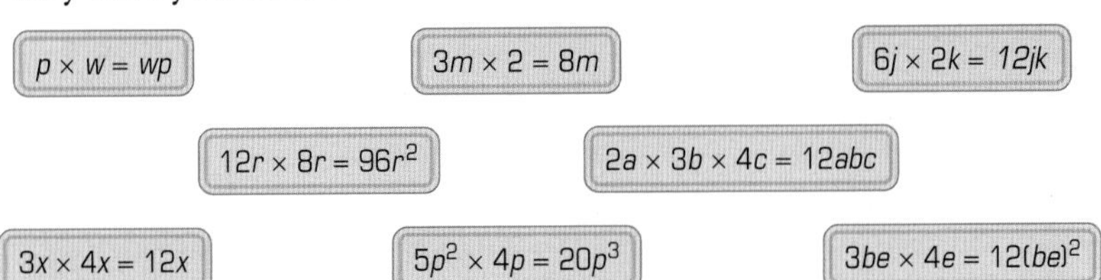

2 Simplify these expressions as fully as possible:

a $9r \times 12s$ **b** $4 \times p + 5 \times p \times q$ **c** $3a + 4b \times 8c$

d $2x \times 5x$ **e** $9ab \times 5cd$ **f** $15p^2 \times 4p$

3 Simplify these expressions as fully as possible:

a $c \div 5$ **b** $15 \div p$ **c** $36a \div 4$

d $\frac{6b}{2}$ **e** $\frac{15a}{3b}$ **f** $\frac{20ab}{4a}$

g $\frac{12abc}{3ac}$ **h** $\frac{90x^2}{3x}$ **i** $\frac{x+2}{2}$

4 Find simplified expressions for the area of these shapes:

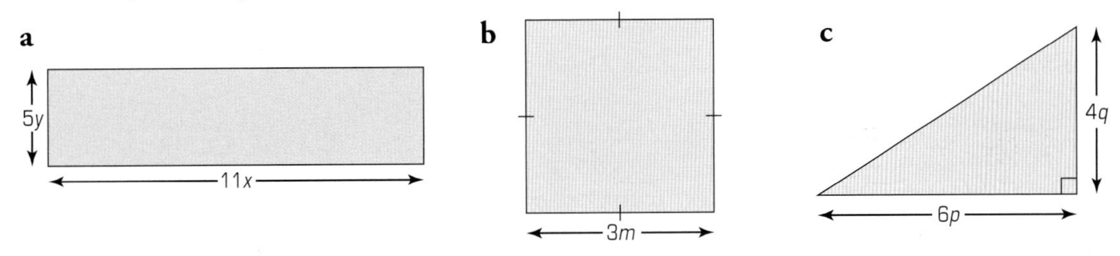

5 Petra saves £m a week. How much does she save in:

a 3 weeks
b k weeks
c a year
d 12 years?

6 A lottery prize comes to £$27abc$

a If 9 people share the prize, how much do they get each?
b If b people share the prize, how much do they get each?
c If $3ac$ people share the prize, how much do they get each?
d If each person gets £9, how many people shared the prize?

A2.5 Using formulae

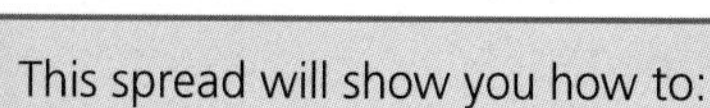

This spread will show you how to:
- Explain the meaning of and substitute integers into formulae.
- Derive algebraic expressions and formulae.

KEYWORDS
Devise
Variable
Formula

Paul's parents calculate his pocket money using this rule:

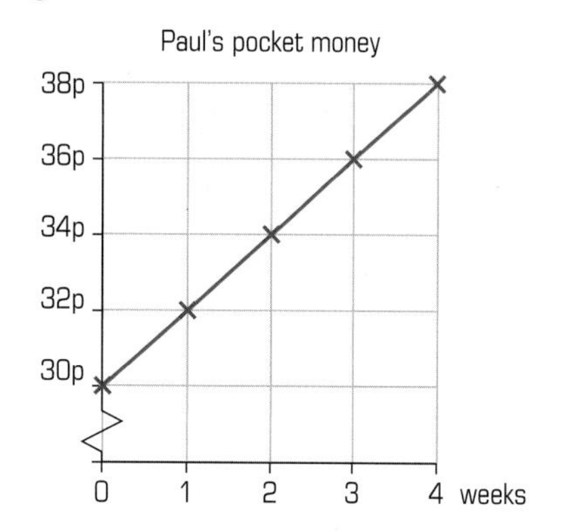

Deepak's parents calculate his pocket money using this rule:

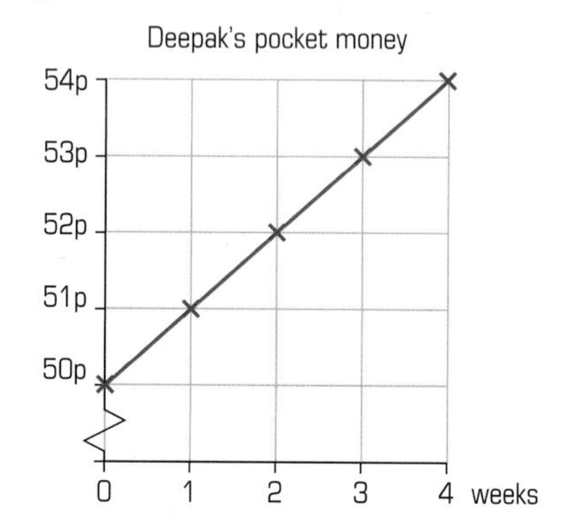

You can write these rules as mathematical formulae.

After 1 week: Money = 30 + 2 × 1
After 2 weeks: Money = 30 + 2 × 2
⋮
After n weeks: Money = 30 + 2 × n

After 1 week: Money = 50 + 1
After 2 weeks: Money = 50 + 2
⋮
After n weeks: Money = 50 + n

The formulae are:

$M = 30 + 2n$ $\qquad\qquad M = 50 + n$

Key: M = amount of money in pence, n = number of weeks

M and n are variables. The key explains what they stand for.

▸ A **formula** describes the relationship between variables.

When a value is known you can substitute it into the formula.

example

Look at the formulae for Paul and Deepak's pocket money.

a How much does each boy receive in week 60?
b How long does it take each boy to earn £3 per week?

---

a After 60 weeks, $n = 60$:
$M = 30 + 2 \times 60 = 150\text{p}$ $\qquad\qquad M = 50 + 60 = 110\text{p}$

b For £3, $M = 300$ (M is measured in pence!)
$300 = 30 + 2n$ $\qquad\qquad 300 = 50 + n$
$2n = 270$ $\qquad\qquad n = 250$

n is 135 weeks $\qquad\qquad$ n is 250 weeks

Exercise A2.5

1 Write formulae to describe these situations.

a The number of days (D) in n weeks.

b Your age in years (Y) when you know your age in months (m).

c The area (A) of a square with length l.

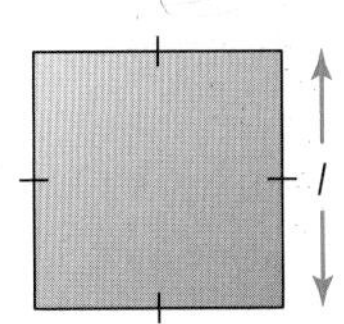

d The perimeter (P) of a rectangle with length l and width w.

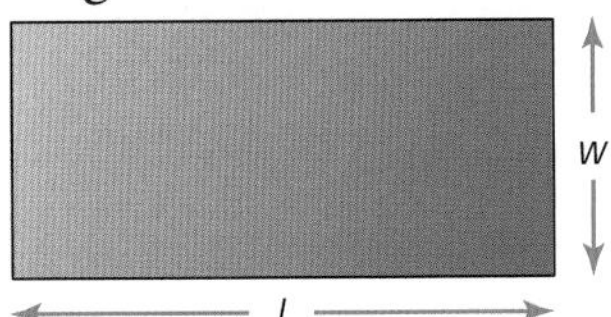

e The cost (C) of p articles when each article costs 30 pence.

f The cost (C) of m articles at 36 pence each and n articles at 49 pence each.

g The amount of tea (T) used in a pot for n people if you need one teaspoon per person plus one extra teaspoon for the pot.

2 An electricity company uses the following formula to calculate bills:

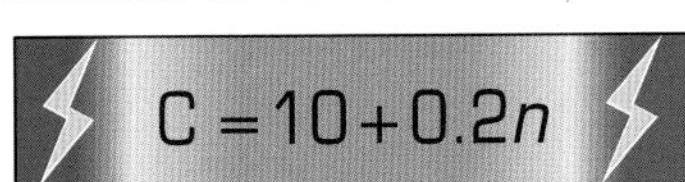

C = cost of bill in pounds
n = Number of units of electricity used.

a Calculate the cost of Mr Potter's bill when he has used 250 units of electricity.

b Mrs Singh's bill came to £17.80. How much electricity had she used?

c The Brown family spent 6 months in America and used no units of electricity. Use the formula to find out how much they must pay. Why is this?

3 Petra says that she always gets 8 birthday presents from her family, and one from each person she invites to her party. Raki says that she only gets 2 presents from her family, but 2 presents from each person she invites to her party.

a Write a formula for each girl's total number of presents (P) when they invite n people to their parties.

b Petra has 15 people to her party, how many gifts does she receive?

c If Raki receives 38 gifts, how many people did she have to her party?

d Who gets most presents? Discuss.

4 Use these formula to calculate the unknown value.

a The formula for converting degrees Fahrenheit into Celsius:

$$C = \frac{5(F - 32)}{9}$$

find 56 °F in Celsius (°C)

b The formula for finding the area of a circle when you know its radius:

$$A = \frac{22r^2}{7}$$

find A when $r = 12$ cm

5 Here are three polygons with diagonals drawn.

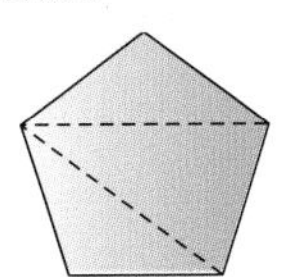
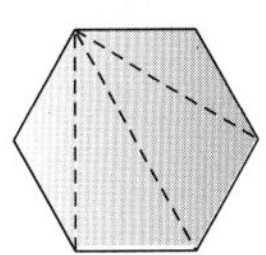
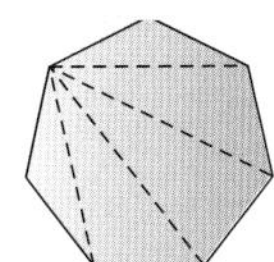

a Find a formula connecting the number of diagonals (D) drawn from one vertex of a polygon with n sides to all of the other vertices.

b Explain why the formula works.

c How many sides has a polygon with 17 diagonals?

A2 Summary

You should know how to ...

1 Know that algebraic conventions follow the same order and conventions as arithmetic operations.

2 Simplify or transform linear expressions by collecting like terms.

3 Substitute integers into simple formulae.

4 Use logical argument to establish the truth of a statement.

Check out

1 a Write these statements using algebra:

i I think of a number, double it, add 5 and divide this by four.

ii I think of a number, square it, subtract 2 and multiply this by 7.

iii I think of a number, multiply it by itself three times and multiply it by 4.

b Pick out expressions which are the same:

xy x^2 $y \div x$ $x \times x$ $y + x$ $x + y$ $\frac{y}{x}$ $2x$ yx

2 Simplify by collecting like terms, if possible:

a i $3p + 4q - 2p + 6q$

ii $3x^2 + 4x - x^2 + 11x$

iii $11ab + 4ba + 12ca$

iv $11x + 9$

Simplify these expressions fully:

b i $3x \times 4y$

ii $9b \div 3$

iii $11a \times 12a$

iv $4 \times p \times 8 \times p \times p$

v $\frac{10w^2}{5w}$

vi $2x \times 3x \times 4x$

3 a If $a = 7$ and $b = {}^{-}2$, evaluate:

i a^2 ii $2a - 4$ iii $5b$

iv $a - b$ v $3a + 4b$

b The area of a shape can be found using the formula $A = 3(L + 6)$, where L is the length of the shape.

i Find the area of a shape of length 5 cm.

ii How long is a shape with area 99 cm^2?

4 a Prove that the sum of three consecutive numbers is always divisible by three.

b Prove that the sum of two consecutive odd numbers is always even.

2 Angles and shapes

This unit will show you how to:

- Understand a proof that the sum of the angles of a triangle is 180° and of a quadrilateral is 360°.
- Identify alternate angles and corresponding angles.
- Classify quadrilaterals by their geometric properties.
- Solve problems and investigate in the context of shape.
- Identify the necessary information to solve a problem.

You can see angles and parallel lines in railway tracks.

Before you start

You should know how to ...

1 Estimate angles in degrees.

2 Know the properties of 2D shapes.

Check in

1 Estimate these angles:

a **b** **c**

2

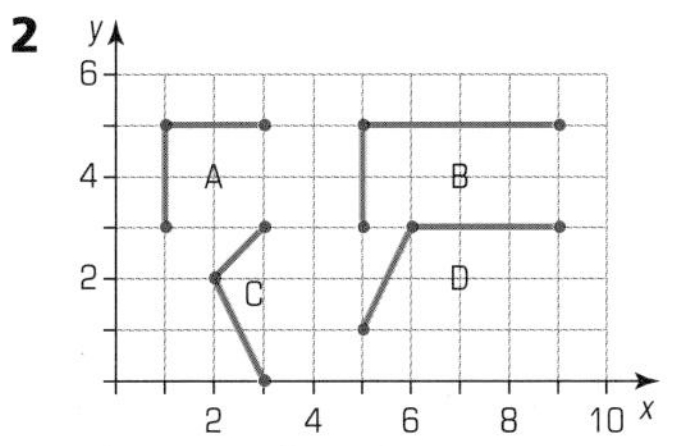

Choose a fourth point:

a to make A a square.

b to make B a rectangle.

c to make C a kite.

d to make D a parallelogram.

S2.1 Finding angles

This spread will show you how to:

- Use accurately the notation for lines, angles and shapes.
- Know the sum of angles at a point, a straight line and a triangle.
- Understand a proof that sum of the angles of a triangle is 180° and of a quadrilateral is 360°.

KEYWORDS

Angle, Acute, Degree, Obtuse, Protractor, Reflex, Point

An angle is a measure of turn. You can measure it in degrees, °, using a protractor.

An angle smaller than 90° is an acute angle.
Angle *AOB* is 55°

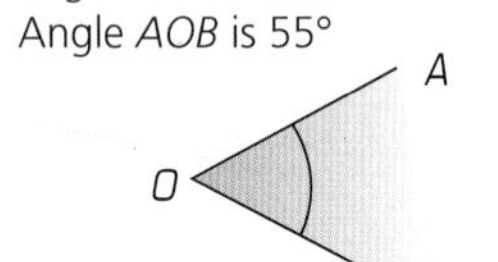

An angle between 90° and 180° is an obtuse angle.
$\angle A$ is 154°

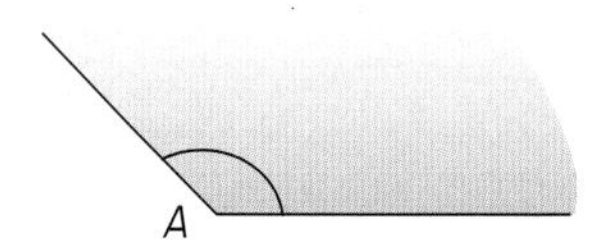

An angle between 180° and 360° is a reflex angle.
$D\hat{E}F$ is 220°

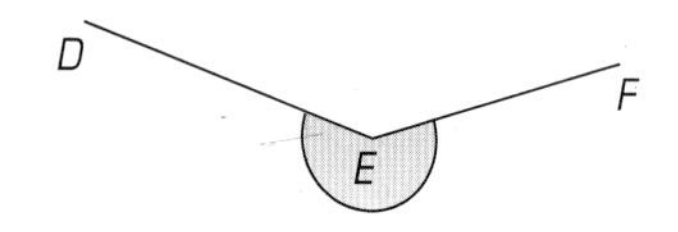

- There are 360° at a point

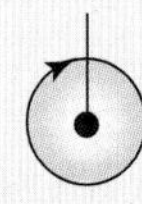

... 180° on a straight line

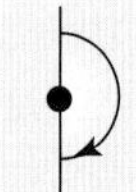

... 90° on a corner

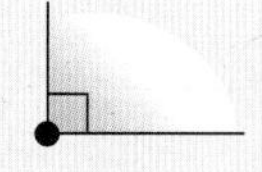

You can use these facts to solve angle problems.

example

Find the missing angles in these diagrams:

a

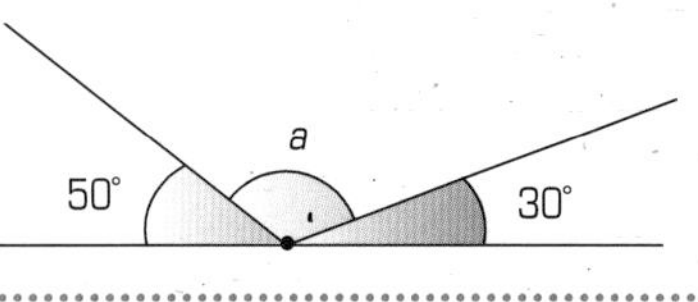

b

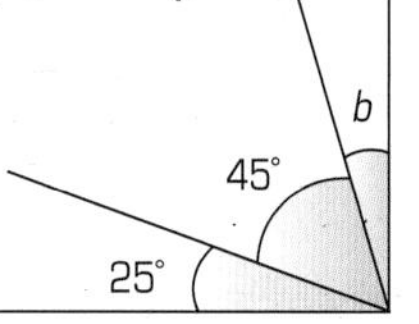

c

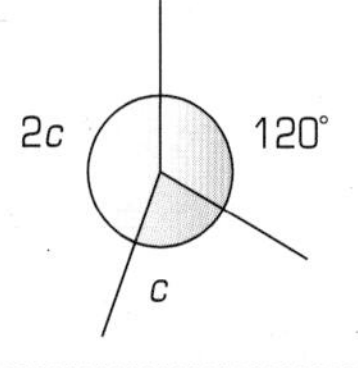

a $a + 50° + 30° = 180°$
$a + 80° = 180°$
so $a = 100°$

b $25° + 45° + b = 90$
$70° + b = 90°$
so $b = 20°$

c $120° + c + 2c = 360°$
$120° + 3c = 360°$
so $3c = 240°$ and $c = 80°$

Activity

Draw any triangle. Cut out the corners. Put them together:

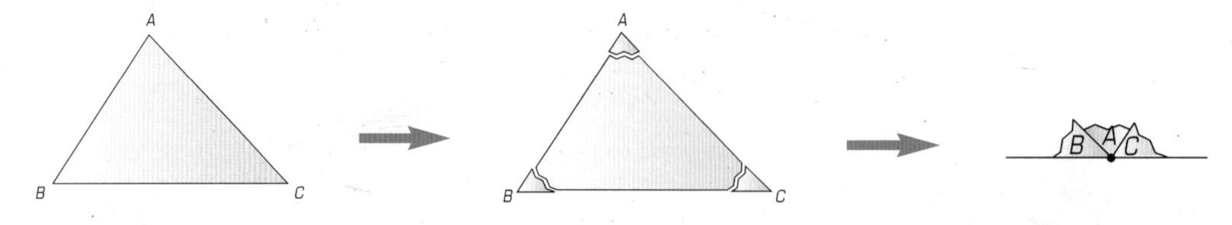

- The angles in a triangle add to 180°. They make a straight line.

Exercise S2.1

Find the value of the letters in questions 1 to 12.

1

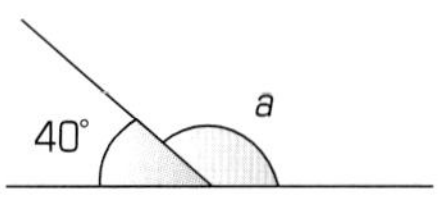

2

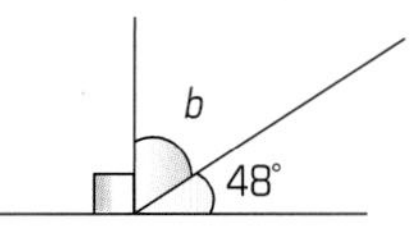

3

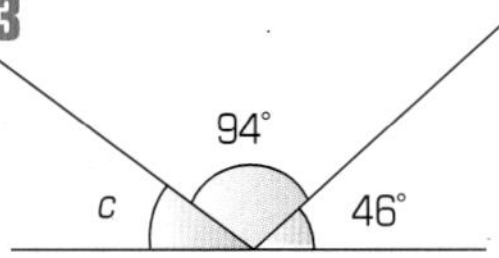

4

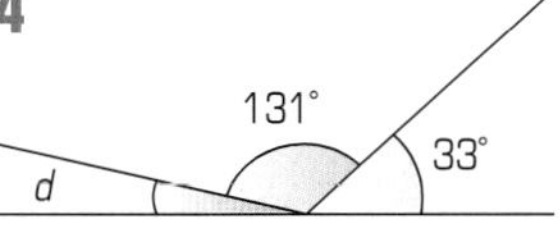

5

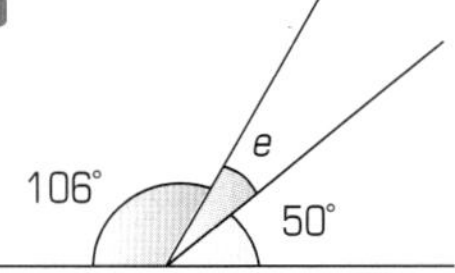

6

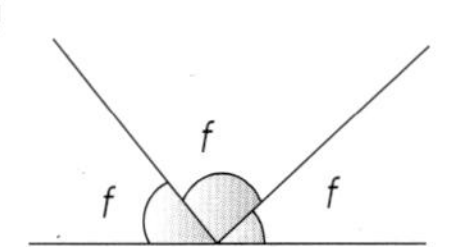

7

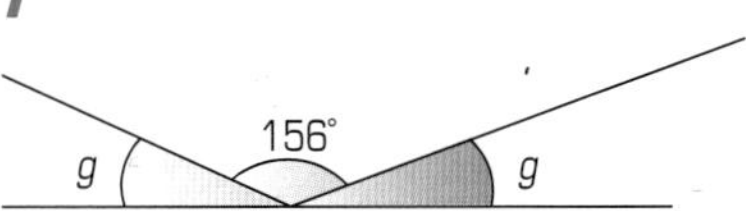

8

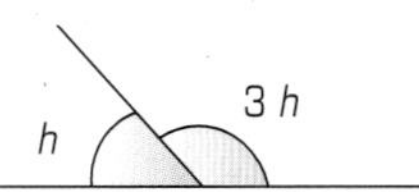

9

$j = 4i$

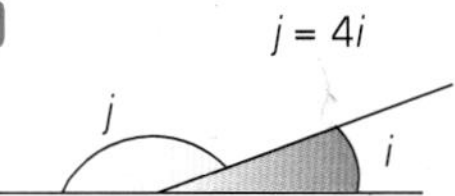

10

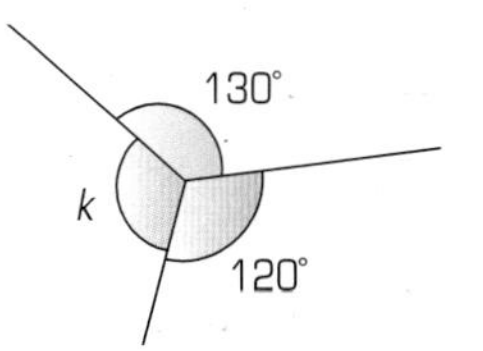

11

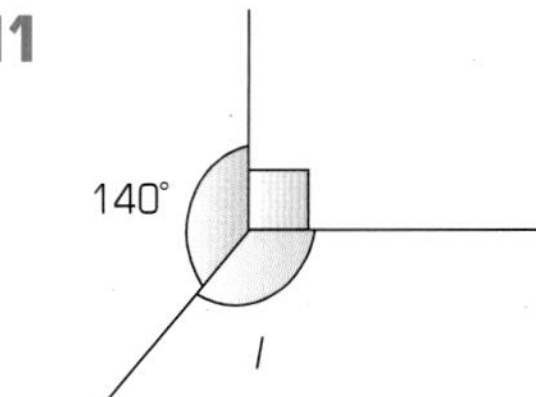

12

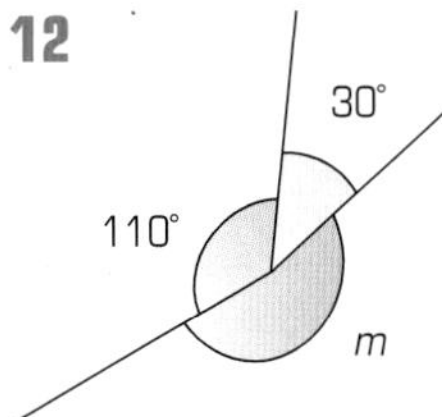

13 **a** Draw any quadrilateral.
Draw in one of its diagonals.

b Copy and complete:

> The sum of the angles of a triangle is ...
> A quadrilateral contains ... triangles.
> The sum of the angles of a quadrilateral is ...

14 **a** Sketch this triangle in your book.

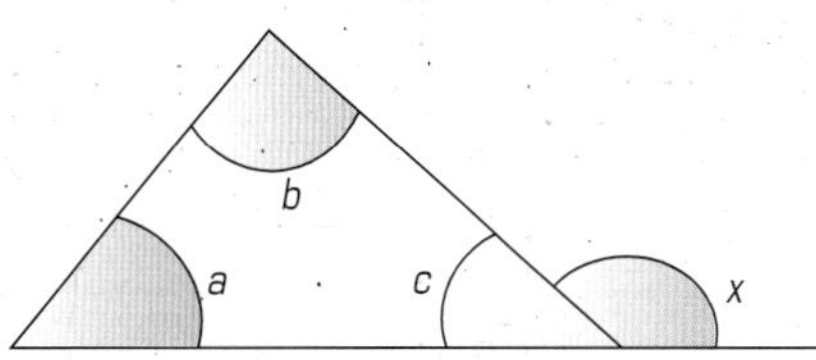

b Copy and complete:

> Angles on a straight line add to 180°:
> $c + x = \cdots$
> So: $c = \cdots - x$
> Angles in a triangle add to 180:
> $\cdots + \cdots + c = 180°$
> So: $\cdots + \cdots + \cdots - x = 180°$
> So: $\cdots + \cdots - x = 0$
> So: $\cdots + \cdots = x$

S2.2 Angles and lines

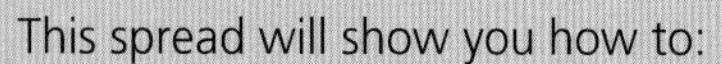

This spread will show you how to:

- Recognise vertically opposite angles.
- Identify alternate and corresponding angles.
- Classify quadrilaterals by their geometric properties.

KEYWORDS

Alternate, Corresponding, Parallel, Equidistant, Perpendicular, Opposite, Intersection

- **Parallel lines || are always equidistant.**
- **Perpendicular lines meet at right angles or 90°.**
- **All other lines meet at an angle other than 90°.**

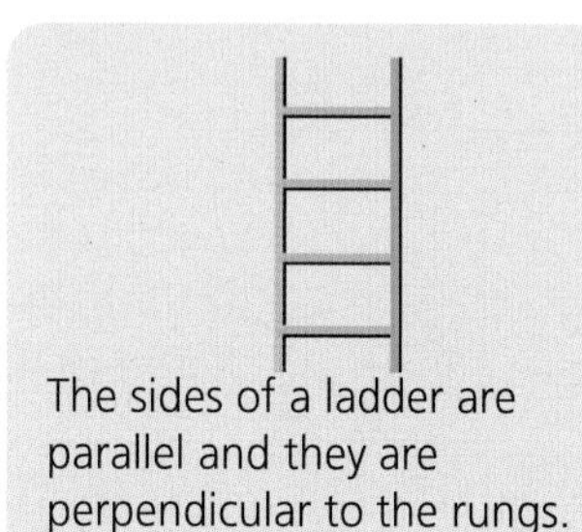

The sides of a ladder are parallel and they are perpendicular to the rungs.

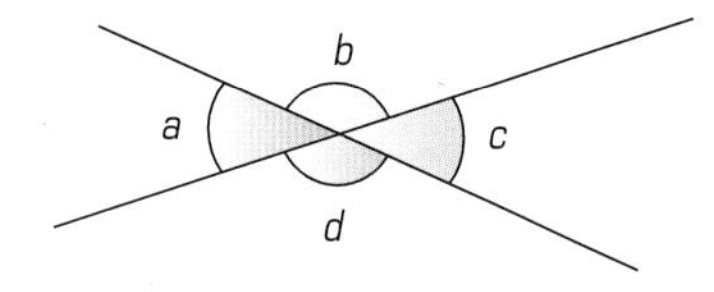

The intersection forms two different angles: a and b make a straight line.

The opposite angles are equal: $a = c$ and $b = d$.

Draw a cross and measure the angles to check.

example

To describe a rectangle you would say:

Adjacent sides (such as *AB* and *AD*) are perpendicular.

Opposite sides *AB* and *CD* (and *AD* and *BC*) are equal in length and are parallel. The marks > and >> show they are parallel.

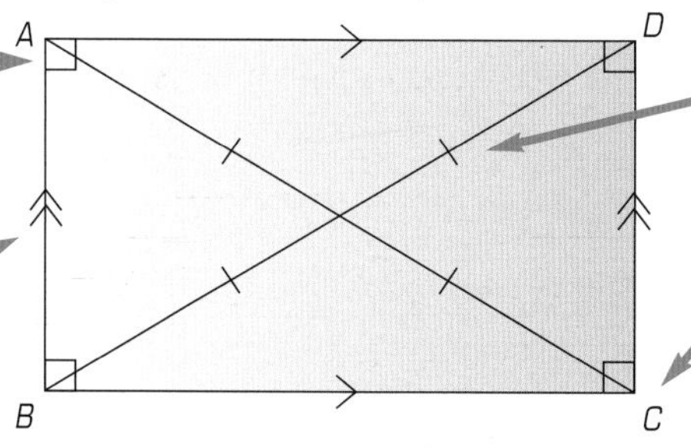

Diagonals *AC* and *BD* are equal in length. These marks show they are equal in length.

All angles are 90° or right angles.

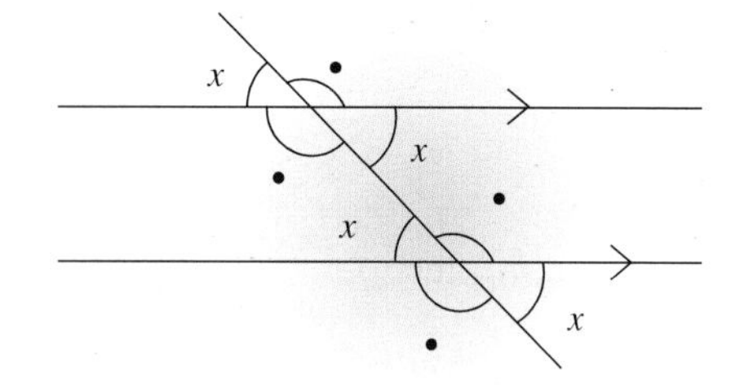

When a straight line crosses a pair of parallel lines 8 angles are formed. In fact there are only two different angles: one acute angle and one obtuse angle.

$x + \bullet = 180°$

These angles are called **corresponding angles.** They are equal.

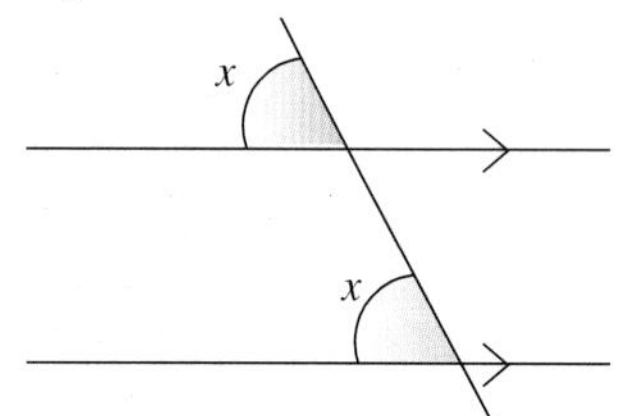

These angles are called **alternate angles.** They are equal.

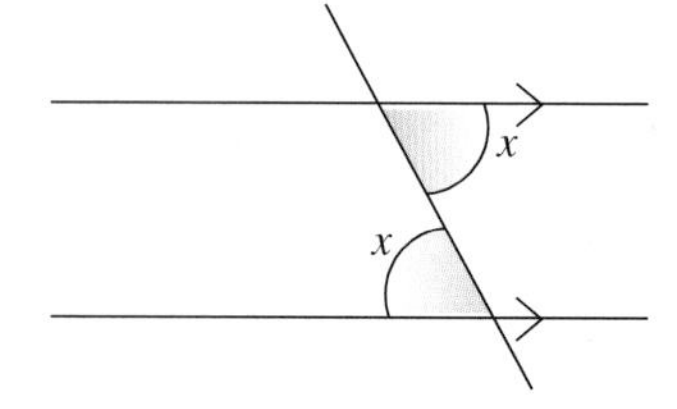

Exercise S2.2

1 For each of these quadrilaterals, describe its geometric properties as accurately as possible. Use words from this list:

Parallel	*Perpendicular*	*Diagonal*	*Bisect*
Adjacent	*Opposite*	*Equal*	*Angles*

a Trapezium

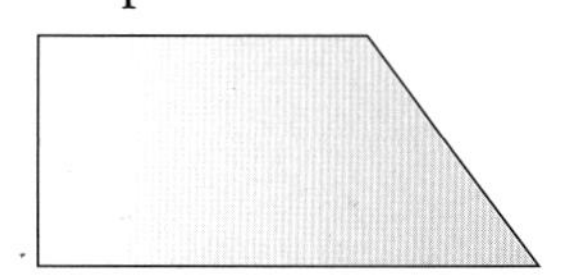

b Parallelogram

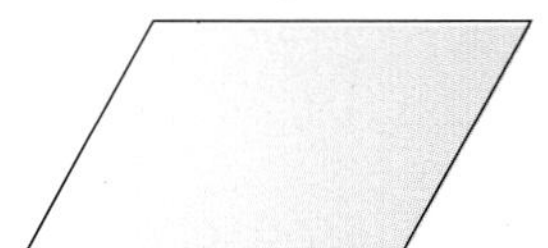

c Square

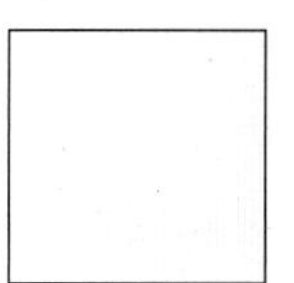

2 **a** Copy this diagram and label all the angles that are *equal* to x.

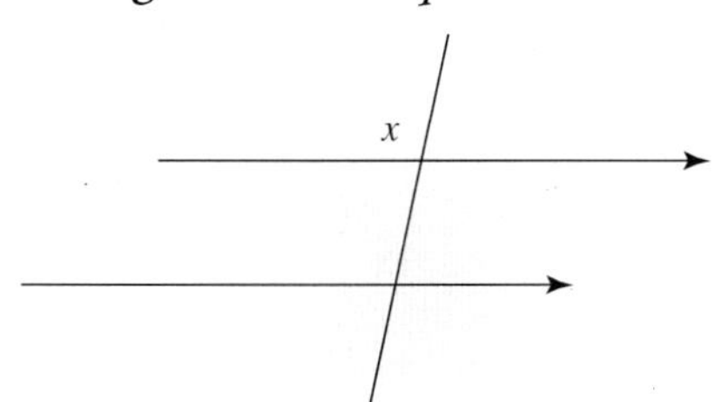

b Copy this diagram and label any angles that *are **not** equal* to a.

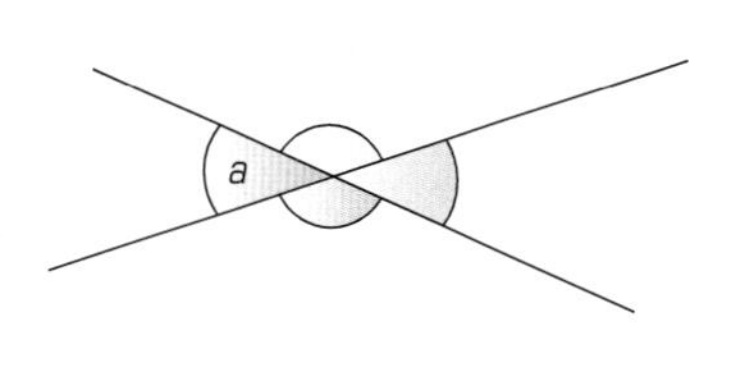

3 Copy the diagrams and label the corresponding angles to those shown.

a

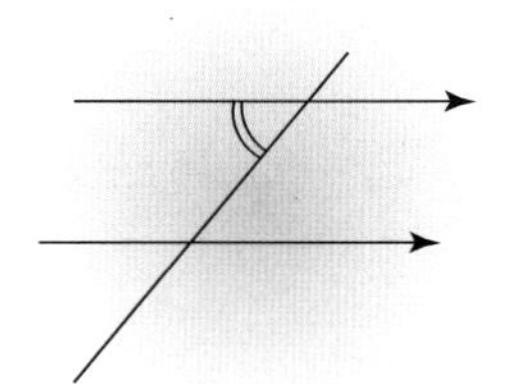

b

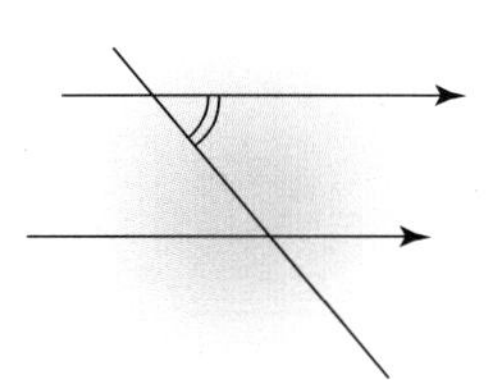

c

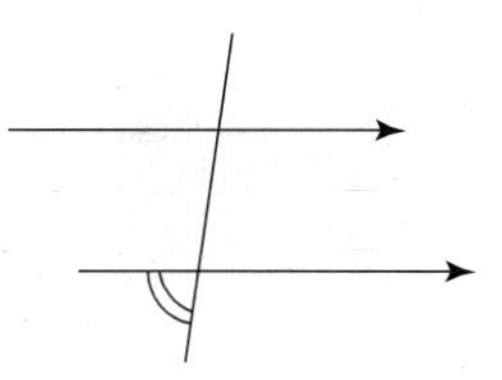

4 Copy the diagrams and label the alternate angles to those shown.

a

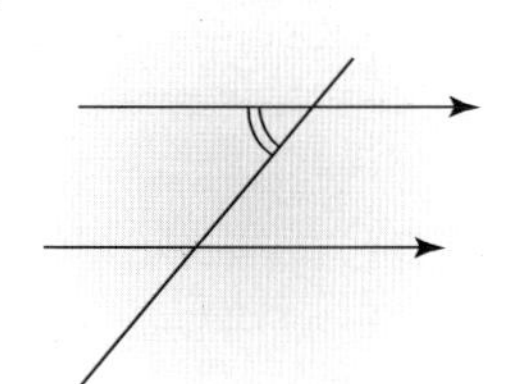

b

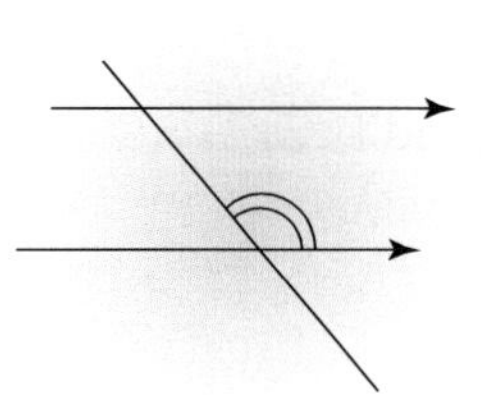

c

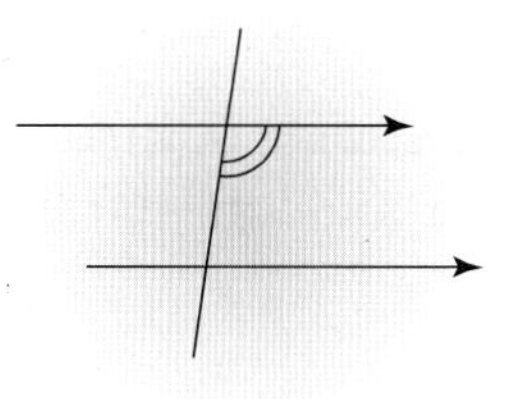

5 **Calculate** the unknown angles:

a

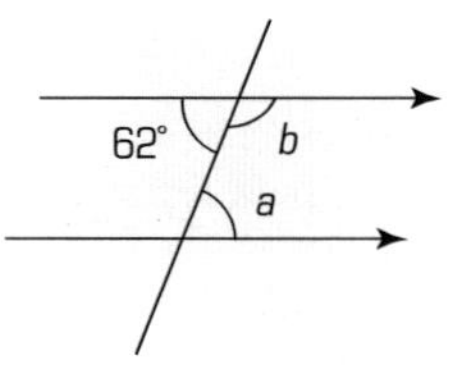

b

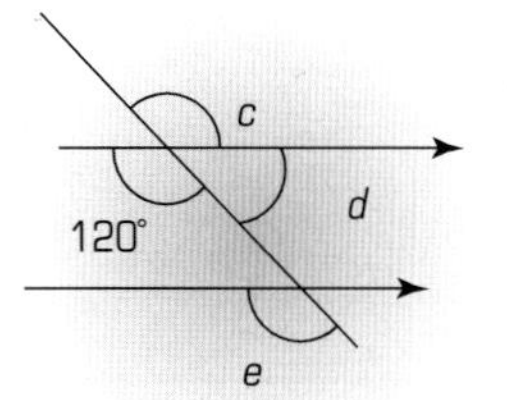

c

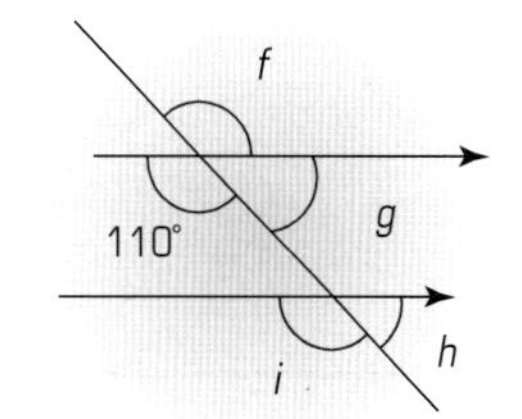

S2.3 Coordinates and shapes

This spread will show you how to:

- Read and plot points in all four quadrants.
- Plot points determined by geometric information.
- Know and use side, angle and symmetry properties of triangles.

KEYWORDS

Isosceles, Rhombus, Equilateral, Trapezium, Parallelogram, Kite

You can plot points on a grid.

- **A grid has two axes that are perpendicular to each other: the *x*-axis and the *y*-axis. The axes split the grid into four quadrants.**

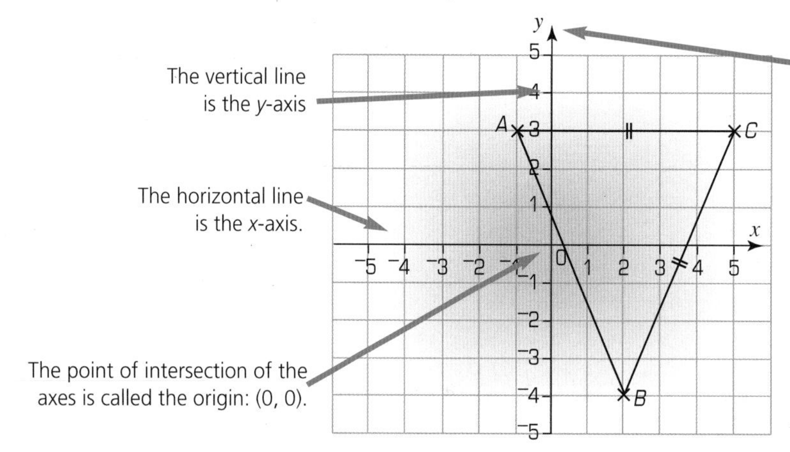

The arrows on the axes show the positive direction.

A coordinate is a pair of numbers: (x, y) that fixes a point on a grid. (5, 3) means 5 along the *x*-axis and 3 up the *y*-axis.

$(^{-}1, 3)$ means 1 backwards along the *x*-axis and 3 up the *y*-axis. This is point *A* on the grid.

$(2, ^{-}4)$ means 2 along the *x*-axis and 4 down the *y*-axis. This is point *B* on the grid.

- **Negative numbers in coordinates mean you go backwards or downwards.**

You can draw shapes on a grid.

example

Two vertices of a triangle are $(^{-}2, 0)$ and $(^{-}2, 2)$
What could the third vertex be if the triangle is:

a right-angled
b isosceles?

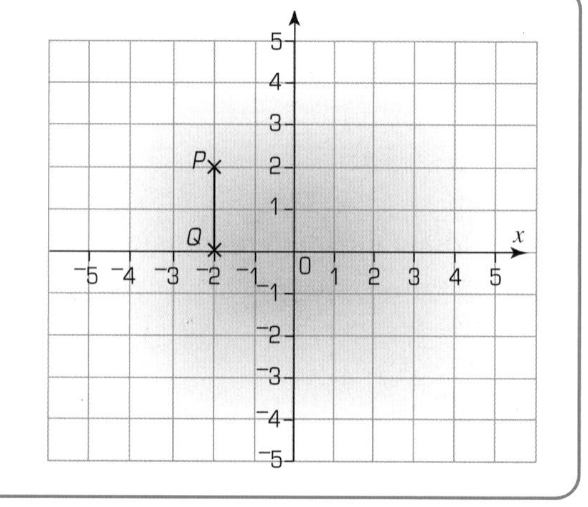

a The third vertex could be at $(^{-}4, 0)$.
b The third vertex could be at (1, 1).

There are lots of different answers. Can you see any more?

Exercise S2.3

1 Copy this grid and plot the points shown.

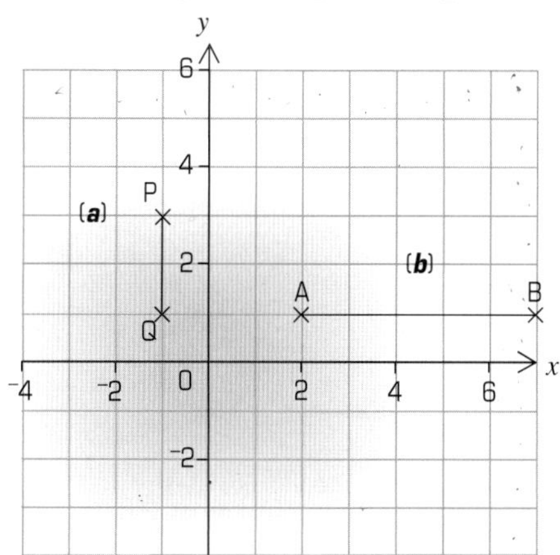

Write down all the possible pairs of points that will make

a rectangles with dimensions 2 units by 3 units, including *PQ* as one side.

b squares with dimensions 5 units, including *AB* as one side.

2 Copy this grid and plot the points shown.

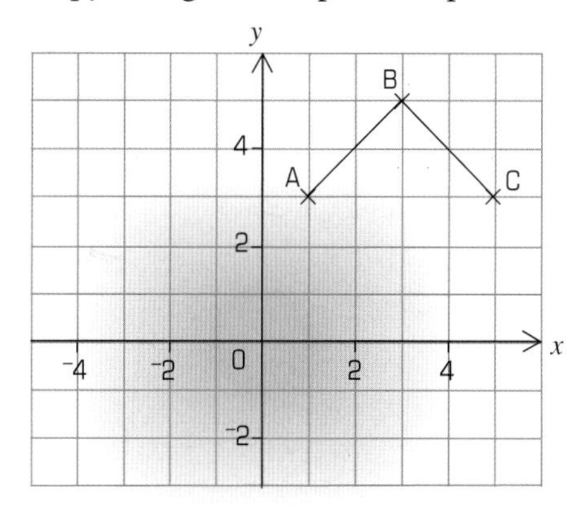

Write down, if possible, the extra point that will make

a a rectangle

b a parallelogram

c a rhombus

d an arrowhead

e a square

Explain your choice.

3 Copy this grid and plot the points shown.

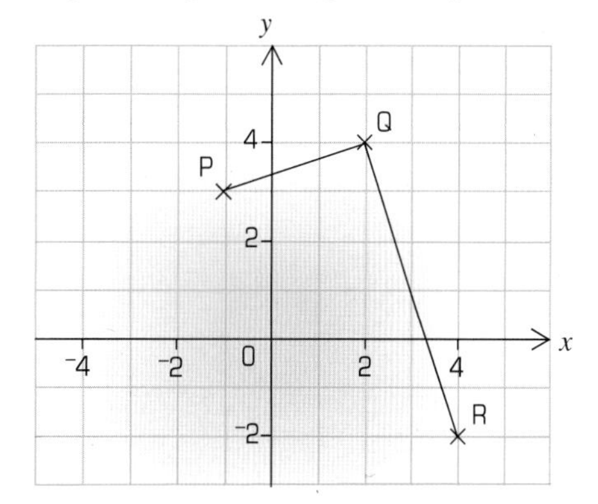

Write down, if possible, the extra point that will make

a a rectangle

b a parallelogram

c a rhombus

d an arrowhead

e a square

Explain your choice.

4 On a grid marked from $^{-}4$ to 4, plot the points A(2, 3) and C($^{-}2$, 3).

Write down points B and D so that ABCD is:

a a square

b a parallelogram

c a kite

Points E and F are at (0, $^{-}4$) and (0, $^{-}1$) respectively.

d How would you describe shape AECF?

e Find the area of the shape AECF in square units. (Hint: find areas of shapes ACF and ACE separately).

Summary

You should know how to ...

1 Identify alternate angles and corresponding angles.

2 Understand a proof that the sum of angles of a triangle is 180° and of a quadrilateral is 360°.

3 Identify the necessary information to solve a problem.

4 Classify quadrilaterals according to their properties.

Check out

1 Find the missing angles:

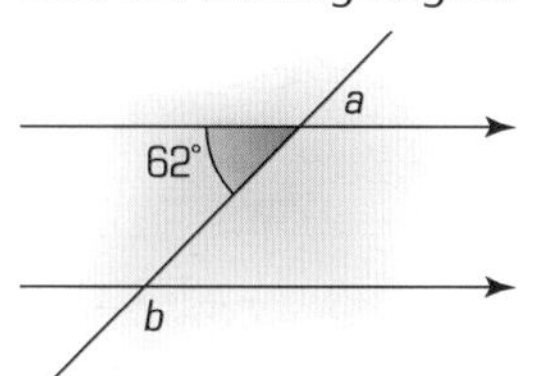

2 a Name this shape:

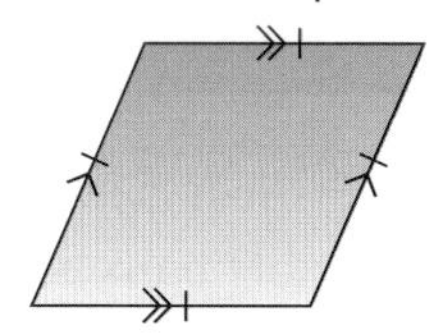

b Copy the shape and draw in one of its diagonals.
Name the triangles you have made.

c What is the angle sum of each triangle?

d What is the angle sum of the quadrilateral?

3 In the shape in question 2, if one of the interior angles is 65° what are the other angles?

4 Plot the points P($^{-}1$, 1) Q(1, 2) and R(4, 1) on a coordinate grid.
What other point would make:

a a rhombus **b** a kite

c a parallelogram?

2 Handling data

This unit will show you how to:

- Discuss a problem that can be addressed by statistical methods and identify related questions to explore.
- Decide which data to collect to answer a question, and the degree of accuracy needed; identify possible sources.
- Plan how to collect the data, including sample size.
- Construct frequency tables with given equal class intervals for sets of continous data.
- Construct diagrams to represent data, including pie charts, bar charts, frequency diagrams and simple line graphs for time series.
- Interpret tables, graphs and diagram for both discrete and continuous data, and draw inferences that relate to the problem being discussed.
- Identify the information necessary to solve a problem.
- Represent problems and interpret solutions in graphical form, using correct notation and appropriate diagrams.

Sport questionnaire

Please take a few minutes to complete this questionnaire

1. Are you: Male ☐ Female ☐

2. How old are you? 11-12 ☐ 13-15 ☐ 16-19 ☐ 20+ ☐

3. Do you enjoy: Yes No

   a) taking part in sport ☐ ☐

   b) watching sport ☐ ☐

If you answered 'yes' to either a or b, go on to question 4.
If not, go directly to question 8.

Questionaires provide an effective way of collecting data.

Before you start

You should know how to ...

1 Use inequalities.

2 Use tally marks.

Check in

1 a Insert $<$ or $>$ between the following pairs of numbers **(i)** 7 5 **(ii)** 3.7 3.64

b If $x \geqslant 4$ could $x = 4$?

2 a Write using tally marks

(i) 4 **(ii)** 7 **(iii)** 12

b What do these tallies show:

(i) III **(ii)** 𝍸 IIII **(iii)** 𝍸 𝍸 𝍸 I

D2.1 Discussing statistical methods

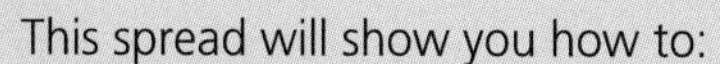

This spread will show you how to:

- Suggest possible answers to a problem that can be addressed by statistical methods.
- Decide which data would be relevant to the inquiry and possible sources.

KEYWORDS

Data
Primary
Secondary
Survey
Experiment

A parent governor of Maypole Primary School is concerned about the fitness of the pupils. She wants to encourage pupils to walk to school.

She decides that she needs to find out:

- the most important concerns of parents
- if pupils would be willing to walk if they walked in a group or with friends
- if parents would help organise a walking bus.

She also wants to send information home to every parent to explain why walking to school would be a good idea.

Once you know what information you need to find out you must think about where to find the answers!

Data you collect yourself is **primary data**. You can collect primary data using a survey or experiment.

Data already collected is **secondary data**. You can find secondary data in books or on the internet.

Exercise D2.1

1 The governors of Maypole Primary School ask Year 7 at a nearby secondary school to identify the issues relevant to pupils walking to and from school.

Make a list of the issues that you think would be important to:

a parents
b pupils.

2 Choose the three issues that you think are most important from each list and give reasons why they are important.

3 You are asked to find out:

- how many pupils currently walk to school
- how many pupils would be willing to walk to school
- how many parents would be willing to help run a walking bus.

How could you collect primary data to answer these questions?

4 Describe how you could use secondary data to find information for parents to explain why a walking bus would be a good idea.

5 Describe how you could use secondary data to compare the results of your survey with other schools in the UK.

6 Make a list of issues that affect traffic outside schools. Which issues would be difficult to collect data on? Explain why it would be difficult.

Collecting data

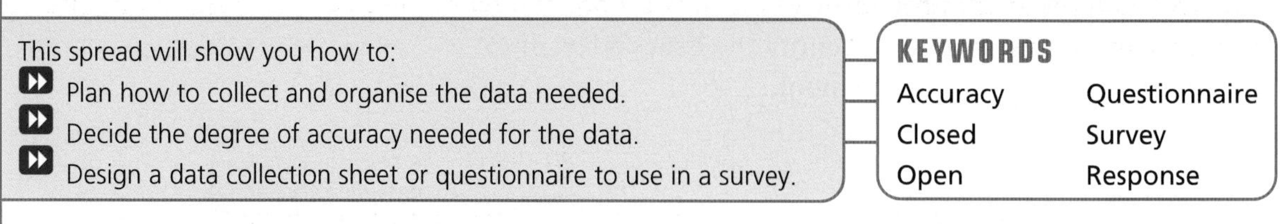

This spread will show you how to:

- Plan how to collect and organise the data needed.
- Decide the degree of accuracy needed for the data.
- Design a data collection sheet or questionnaire to use in a survey.

KEYWORDS

Accuracy, Questionnaire, Closed, Survey, Open, Response

To decide whether pupils and parents at Maypole Primary School would support a walking bus, the parent governor decided to ask the pupils:

Do you walk to school most days?

This question has yes or no answers.

How many of your friends live in the same street as you?
How far do you live from school?
How long does your journey usually take?

These questions have numerical answers.

How do you think car journeys affect the environment?
In which direction do you travel to school?

These questions have many different answers.

▶ **Questions that have particular answers are closed questions. You can use tick boxes to collect this data.**

▶ **Questions that have many different answers are open questions. They can include answers that you may not have thought of.**

A questionnaire is developed to give to the pupils.

Walking Bus Questionnaire

Do you walk to school most days? Yes ☐ No ☐

Yes/No answers give limited information, but the data is easy to collect.

How many of your friends live in the same street as you?
0 ☐ 1 ☐ 2 ☐ 3 ☐

This question has answers that are specific values.

How far do you live from school?
Less than 1 km ☐ 1 – 2 km ☐ over 2 km ☐

How long does your journey usually take?
< 5 mins ☐ $5 \leqslant$ mins $\leqslant 10$ ☐ > 10 mins ☐

These questions have answers organised in groups. You need to decide how accurate you want answers to be.

How do you think car journeys affect the environment?

In which direction do you travel to school?

These questions are open. Responses may be difficult to categorise. You can make it closed by giving a choice of typical responses.

▶ **You can record answers given to a questionnaire in a table. This is called a data collection sheet.**

Exercise D2.2

Gina and Sam decide to find out who walks to school and why. They develop a questionnaire with these questions.

Walk to School Questionnaire

1. Do you walk to school? Every day ☐ Sometimes ☐ Never ☐

2. How long does it take you to get to school? __________

3. How far do you live from school?

C B A School ×

4. Would you use a walking bus to get to school?

Comment on the following for each question:

1 If pupils answer 'Sometimes' or 'Never' would it be useful for them to continue with the rest of the questionnaire?
If they were not to continue what could you write before question 2? (Hint: read the whole questionnaire before answering).

2 How could you improve on the way answers are given?
Give a reason for your suggestion.
Write out the question and include your improvement for answers.

3 What is missing from the diagram in the questionnaire?
How else could you use the diagram to find other useful information to set up a walking bus to school?

4 Should this question be the last one on the questionnaire? Give a reason for your answer.

5 Gina and Sam want to collect data on the students at Maypole High, but the school contains about a thousand students.
They decide to collect information on 60 students, chosen throughout the school.
Suggest a way in which they could do this fairly.

Organising the data

This spread will show you how to:

- Construct frequency tables for discrete data and for continuous data, grouped where appropriate in equal class intervals.

KEYWORDS
Class interval Discrete
Continuous Frequency

- Numeric data can be **discrete** or **continuous**.

Discrete data is countable, it can only be specific values.
Continuous data is measurable, it can be any value within a given interval.

Discrete data and continuous data can be collected in a frequency table.

This question has responses that are specific values. The data you collect is discrete.

How many of your friends live in the same street as you?
0 ☐ 1 ☐ 2 ☐ 3 ☐

You could use a tally chart to collect the data.

Number of friends	Tally	Frequency
0		
1		

Using tallies makes the data easier to collect.

Frequency is the total.

This question has responses that are not exact. The responses are given in class intervals. The data you collect is continuous.

How long does it take you to walk to school?	Tally	Frequency
Less than 5 minutes		
$5 \leqslant$ minutes < 10		
$10 \leqslant$ minutes < 15		
$15 \leqslant$ minutes < 20		

Remember there should be no gaps between each class interval and they should not overlap.

- This table has equal class intervals.
  This makes it easier to compare the frequencies in each of the classes.

You can organise discrete data into class intervals.
If pupils had between 0 and 30 friends living in the same street then it may be easier to group the number of friends, 0–4, 5–9, 10–14, ...

- When you group data into equal class intervals you should have between 4 and 10 intervals.

Exercise D2.3

1 Gina collects data on how long it takes a class of Year 6 pupils to walk to school. The data is given to the nearest minute.

6	4	6	9	1	22	24	7	17	25
5	28	26	13	18	1	2	11	8	12
22	15	19	14	12	3	19	15	7	23

Use the frequency table on the opposite page to collect these data. (*You may need to extend the table and add more class intervals*).

2 The lifetime, in hours, of 40 batteries was recorded as follows:

34	61	48	78	76	65	56	43	57	47
62	57	31	43	72	64	64	59	58	51
75	78	53	32	59	70	43	67	44	64
67	39	78	49	64	32	55	39	45	54

Copy and complete the following grouped frequency table.

Lifetime (hours)	Tally	Frequency
30–39		
40–49		
50–59		
60–69		
70–79		

3 The times taken, in seconds, for a class of children to swim 25 metres front crawl were:

28.4	34.1	20.7	28.0	29.9	24.6	32.3	28.5	39.2	36.3
22.7	31.5	34.6	27.9	35.0	21.9	29.2	29.9	31.1	32.0

Copy and complete the following grouped frequency table.

Time (seconds)	Tally	Frequency
$20 \leqslant$ seconds < 24		
$24 \leqslant$ seconds < 28		
$28 \leqslant$ seconds < 32		
$32 \leqslant$ seconds < 36		
$36 \leqslant$ seconds < 40		

D2.4 Displaying your results

This spread will show you how to:

- Construct graphs and diagrams to represent data on paper and using ICT.

KEYWORDS

Bar chart
Time series
Category
Pie chart

The parent governor of Maypole Primary School wants to get sponsorship from local businesses to buy fluorescent tabards for the children to wear on their walking bus.

To illustrate her findings she wants to use graphs and diagrams to show the data. She must choose the most appropriate diagram for her data.

- **Qualitative or discrete data can be represented using a pie chart or a bar chart.**

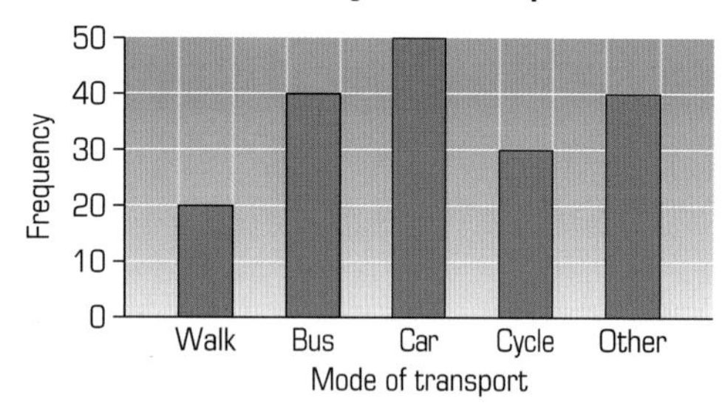

Each bar represents a category (or class). The height of each bar represents the frequency of each category.
Gaps between each bar highlight the fact that the data is discrete.

- **A pie chart represents data on a circle. Each sector represents a category or class.**

- **To calculate the size of each sector, divide 360° by the total frequency then multiply by the frequency of the class.**

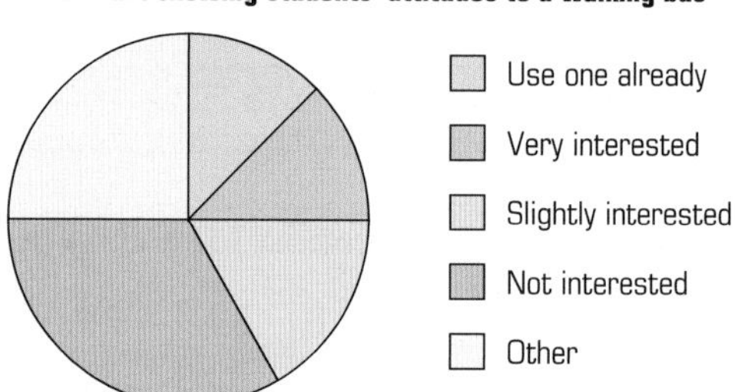

- **Continuous data can be represented using a frequency diagram or a time series graph.**

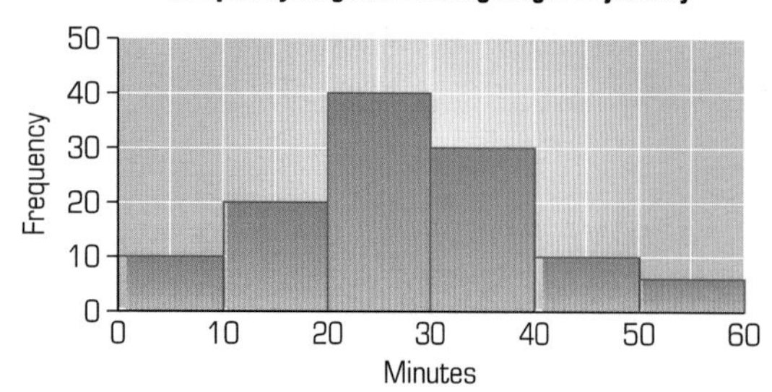

The axes must be scaled.
There are no gaps between the bars.

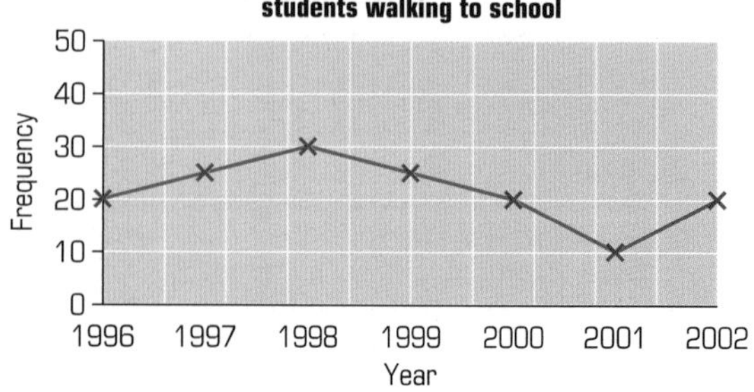

Points on the graph can be joined in order to compare trends over time

Exercise D2.4

1 For the data given in Exercise D2.3 question 1, Gina states that it would be best represented on a pie chart or a frequency diagram.
Give a reason for Gina to make this statement.

Use the data collected in the frequency table to draw:

a a pie chart

b a frequency diagram to represent the data.

2 Samantha asked some pupils at Maypole Primary School how many other pupils in their school lived in their street.
She put her findings in a frequency table.

Number of pupils	Frequency
0	2
1	5
2	7
3	12
4	8
5	3

Draw a bar-line graph to represent these data.

3 Elena collected data over two weeks on how many children walked to school at Maypole Primary School.

Day	Mon	Tue	Wed	Thur	Fri	Mon	Tue	Wed	Thur	Fri
Frequency	46	44	112	96	80	62	16	40	44	48

Draw a time series graph to represent these data.

4 **a** Draw a frequency diagram to represent the grouped data given in Exercise D2.3 question 3 showing the time taken for a group of children to swim 25 metres front crawl.
(Hint: Use your completed frequency table to draw the frequency diagram.)

b Describe briefly what your frequency diagram shows.

D2.5 Interpreting your diagrams

This spread will show you how to:

- Interpret diagrams, graphs and charts, and draw simple conclusions based on the shape of the graphs.
- Relate the conclusions to the initial problem.

KEYWORDS
Infer
Interpret

It is easy to generate lots of diagrams, but it is more important to use them to comment on what they show.
Different people may find different ways to interpret a graph.

▸ **When you interpret a diagram you make a conclusion about what it shows.**
The conclusion should relate to the reason that you collected the data.
The conclusion may lead you to ask other questions and collect more data.

These graphs were drawn to represent the data collected at Maypole Primary School.

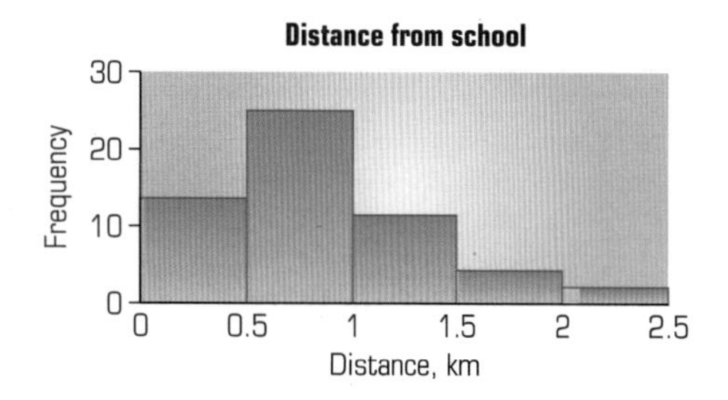

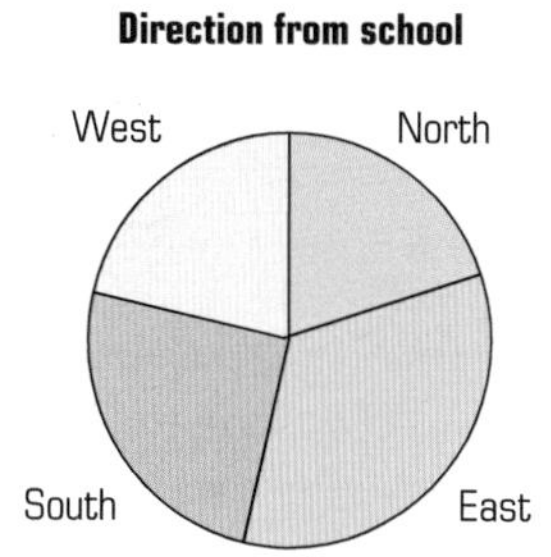

The frequency diagram shows that most children live 0.5–1 km from the school.
The parent governor decided that this distance was not too far to walk each morning.
The pie chart shows that the pupils travel from several different directions each day.

This raised an issue that there may be a need for more than one walking bus.
The parent governor needed to do more research; she must start the cycle again.

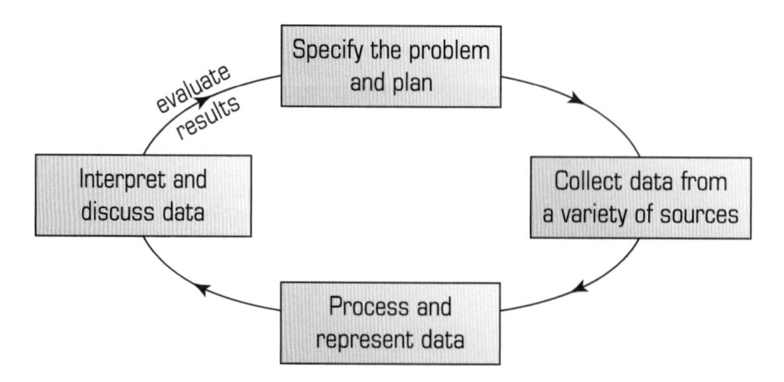

Exercise D2.5

1 In Exercise 2.4 question 2, Samantha asked some pupils at Maypole Primary School how many other pupils in their school lived in their street.
This bar-line graph shows her results.
Using the diagram state:

a How many pupils at Maypole Primary School did Samantha ask?

b How many pupils did not know anyone at their school that lived in their street?

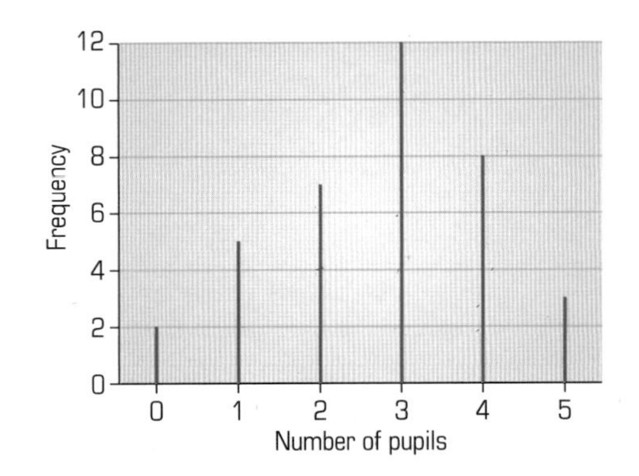

2 In Exercise 2.4 question 3, Elena collected data over two weeks on how many children walked to school at Maypole Primary School.
This time-series graph shows her results.

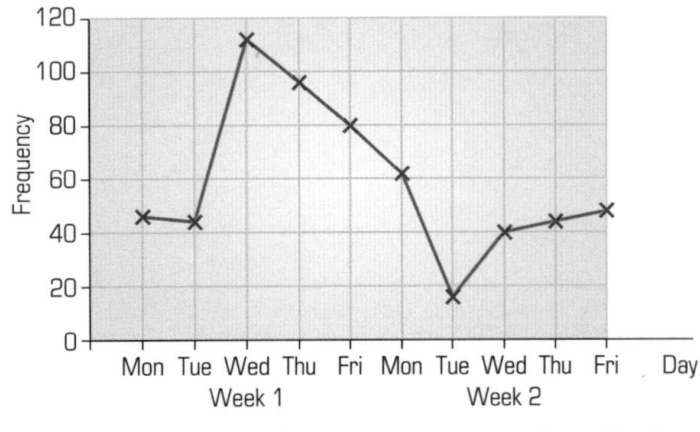

a On one of these days it was 'national walk to school day'.
Which day do you think that was?
Give a reason for your answer.

b Do you think that 'national walk to school day' affected whether or not pupils walked to school? Explain your answer.

c On the second Tuesday the number of pupils that walked to school was much smaller than any other day in those two weeks. Why do you think the number was smaller?

3 The scatter graph shows the distance from school and the time taken to travel to school for a group of 10 pupils at Maypole Primary School.

a Decide whether these statements are true or false:

i The greater the distance from school, the longer the journey time.

ii Most pupils live more than 1 km away from school.
Explain your answers.

b How many pupils:

i live 2 km from school

ii take 25 minutes to travel to school?

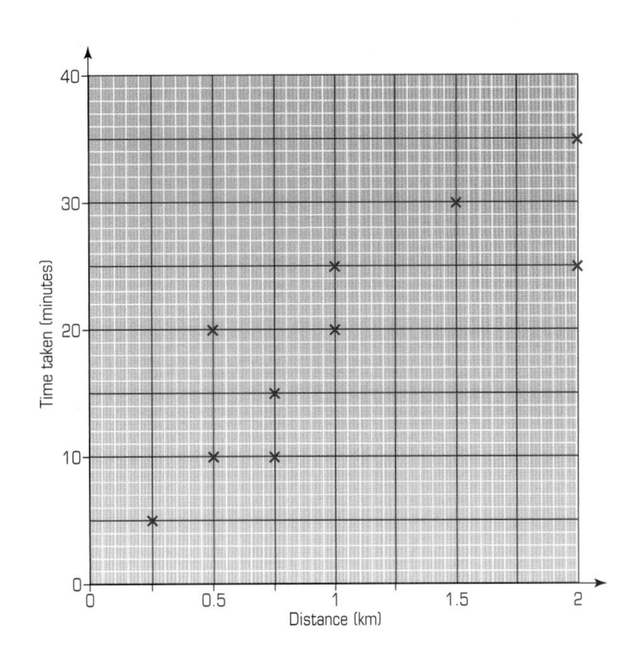

D2 Summary

You should know how to ...

1 Identify the necessary information to solve a problem.

2 Represent problems and interpret solutions in graphical form.

Check out

Is there any difference in the average amount of time boys and girls in Year 7 spend on homework?

1 a Write two relevant questions for a questionnaire to collect this data.

b Design a data collection sheet to collect responses to your questions.

2 A survey of the time spent on homework of 20 boys and 20 girls was carried out.

a The pie chart represent the result of the boys' survey.

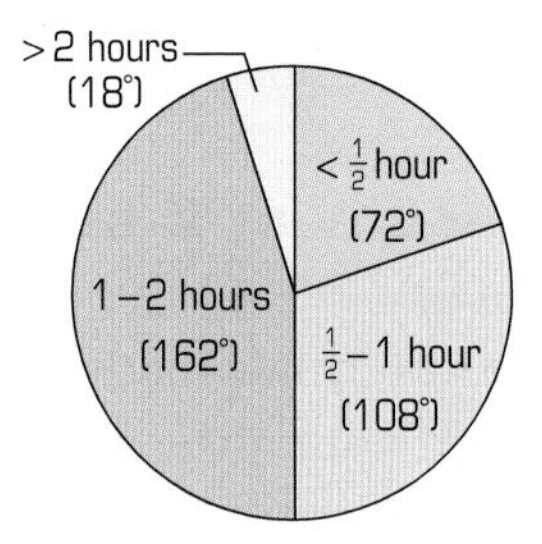

What angle is represented by 1 boy?
How many boys spent **(i)** more that 2 hours **(ii)** less than 1 hour on their homework?

b This frequency polygon was drawn to show the results of the survey of the boys and girls.

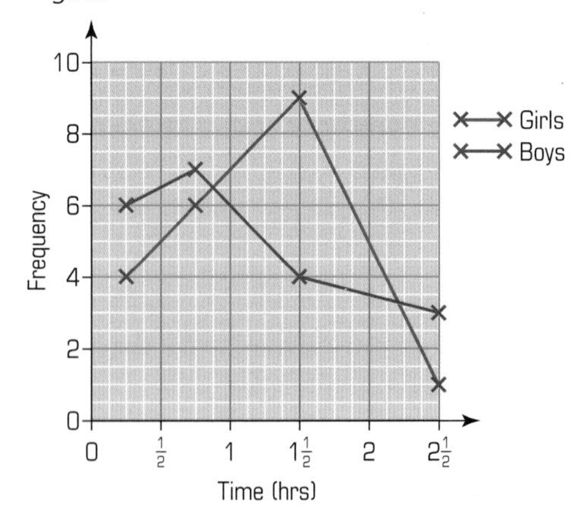

Comment on the similarities in the homework habits of the boys and girls in this survey.

N3 Multiplication and division

This unit will show you how to:

- Convert one metric unit to another.
- Multiply and divide integers and decimals by 0.1 and 0.01.
- Use the order of operations, including brackets, with complex calculations.
- Consolidate and extend mental methods of calculation, working with decimals.
- Carry out calculations on a calculator using the function key for powers.
- Enter numbers and interpret the display in different contexts.
- Use standard column procedures for multiplication and division of integers and decimals.
- Check a result by considering whether it is of the right order of magnitude.
- Use index notation for small positive integer powers.
- Solve word problems and investigate in the context of number.
- Solve complex problems by breaking them into smaller steps or tasks.

Design and technology involves calculating with decimals.

Before you start

You should know how to …

1 Work out simple order of operations.

2 Use mental strategies of multiplication with whole numbers.

Check in

1 Work out

a $3 + 4 \times 5$ **b** $\frac{9-3}{2}$

c $9 - 2 \times 3 + 1$ **d** $5 - 8 \div 2 + 1$

2 Work out mentally:

a 30×12 **b** 19×17

c 42×15 **d** 25×18

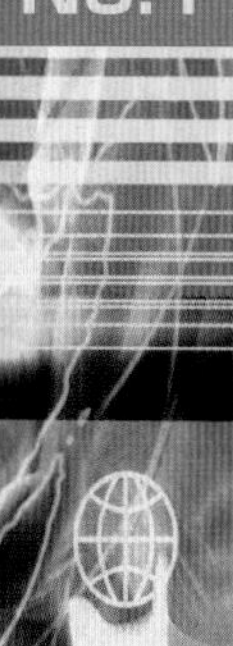

N3.1 Number and measures

This spread will show you how to:
- Multiply and divide numbers by 0.1 and 0.01.
- Convert between metric units in length and area.

KEYWORDS
Convert
Measure
Metric
Place value
Zero

The numbers that you use are based on the decimal system.
This system makes it easy to multiply by 0.1 and 0.01.

	Th	H	T	U	.			
			4	3	.	5		
$43.5 \times 0.1 =$				4	.	3	5	
$43.5 \times 0.01 =$				0	.	4	3	5
$43.5 \div 0.1 =$		4	3	5	.			
$43.5 \div 0.01 =$	4	3	5	0	.			

When you multiply by 0.1 each digit moves one place to the right.

When you multiply by 0.01 each digit moves two places to the right.

When you divide by 0.1 each digit moves one place to the left.

When you divide by 0.01 each digit moves two places to the left.

- **Multiplying a number by 0.1 is the same as dividing it by 10:** $43.5 \times 0.1 = 43.5 \div 10$.
- **Dividing a number by 0.1 is the same as multiplying it by 10:** $43.5 \div 0.1 = 43.5 \times 10$.

Metric measures are based on the decimal system.

Remember these metric lengths:

- 1 km = 1000 m
- 1 m = 100 cm
- 1 cm = 10 mm

Remember these metric areas:

- $1 \text{ cm}^2 = 100 \text{ mm}^2$
- $1 \text{ m}^2 = 10\,000 \text{ cm}^2$
- $1 \text{ km}^2 = 1\,000\,000 \text{ m}^2$

example

Convert:

a 55 mm to cm
1 cm = 10 mm
55 mm = (55 ÷ 10) cm
= 5.5 cm

b 6.8 m to cm
1 m = 100 cm
6.8 m = (6.8 × 100) cm
= 680 cm

c $72\,000 \text{ cm}^2$ to m^2
$1 \text{ m}^2 = 10\,000 \text{ cm}^2$
$72\,000 \text{ cm}^2 =$
$(72\,000 \div 10\,0000) \text{ m}^2$
$= 7.2 \text{ m}^2$

- **To convert from one unit of measure to another:**
  large to small means **more** so **multiply**
  km → m
  small to large means **less** so **divide**
  $\text{cm}^2 \rightarrow \text{m}^2$

Exercise N3.1

1 Copy and complete:

a $36 \times 10 = \_\_$ **b** $7 \div 10 = \_\_$

c $\_\_ \times 100 = 145$ **d** $67 \div 100 = \_\_$

2

> Multiplying by 10 moves all the digits one place to the left.

Describe the effect these have:

a $\times 100$ **b** $\div 10$ **c** $\div 100$

3 Convert the following measurements as stated:

a 26.7 cm into mm

b 5600 mm into km

c 90 cm into mm

4 Use a place value table to answer these questions. Allow two rows for each question.

Example: $4 \times 0.1 = 0.4$

H	T	U	.	t	h	th
		4				
		0	.	4		

a 6×0.1 **b** $6 \div 0.01$

c 3×0.01 **d** $9 \div 0.1$

e 6.3×0.01 **f** $4.56 \div 0.1$

g $8.1 \div 0.01$ **h** 3.3×0.01

5 **Investigation**

Work out the following:

a 6×100 **b** $3 \div 100$

c $6.3 \div 100$ **d** 8.1×100

Compare your answers to question 4 parts **b**, **c**, **e**, and **g**. What do you notice?

Copy and complete:

- Multiplying a number by 0.01 is the same as dividing it by__
- Dividing a number by 0.01 is the same as multiplying it by__

6 The area of this square is 640 000 cm^2.

What is the length of each side

a in centimetres

b in metres?

7 Convert these areas into the given units:

a 40 000 cm^2 (into m^2)

b 3 m^2 (into cm^2)

c 600 mm^2 (into cm^2)

8 **Investigation**

Here is a cube of side 2 m.

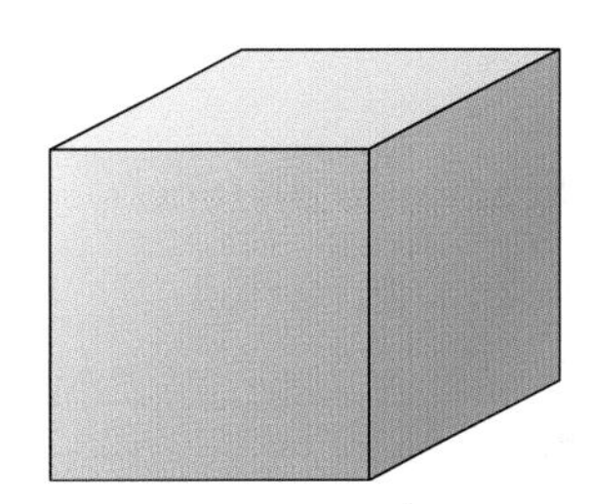

a What is its volume in cubic metres (m^3)?

b Find the length of its sides in cm.

c Find its volume in cubic centimetres (cm^3).

d Copy and complete:

$8\ m^3 = \_\_\_\ cm^3$

$1\ m^3 = \_\_\_\ cm^3$

9 Copy and complete:

a $0.46 \times 0.1 =$

b $0.6 \div 0.01 =$

c $\_\_\_ \times 0.01 = 14.5$

d $0.67 \div 0.1 =$

e $0.201 \div 0.001 =$

f $0.04 \div \_\_\_ = 4$

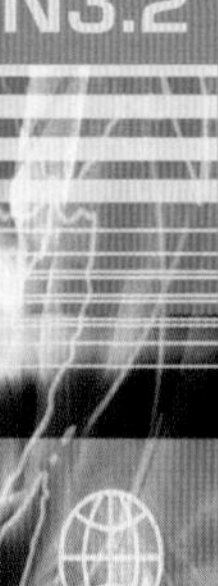

N3.2 Powers and operations

This spread will show you how to:

- Know and use the order of operations, including bracket keys on a calculator.

KEYWORDS

Base	Index, indices
Brackets	Operation
Convention	Power

A power or index is an **operator**, like +, −, × and ÷.

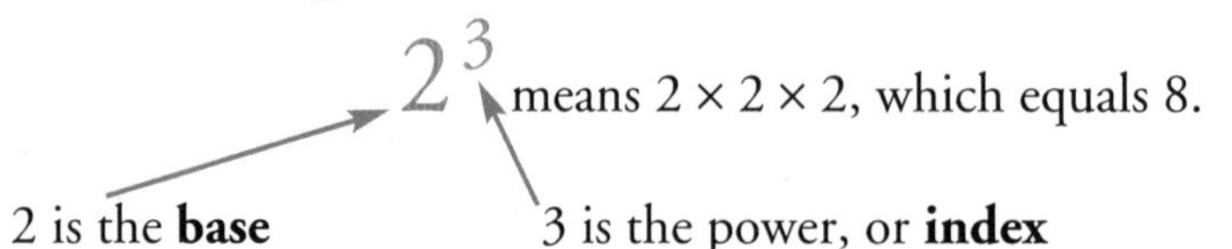

2^3 means 2 × 2 × 2, which equals 8.

2 is the **base** 3 is the power, or **index**

Note: 2^3 is not the same as 2 × 3, which is 6.

When there are a lot of different operations you must do them in this order:

example

Work out $\dfrac{(5+6)^2}{2} + 20 - 3.5$

- Brackets $\dfrac{11^2}{2} + 20 - 3.5$
- Powers $\dfrac{121}{2} + 20 - 3.5$
- Multiplication or division $60.5 + 20 - 3.5$
- Addition or subtraction $80.5 - 3.5$

So $\dfrac{(5+6)^2}{2} + 20 - 3.5 = 77$

The next example is best worked out on a calculator.

example

Work out $\dfrac{^-4 \times (^-7.3 + 4.8)^3}{21.24 + 10.1}$

Use the **bracket** keys: (and)	$= \dfrac{^-4 \times {^-2.5}^3}{21.24 + 10.1}$
Use the **index** key. On a Casio it looks like: x^y	$= \dfrac{^-4 \times {^-15.625}}{21.24 + 10.1}$
Work out the numerator	$= \dfrac{62.5}{21.24 + 10.1}$
Work out the denominator	$= \dfrac{62.5}{31.34}$
Divide and press the = key	= 1.9942565 = 1.99 to 2 decimal places.

Exercise N3.2

1 Work out the answer using either a mental or a written method.
In each case, first estimate the answer.

a $32 \div 8 \times 2$
b $13 \times 7 - 18 \times 9$
c $(3 + 4)^2 - 15 \times 9$
d $\frac{(4+1)^2}{(13-8)^2}$
e $3 \times 4^2 \div (3^2 \times 2 + 6) + 2$
f $5^2 - \frac{6}{2^2} - 1$

2 Work these out using either a written or a calculator method.
First estimate the answer.

a $4.96 + 7.4$
b $7.2 - (3.6 - 1.03)$
c $5 \times (4 - 3.7)$
d $5.74 - 3 \times (4.3 - 1.2)$
e $9.4 + \frac{7.5}{4.7} + 5.1$
f $6.3 \frac{(2.3 - 1.1)^2}{7.9} - 4.3$

3 Work these out using a calculator.
Give your answer to 2 decimal places.
Estimate first.

a $(4.9)^2 + 8.3 \times 4.1$
b $(7.4 + 8.1)^3$
c $7\frac{3}{4} \times (6 + 9.7)$
d $\frac{(5.4 + 2)}{7}$
e $8 + \frac{5}{13} - 2$
f $\frac{25}{6} \times 92$

4 Use your calculator to give the decimal equivalent of the following fractions.
Give your answer to 2 d.p.

a $\frac{7}{9}$ **b** $1\frac{7}{8}$ **c** $\frac{43}{9}$

5 **a** Work out the value of $\sqrt{6}$ on your calculator. (You can use the $\sqrt{\ }$ key. This is called the **square root**.)
Write the display in full.
b **Square** your answer to **a**.
c Copy and complete:
The opposite of square is ___ ___.

6 **Activity**

a Make as many different answers as possible by putting brackets into the expression
$2 \times 6 + 4 - 3 \times 5 + 2$
b **Extension:** Repeat part **a**, but now use **nested brackets** (brackets inside brackets).
Example: $2 \times (6 + (4 - 3 \times 5) + 2)$
The answer is $^{-}6$. Check that you can see why.

7 **a** Write down the key sequence for the expression $4 \times (\frac{19}{20})^{\frac{1}{2}}$
b Write down the output display.
c **Investigate** the function $x^{\frac{1}{2}}$. Can you find an equivalent function?

8 **Countdown**

In this activity you must use numbers from Box A, along with any operations, to make the specified number.

Box A

20	4	11	45
	7	2	

All answers should be accurate to 2 decimal places.

a Use 2 numbers to make 4.09
b Use 3 numbers to make 6.29
c Use 4 numbers to make 5.87
d **Challenge**
Use 5 numbers to make 4.90

N3.3 Mental methods

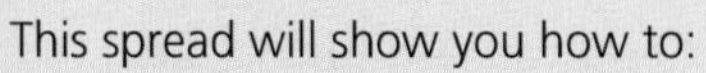

This spread will show you how to:

- Consolidate and extend mental methods of multiplication and division.

KEYWORDS

Compensate Place value
Partition Product

You can work out most multiplications in your head. Here are three useful strategies:

Partitioning

You can **partition** either number of a product by splitting it up into smaller pieces.

example

Use partitioning to work out 6.4×21.

$$\begin{aligned} 6.4 \times 21 &= 6.4 \times 20 + 6.4 \times 1 \\ &= 6.4 \times 10 \times 2 + 6.4 \times 1 \\ &= 64 \times 2 + 6.4 \\ &= 128 + 6.4 \\ &= 134.4 \end{aligned}$$

You can use jottings to help you multiply or divide mentally.

Place value

You can use place value to multiply or divide mentally any number by 0.1 or 0.01.

example

Work out mentally

a 6.7×0.1 **b** $0.35 \div 0.01$ **c** 0.6×0.9

a $6.7 \times 0.1 = 6.7 \div 10 = 0.67$

b $0.35 \div 0.01 = 0.35 \times 100 = 35$

c $$\begin{aligned} 0.6 \times 0.9 &= 6 \div 10 \times 9 \div 10 \\ &= 6 \times 9 \div 10 \div 10 \\ &= 54 \div 100 \\ &= 0.54 \end{aligned}$$

Doubling and halving

Some calculations are easier to do if you first double one of the numbers. You later compensate by halving the answer.

example

Work out mentally: **a** 12×1.5 **b** 5.2×4.5

a First double 1.5: $12 \times 3 = 36$
Then halve the answer: $36 \div 2 = 18$

b First double 4.5: $5.2 \times 9 =$
Partition: $5.2 \times 10 - 5.2 \times 1 = 52 - 5.2 = 46.8$
Halve the answer: $46.8 \div 2 = 23.4$

Exercise N3.3

1 Use partitioning to calculate these mentally:

a 7.4×11
b 16.5×9
c 3.7×19
d 170×21

2 Choose a strategy to do each of these multiplications mentally:

a 62×5 **b** 6.2×5
c 12.4×10 **d** 1.24×0.5
e 3.7×0.1 **f** 98.3×0.01

3 Game

You will need two 9-sided dice or two sets of digit cards.
Use labels to number the faces 0.1, 0.2, ... 0.9.

- Player 1 rolls the dice twice and works out the score by multiplying the two values together.
  For example, if she rolls 0.3 and 0.7, the score is $0.3 \times 0.7 = 0.21$
- Player 1 makes a note of her score and passes the dice to player 2.
- The players keep a running total of their scores. If you throw a double you get another turn.
- The aim of the game is to make 3 exactly.

4 Choose a strategy to solve each of these divisions mentally:

a $88 \div 4$ **b** $88 \div 0.5$
c $22 \div 4$ **d** $440 \div 4$
e ${}^{-}2.6 \div 0.1$ **f** $0.27 \div 0.01$

5 Use your knowledge of place value to answer these questions mentally.

a 1.7×3 **b** 0.3×0.4
c 0.7×0.05 **d** 75×0.04
e $20 \div 0.5$ **f** ${}^{-}40 \div {}^{-}0.2$

6 Work these out by partitioning:

a 320×2.4 **b** 7.2×1.9
c 4.3×2.1 **d** 6.7×190

7 Reena weighs 20 sweets in a science experiment, and the total weight is 45 g.
What is the mean weight of each sweet?

8 Callum has 50 pieces of cotton, each measuring 0.05 m.
He needs three times this amount.
What total length of cotton does Callum need?

9 Imagine you are working at NASA.
The time is ${}^{-}30.5$ minutes before lift off.
You want to share the time at the control panel equally between two colleagues and yourself.
How long would you each get?

10 Puzzle

Use the numbers in boxes A, B and C to make six different divisions.
You must use one number from Box A divided by one number from Box B to make an answer that is in Box C:

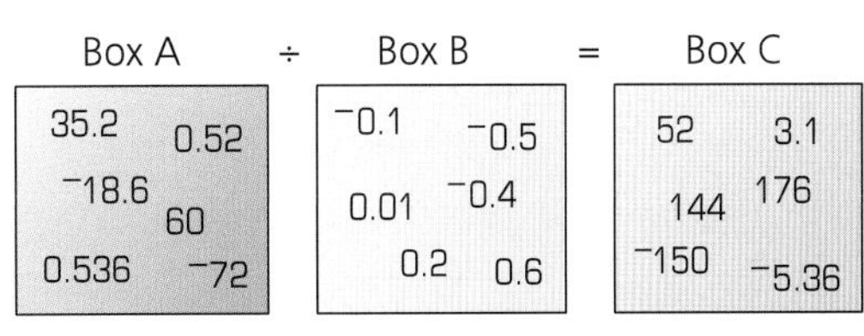

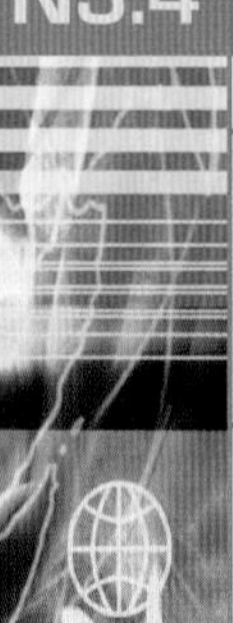

N3.4 Multiplying on paper

This spread will show you how to:

- Refine written methods of multiplication and extend to decimals.
- Check a result by considering whether it is of the right magnitude.

KEYWORDS
Estimate
Partition

When numbers are too difficult to multiply in your head you need to use a written method.
Here are two written methods for multiplying numbers together.

Partitioning and the grid method

You can split up difficult multiplications by partitioning.

example

Work out 3.4×2.7

- First estimate: $3 \times 3 \approx 9$
- Partition the numbers
  - 3.4 can be split into 3 and 0.4
  - 2.7 can be split into 2 and 0.7
- Multiply the pairs of numbers together
- Now add them up: $6 + 0.8 + 2.1 + 0.28 = 9.18$

Using a grid can help:

	2	0.7
3	6	2.1
0.4	0.8	0.28

Note: This is close to the estimate of 9 so it is likely to be correct.

Standard method

Partitioning takes up a lot of space. Instead you could use the standard method.

example

Work out 3.47×27 using the **a** grid method and **b** the standard method.

First estimate: $3 \times 30 = 90$

Grid method

	20	7
3	60	21
0.4	8	2.8
0.07	1.4	0.49

$$60 + 21 + 8 + 2.8 + 1.4 + 0.49$$
$$= 89 + 4.2 + 0.49$$
$$= 93.2 + 0.49$$
$$= 93.69$$

Standard method

$3.47 \times 27 = 347 \div 100 \times 27$
$= 347 \times 27 \div 100$

$$\begin{array}{r} 347 \\ \times\ 27 \\ \hline 2429 \\ 6940 \\ \hline 9369 \end{array}$$

$9369 \div 100 = 93.69$
So $3.47 \times 27 = 93.69$

Exercise N3.4

1 Work these out using a mental or written method.

a 23×30 **b** 45×9
c 4×120 **d** 300×46
e 620×3 **f** 366×6

2 Fill in the missing digits.

a

```
   405
 × 7□
 -----
  1215
 □□□□□
 -----
 □□□□□
 -----
```

b

```
   62□
 × 43
 -----
  1872
 □□□□□
 -----
 □□□□□
 -----
```

3 Use a written method to calculate:

a 4.81×7 **b** 12.13×6
c 2.34×9 **d** 15.8×8

4 Find two consecutive whole numbers with a product of:

a 552 **b** 1406

Hint: **Consecutive** means they have a difference of 1.

5 Investigation

Many products can be made with the digits 1, 2 and 3.

For example: 1×23 and 12×3

a What is the largest product you can make?

b What is the smallest product you can make?

c Investigate using four digits: 1, 2, 3, 4.

d Extend to five digits: 1, 2, 3, 4, 5.

6 Use the grid method to work out:

a 23.4×4.5 **b** 3.5×7.8
c 94.2×28.1 **d** 3.19×4.8

7 Use a written method to calculate:

a 6.25×13 **b** 1.79×24
c 13.4×27 **d** 25.74×12

8 Use the facts in brackets to help you to work out the answers to these questions:

a $16.2 \div 20$ $(162 \div 8.1 = 20)$

b $\_\_ \times 20 = 3.24$ $(3.24 \times 10 = 32.4)$

c $6.93 \div \_\_ = 2.31$ $(231 \times 3 = 693)$

d $\_\_ \times 5 = 355$ $(3.55 \div 5 = 0.71)$

9 Using your knowledge of place value, work out what number you must multiply 0.02 by to make it equal to:

a 1 **b** 10 **c** 100

10 Show how you could use the grid method to work out a problem with 3 decimal places, for example 3.145×6.784.

11 Explain why 0.02×0.04 is not equal to 0.08.

12 A rectangle measures 0.78 m by 0.12 m.

0.78 m

0.12 m

What would be its area

a in square metres

b in square centimetres?

13 Challenge

When is the result of a multiplication smaller than the number you multiplied?

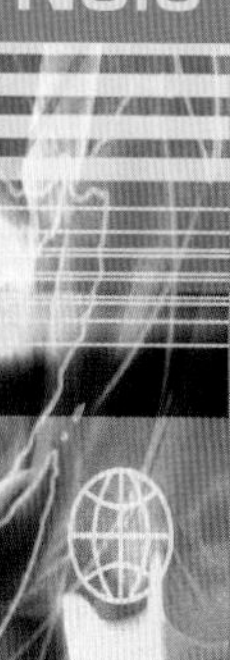

N3.5 Multiplying decimals

This spread will show you how to:

- Refine written methods of multiplication of whole numbers to ensure efficiency, and extend to decimals.
- Round decimals to the nearest whole number or to one decimal place.

KEYWORDS
Estimate
Decimal place
Round

When you calculate with decimals you need to understand about **rounding**.

example

Round 4.378 to **a** the nearest whole number **b** one decimal place.

a The whole number is 4 so 4.378 is between 4 and 5. It is nearer 4 so 4.378 is 4 to the nearest whole number.

b The first decimal place is 3 so 4.378 is between 4.3 and 4.4. It is nearer 4.4 so 4.378 is 4.4 to one decimal place.

A number line will help:

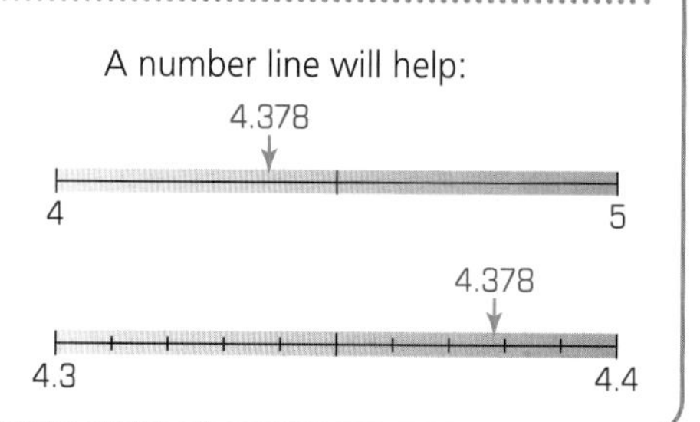

- To round to one decimal place, look at the first decimal place. If the second decimal place is 5 or more you round up.

You can use the standard method to multiply decimals.

example

Work out 32.4×4.5 and give your answer to the nearest whole number.

First estimate: $30 \times 5 = 150$

32.4×4.5

$= 324 \div 10 \times 45 \div 10$

$= 324 \times 45 \div 10 \div 10$

$= 324 \times 45 \div 100$

Now use the standard method:

$$\begin{array}{r} 324 \\ \times\ 45 \\ \hline 1620 \\ 12960 \\ \hline 14580 \end{array}$$

$14\,580 \div 100 = 145.8$

The answer is 146, to the nearest whole number.

Exercise N3.5

1 Calculate the following. Show your working and estimate the answer before each calculation. Set out your work like this:

Work out 354×18.

Estimate is

$356 \times 20 \approx 350 \times 10 \times 2 = 7000$

$$\begin{array}{r} 354 \\ \times 18 \\ \hline 2832 \\ 3540 \\ \hline \end{array}$$

Answer 6372

a 97×58 **b** 36×58
c 78×16 **d** 70×32
e 17×21 **f** 391×93
g 446×27 **h** 276×68
i 479×18 **j** 242×76

2 Use the standard method to answer these questions. Estimate first.

a If a school can fill 16 coaches that each seat 35 pupils, how many pupils does the school have?

b What would be the total number of pennies in 23 piggy banks if each piggy bank contained exactly 431 pennies?

c If petrol costs 78p per litre, how much would 560 litres cost?

d A shop has 329 gift vouchers left in stock. If each gift voucher is worth £25 how much are the gift vouchers worth altogether?

3 Use the standard method to multiply these decimals, estimating first.

a 9.47×43 **b** 1.78×16
c 13.8×65 **d** 6.46×55
e 7.37×18 **f** 8.24×9.1
g 5.16×7.4 **h** 6.41×1.3

4 Use the standard method to multiply the decimals, estimating first. Give your answers correct to 1 d.p.

a 8.47×0.46 **b** 65.4×0.85
c 5.73×0.46 **d** 5.15×0.48
e 8.13×0.25 **f** 57.1×0.43
g 7.43×0.45 **h** 3.46×0.84
i 6.37×0.38 **j** 2.21×0.46

5 **Challenge**

Use the standard method to work out 0.00056×0.769.

Give your answer as an exact decimal.

6 Here is a diagram of a patio:

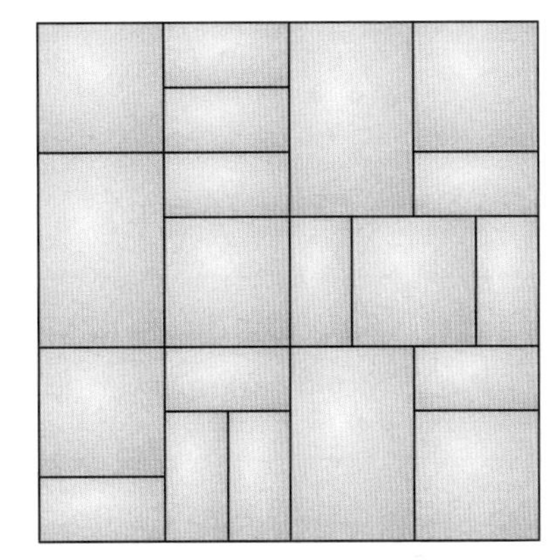

The patio flags come in three sizes:

1.26 m by 0.84 m (large)
0.84 m by 0.84 m (medium)
0.84 m by 0.42 m (small)

a What is the area of each size of flags?

b What is the:

i area in m^2, of the whole patio?

ii total area of the medium sized flags used?

iii area in cm^2, of the whole patio?

c **Investigation**

Can the whole patio be filled exactly with only one size of flag?

Show your working.

N3.6 Dividing on paper

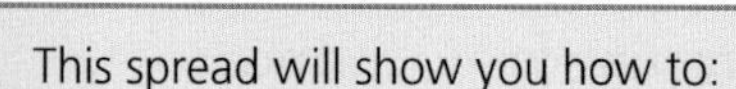

This spread will show you how to:

- Refine written methods of division and extend to decimals.
- Check a result by considering whether it is of the right magnitude.

KEYWORDS
Decimal
Divide
Estimate

It can be difficult to divide decimals in your head.
You can use the standard method on paper.

example

Work out $70.4 \div 16$.
Give an estimate first and show your workings clearly on paper.

First estimate: $80 \div 20 = 4$

Set out your working like this:

$$\begin{array}{rl} 16\overline{)70.4} & \\ \underline{64} & 16 \times 4 \\ 6.4 & \\ \underline{6.4} & 16 \times 0.4 \\ 0 & \end{array}$$

$16 \times 4.4 = 70.4$

Check: 4.4 is close to the estimate of 4 so it is likely to be correct.

$16 \times 4.4 = 70.4$ so $70.4 \div 16 = 4.4$

- Often when you divide two whole numbers you get a decimal answer. You may have to round your answer up or down.

example

Work out 6325 divided by 8 to the nearest whole number.

First estimate: $6400 \div 8 = 800$

Now set out your working:

$$\begin{array}{rl} 8\overline{)6325.0} & \\ \underline{5600.0} & 8 \times 700 \\ 725.0 & \\ \underline{720.0} & 8 \times 90 \\ 5.0 & \\ \underline{4.8} & 8 \times 0.6 \\ 0.2 & \\ & \overline{8 \times 790.6} \end{array}$$

Estimate with 6400 rather than 6300 because you can work it out in your head.

All workings are to 1 decimal place.

Stop when you reach a number to one decimal place.

So $6325 = 791$ to the nearest whole number.

- If your answer is to be rounded to the nearest whole number, you should work to one decimal place throughout.

Exercise N3.6

1 Calculate these mentally.

a 68 ÷ 4 **b** 64 ÷ 8
c 72 ÷ 8 **d** 15 ÷ 5
e 96 ÷ 6 **f** 49 ÷ 7
g 39 ÷ 3 **h** 88 ÷ 4

2 Calculate these using a mental or written method.

a 325 ÷ 5 **b** 306 ÷ 9
c 488 ÷ 8 **d** 198 ÷ 6
e 261 ÷ 3 **f** 548 ÷ 2
g 236 ÷ 4 **g** 189 ÷ 7

3 Calculate these using a mental or written method.

a 9.5 ÷ 5 **b** 7.4 ÷ 2
c 4.5 ÷ 9 **d** 3.5 ÷ 7
e 7.8 ÷ 6 **f** 7.2 ÷ 8
g 6.3 ÷ 3 **h** 5.2 ÷ 4

4 Calculate these using a written method.

a 13.6 ÷ 4 **b** 24.4 ÷ 8
c 17.5 ÷ 7 **d** 31.5 ÷ 5
e 9.66 ÷ 6 **f** 4.59 ÷ 9
g 5.37 ÷ 3 **h** 7.28 ÷ 4

5 Harry has 8 friends. He wants to share out some money equally between them. What amounts between £8 and £9 can be shared out exactly without any money left over?

For example, £8.40 can be shared out exactly:

8.40 ÷ 8 = 1.05

Each person would get £1.05.

6 Complete the table below dividing the number on the top by the number down the side. One is already done for you:

÷	0.72	3.60	0.0288
6			0.0048
4			
3			

7 Zog the alien spent most of his time on a planet where 7 pence coins were in circulation. He worked out that using 7p coins he could make exactly £6.14.

a Is he correct? Show your working out.

b What amounts between £6 and £7 can he make?

8 **Investigation**

In a game:

blue counters = 13 points
red counters = 14 points
yellow counters = 15 points

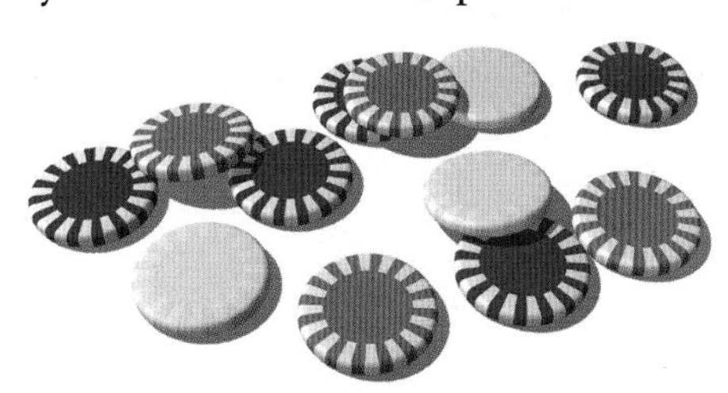

a What scores between 100 and 120 can you make exactly?

b Would your answer to part **a** change if each counter's value was doubled?

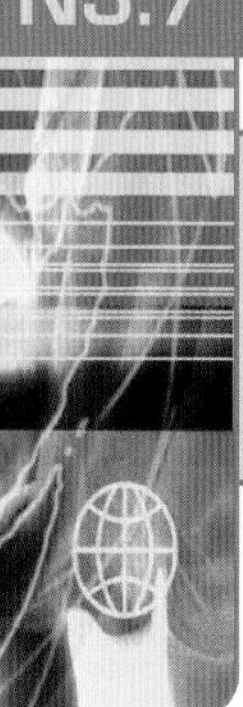

N3.7 Dividing with remainders

This spread will show you how to:

- Refine written methods of division of decimals.
- Round decimals to the nearest whole number or to one d.p.
- Check a result by considering whether it is of the right magnitude.

KEYWORDS

Decimal, Estimate, Divide, Multiply, Divisor

Now you can divide:

- Whole numbers by whole numbers, for example $6325 \div 8$.
- Decimals by whole numbers, for example $70.4 \div 16$.

> **When you divide by a decimal on paper, first turn the decimal into a whole number by multiplying.**

example

Calculate $467 \div 5.4$ giving your answer to 1 decimal place.

Estimate first: $450 \div 5 = 90$

Now change the divisor to a whole number.

You can do this by multiplying by 10. $467 \div 5.4 = 4670 \div 54$

Now use the standard written method:

$$
\begin{array}{r l}
54\overline{)4670.00} & \\
\underline{4320.0} & 54 \times 80 \\
350.00 & \\
\underline{324} & 54 \times 6 \\
26.00 & \\
\underline{21.6} & 54 \times 0.4 \quad \leftarrow 54 \times 4 = 216 \\
4.40 & \\
\underline{4.32} & \underline{54 \times 0.08} \quad \leftarrow 54 \times 8 = 432 \\
0.08 & \\
 & \underline{54 \times 86.48}
\end{array}
$$

Stop at 2 decimal places.

So $467 \div 5.4 = 86.48$

$= 86.5$ to 1 decimal place.

The answer is close to the estimate.

The remainder after a division can be expressed as a fraction or a decimal.
For example:

$$193.6 \div 7 = 1936 \div 70 = \frac{1936}{70} = 27\tfrac{46}{70} = 27\tfrac{23}{35}$$

or $193.6 \div 7 = 27.65714 .... = 27.7$ (to 1 d.p.)

Remember that division is related to fractions: $\frac{24}{3} = 24 \div 3 = 8$

Exercise N3.7

1 Calculate the following.

a 184 ÷ 8 b 243 ÷ 9
c 59.8 ÷ 13 d 47.6 ÷ 17
e 66.7 ÷ 23 f 231.2 ÷ 34

2 Calculate the following.

a 129.6 ÷ 18
b 327 ÷ 23 (leave the remainder as a fraction)
c 285 ÷ 16 (whole number remainder)
d 1050.8 ÷ 71
e 876 ÷ 28 correct to 1 d.p.
f 995.4 ÷ 42

3 a Calculate the following to the nearest whole number.

i 1873 ÷ 8 ii 2743 ÷ 11
iii 6943 ÷ 13 iv 4876 ÷ 27

b Calculate the above divisions to 2 d.p.

4 Calculate the following to 1 d.p.

a 435 ÷ 2.8 b 573 ÷ 4.6
c 265 ÷ 3.3 d 469 ÷ 7.2
e 74 ÷ 2.6 f 384 ÷ 6.3

5 a A piece of wood 504 cm is to be cut into pieces of length 3.6 cm. How many pieces will there be?

b Cartmen has saved $803 so far this year. If he saves $2.20 per day for how many days has he been saving?

6 Karen has earned £9338.77 so far this year. If she gets paid £6.23 per hour how many hours has she worked?

7 Calculate the missing numbers.

a 5.6 × ____ = 95.2
b ____ × 34 = 486.2
c 12.7 × ____ = 546.1
d ____ × 23.3 = 955.3

8 In these questions you will need to use a mental or written method and decide what to do with the remainder.

a 118 students and staff are going on a school trip to the zoo. How many 36-seat coaches need to be booked?
b How many crates can be completely filled from 226 bottles, if each crate can hold 18 bottles?
c Sanita is saving up for a computer game costing £58.65. She can save £4.30 per week.
How many weeks will it take before Sanita can afford the game?
d A plank of wood is 3.8 m long. Chanelle needs to cut it into equal lengths of 5.6 cm. How much will be left over?

9 If you know that 2.34 ÷ 0.4 = 5.85 then use this to calculate

a 23.4 ÷ 0.4 b 234 ÷ 0.4
c 2.34 ÷ 0.8 d 23.4 ÷ 40
e Copy and complete this diagram.

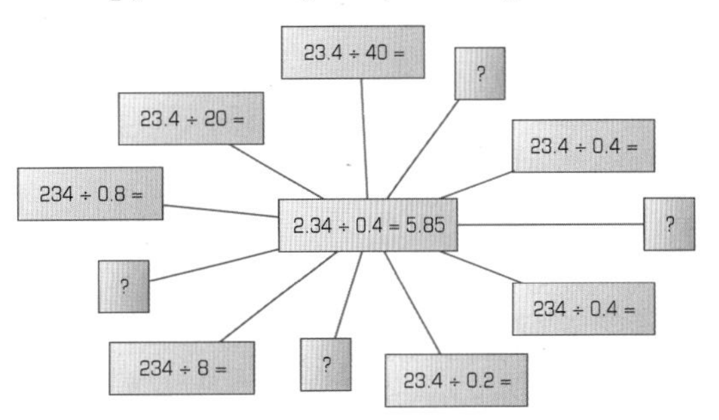

10 **Extension:**
Here is a rectangular paving slab.
Area = 0.4859 m^2
Length = 1.13 m
Calculate the width of the paving slab.

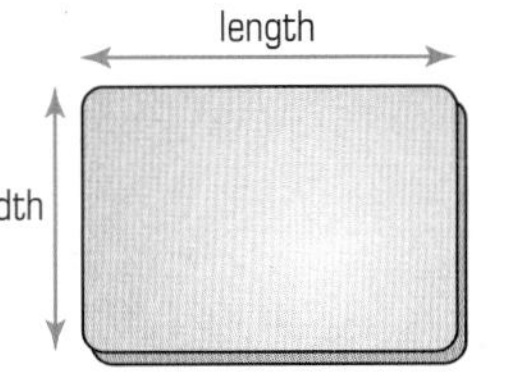

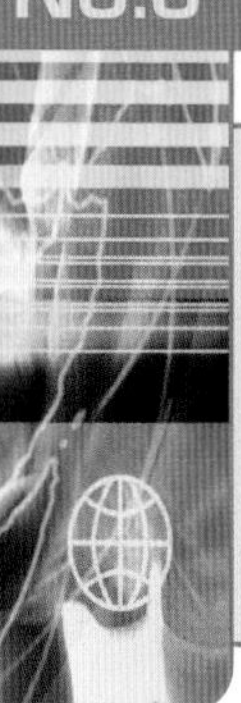

N3.8 Calculator methods

This spread will show you how to:

- Carry out more complicated calculations using the facilities on a calculator.
- Interpret the display on a calculator in different contexts.
- Check a result by considering whether it is of the right order of magnitude.

KEYWORDS

Accuracy
Brackets
Interpret
Operation
Memory

You can use a calculator to do difficult calculations but always estimate the answer first.
Use these examples to check you know how your calculator works.
You should find out what order of operations your calculator uses.

example

Work out $\dfrac{3.24 \times 5^2}{6.08 \times (34 - 3.7)}$ accurate to 2 decimal places.

First estimate: $(3 \times 25) \div (5 \times 30) = 75 \div 150 = 0.5$

This really means: $(3.24 \times 5^2) \div [6.08 \times (34 - 30.7)]$
Work out the two main brackets first, starting with the one on the left:

$(3.24 \times 5^2) = 3.24 \times 25$
$= 81$

Powers go before multiplication, and you can use the x^2 key for this. Store the answer in the calculator's memory, often [Min].

$[6.08 \times (34 - 3.7)] = 6.08 \times 30.3$
$= 184.224$

Brackets go before multiplication.

Now divide the contents of one bracket by the other.
You can recall the memory:
MR or $81 \div 184.224 = 0.43968212$

This is the calculator answer. You should round it off to the chosen degree of accuracy.

$\dfrac{3.24 \times 5^2}{6.08 \times (34 - 3.7)} = 0.44$ to 2 d.p.

This is close to the estimate of 0.5.

▸ **You need to be able to interpret your calculator answer in the context of the original problem.**

example

A calculator gives an answer as 6.15, measured in hours.
How many hours and minutes is this?

Note: the answer is not 6 hours and 15 minutes.

$0.15 = \frac{1}{10} + \frac{5}{100} = \frac{15}{100}$
$\frac{15}{100}$ of an hour $= \frac{15}{100} \times 60$ minutes $= 15 \div 100 \times 60 = 9$ minutes
So 6.15 hours = 6 hours and 9 minutes

Exercise N3.8

1 Use your calculator to work out the following questions. Do a mental check first and show your working for the mental approximation.

a $\dfrac{6.78 \times 3.4}{34}$ **b** $\dfrac{3.6 \times 8.86}{26}$

c $\dfrac{4.5^2 \times 6}{5^2}$ **d** $\dfrac{4^3 \times 5^2}{3 + 6}$

e 2^7

2 If you worked out £10 ÷ 4 on your calculator, what would:

a the calculator display read?

b the answer mean in money?

3 Write in words what these calculator displays mean:

a 6.77777777777778

b 9.6 (when the answer is in £)

c 10.25 (when the answer is in minutes)

4 Interpret the following calculator displays giving your answers in minutes and seconds:

a 4.5 minutes **b** 6.2 minutes

c 2.25 minutes **d** 0.1 minutes

5 **Investigation**

The square root of 9 is 3, or $\sqrt{9} = 3$ because $3 \times 3 = 9$.

Use a calculator to work out the answers to these square roots.

a $\sqrt{10}$ **b** $\sqrt{20}$ **c** $\sqrt{50}$

d $\sqrt{60}$ **e** $\sqrt{70}$

Consider your answers to questions **a** to **e** and write down the key sequence you would have to press to 'get back to' the numbers 10, 20, 30, 40 50. Write down what you have noticed.

6 Use your calculator to help you order these fractions (by converting them to decimals): $\frac{9}{16}$, $\frac{2}{3}$, $\frac{4}{7}$, $\frac{5}{8}$, $\frac{9}{14}$, $\frac{5}{9}$, $\frac{7}{10}$, $\frac{22}{30}$

7 Using the fraction key $a\frac{b}{c}$ convert each of the following times to hours as decimals, for example 33 minutes = $\frac{33}{60}$ $= 33 \div 60 = 0.55$ hours

a 38 min **b** 1 h 41 min

c 2 h 38 min **d** 3 h 23 min

e 1 min **f** 4 h 3 min

8 **a** Estimate the answer to $(3.8 \times 9)^2 \times 9.9^2$

b Write down the key sequence you would use to work out part **a** on your calculator.

c Do the calculation and write down the answer you get.

9 What calculator keys would you press to work out these calculations? Write down

a the key sequence

b the answer you get.

i $3^{\frac{1}{3}}$ **ii** $3\sqrt{5}$

10 Interpret the following calculator displays giving your answers in seconds:

a 4.3333333 minutes

b 6.6666666 minutes

c 2.15 minutes

d 0.125 minutes

11 Use your calculator to work out the following.

a $\dfrac{(7 \times 8)^2}{7 \times 4}$ **b** $\dfrac{7 \times 8^2}{7 \times 4}$

c $-3.4^2 + 8$ **d** $(-3.4)^2 + 8$.

N3 Summary

You should know how to ...

1 Convert one metric unit to another.

2 Use standard column procedures for multiplication and division of integers and decimals.

3 Understand where to position the decimal point by considering equivalent calculations.

4 Interpret the calculator display in different contexts.

Check out

1 Convert:

a 35 m to cm b 35 m to km

c 35 m^2 to cm^2 d 35 m^2 to km^2

2 Showing all your working calculate:

a 4.96 × 16 b 19.61 × 23

c 3.14 × 0.52 d 6.4 × 0.31

e 196 ÷ 7 f 161.7 ÷ 7

g 56 ÷ 3.2 h 442.8 ÷ 3.6

3 Given that 57 × 3 = 171 and 57 ÷ 3 = 19

Work out mentally:

a 5.7 × 0.01 b 5.7 × 0.03

c 0.57 × 0.01 d 5.7 ÷ 0.1

e 5.7 ÷ 0.3 f 5.7 ÷ 0.03

4 The square root button on your calculator finds a number x such that x^2 is the number you input:

For example, $\sqrt{4} = 2$ so $2^2 = 4$.

Find the square root of all the integers up to 30.

Which of the square roots can be written exactly? What can you say about these numbers?

A3 Functions and graphs

This unit will show you how to:

- Recognise and use multiples, factors, common factor, highest common factor, lowest common multiple and primes.
- Find the prime factor decomposition of a number.
- Know the meaning of the words formula and function.
- Use linear expressions to describe the nth term of an arithmetic sequence, justifying its form.
- Generate points in all four quadrants and plot the graphs of linear function.
- Recognise that equations of the form $y = mx + c$ correspond to straight-line graphs.
- Solve word problems and investigate in the context of algebra.
- Identify exceptional cases or counter-examples.

The output of a computer depends on the data that you input.

Before you start

You should know how to ...

1 Plot points on a coordinate grid.

2 Substitute integers into simple formulae.

3 Use index notation for small positive powers.

Check in

1 Draw a set of axes labelled $^{-}5$ to 5.
Plot these points on your grid and join them up.
(3, 1), (3, 4), (3, $^{-}1$), (3, $^{-}4$), (3, 2)
What do you notice? Explain any pattern.

2 Find the value of the required variable.
- **a** Find y when $x = 3$ and $y = 2x + 1$
- **b** Find T when $n = 6$ and $T = 6n - 4$
- **c** Find w when $z = 3$ and $w = 4z^2$

3 **a** Evaluate:
i 3^3 **ii** 13^2 **iii** 2^5 **iv** $(^{-}4)^2$ **v** 10^3

b Write using index notation:
- **i** $3 \times 3 \times 3 \times 3 \times 4 \times 4 \times 4$
- **ii** $x \times x \times x \times x \times x \times x$
- **iii** $5 \times 5 \times 5 \times y \times y$

A3.1 Factors, multiples and primes

This spread will show you how to:
- Recognise and use factors and primes.
- Find the prime factor decomposition of a number.
- Recognise multiples and find lowest common multiples.

KEYWORDS
Divisible, Prime, Factor, Product, Multiple, Square

- The **factors** of a number are numbers that divide into it without leaving a remainder.

example

Find the factors of 84.

First list the products that make 84: 1×84 2×42 3×28 4×21 6×14 7×12
Now list the factors: 1, 2, 3, 4, 6, 7, 12, 14, 21, 28, 42, 84

- If a number has only two factors it is a **prime** number.

- All numbers can be written as a product of their **prime factors**.

Note: The sequence of prime numbers goes:
2, 3, 5, 7, 11, 13, 17, ...
You should learn the prime numbers up to 30.

example

Write in terms of their prime factors:

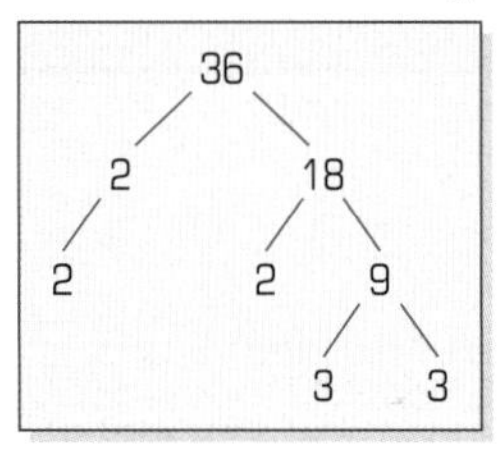

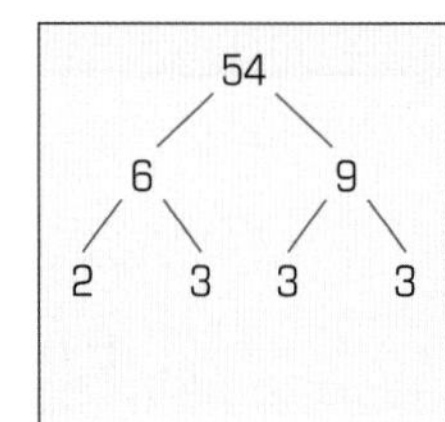

a 36
$36 = 2 \times 2 \times 3 \times 3$
$= 2^2 \times 3^2$

b 54
$54 = 2 \times 3 \times 3 \times 3$
$= 2 \times 3^3$

Note:
2 and 3 are **common factors** of 36 and 54.
$36 = 2 \times 2 \times 3 \times 3$
$54 = 2 \times 3 \times 3 \times 3$
$2 \times 3 \times 3 = 18$ is the **highest** common factor.

- A **multiple** of a number N is $N \times$ any whole number.

example

a List the first eight multiples of 8 and 3.
b Identify the first number that appears in both lists of multiples.

a multiples of 8: 8, 16, 24, 32, 40, 48, 56, 64, . . .
multiples of 3: 3, 6, 9, 12, 15, 18, 21, 24, ...
b 24 appears in both lists so it is the **lowest common multiple** of 3 and 8.

Exercise A3.1

1 Using pairing and deciding when to stop, list all the factors of:
a 64 **b** 126

2 **a** List all the factors of 36 and of 90.
b List all the common factors of 36 and 90.
c State the highest common factor of 36 and 90.
d Cancel the fraction $\frac{36}{90}$ to its simplest form.

3 True or false? 9 is a factor of 1342.

4 Have a look at these calculations. They can be worked out quickly using factors:

$$\begin{aligned} 35 \times 24 &= 5 \times 7 \times 2 \times 12 \\ &= 5 \times 2 \times 7 \times 12 \\ &= 10 \times 84 \\ &= 840 \end{aligned}$$

$$\begin{aligned} 300 \div 12 &= \frac{3 \times 100}{3 \times 4} \\ &= \frac{100}{4} \\ &= 25 \end{aligned}$$

Try these calculations using factors:
a 18×45 **b** $630 \div 14$

5 **a** Find all the numbers, below 100, with only 3 factors.
b Find the number, below 100, which has the most factors.

6 **a** Write down the multiples of 6 and of 13 up to 100.
b Write down the lowest common multiple of 6 and 13.
c Use your answer to **b** to decide if $\frac{5}{6}$ or $\frac{11}{13}$ is bigger.

7 **a** Using a factor tree, decompose both 324 and 468 into their prime factors. Write your answer using indices.
b Use your answers to find out:
i if 324 is a square number and, if so, write its square root
ii the highest common factor of 324 and 468
iii five factors of 468 over 20

8 The product of a married couple's ages is 2106. If they are both at least 25 and both under 60, how old are they?

9 Choosing numbers between 40 and 60, find:
a a multiple of 6 which is also a factor of 162
b a prime number which is also a factor of 590
c a factor of 147 which is also a square number
d a number with just one prime factor

10 True or false? A prime number cannot be a multiple.

11 **Investigation**
Investigate this statement:
All square numbers have an odd number of factors.

12 **True or false:**
a A square number can never be prime.
b A multiple of 6 must be a multiple of 2 and of 3.
c The smallest prime number over 300 is 301.

Explain your answers.

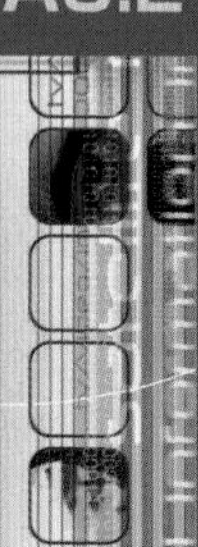

A3.2 Patterns in numbers

This spread will show you how to:

- Generate terms of a sequence, given a rule.
- Use linear expressions to describe the nth term of an arithmetic sequence.

KEYWORDS

Formula, Sequence, Generate, General term, Linear

Many number patterns can be described as **linear sequences.**

- **In a linear sequence, the difference between successive terms is the same.**

For example, the 3 times table is linear: 3, 6, 9, 12, 15, 18, ...
The difference between terms is 3.

- Multiples of a number always follow a linear sequence.

Here is a linear sequence that does not consist of multiples: 4, 7, 10, 13, 16, ...
If you wanted to find the hundredth term of the sequence, you could either:

- write out the first 100 terms, or
- find a general rule.

To find a rule, first look at the differences: in each case the difference is 3.
The sequence behaves like the 3-times table, so there must be a connection.

Term number, n	$3 \times$ table	Term $T(n)$
1	3	4
2	6	7
3	9	10
4	12	13
5	15	16

To get from the term number, n, to the term, $T(n)$, you must $\times$ 3 and then +1.

As a function:

$n \rightarrow 3n + 1$

The **general term** is $T(n) = 3n + 1$

The hundredth term is where $n = 100$
$3 \times 100 + 1 = 301$
So the hundredth term is 301.

- **You can calculate any term in a sequence if you know the general term.**

You can describe the general term in words or using a formula.

In reverse, if you know the formula, you can generate the sequence.

example

Generate the first five terms of the sequence given by the general term $T(n) = 6n - 4$

$N = 1$ $T(1) = 6 \times 1 - 4 = 2$
$N = 2$ $T(2) = 6 \times 2 - 4 = 8$
$N = 3$ $T(3) = 6 \times 3 - 4 = 14$
$N = 4$ $T(4) = 6 \times 4 - 4 = 20$
$N = 5$ $T(5) = 6 \times 5 - 4 = 26$

The sequence is 2, 8, 14, 20, 26, ...

Exercise A3.2

1 Write out the first five terms of the sequences given by:

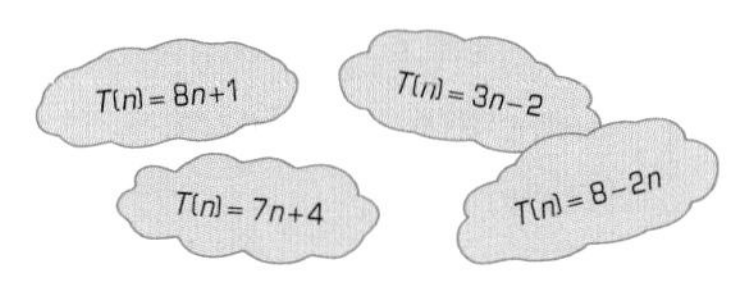

2 Find a formula for $T(n)$, the general term of each sequence, in terms of n, the term number.

a 7, 11, 15, 19, 23, ...
b 1, 10, 19, 28, 37, ...
c −1, 6, 13, 20, 27, ...
d 20, 15, 10, 5, 0, ...

3 **a** Write out the first 6 terms of the sequence $T(n) = 3n$ and also $T(n) = 4n$.
b Which is the first term the sequences have in common? Why?
c Without writing out the sequence, were do $7n$ and $8n$ meet?
d Where do $T(n) = 4n$ and $T(n) = 6n$ meet?
e Where do $T(n) = 5n - 1$ and $T(n) = 7n + 1$ meet?

4 The twentieth, twenty-first and twenty-second terms of a linear sequence are 145, 152 and 159. Write down the first five terms of the sequence.

Hint: try to generate a formula.

5 Generate the sequences described in these boxes. Decide if each sequence is linear or not.

a 1 2 3 4 5 → $6n$

b 1 2 3 4 5 → $7n-2$

c 1 2 3 4 5 → n^2+1

d 1 2 3 4 5 → $3-n$

6 Find the missing function from these boxes:

a

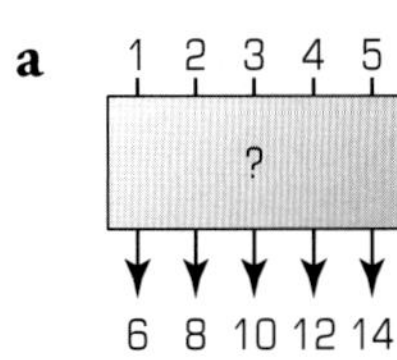

b

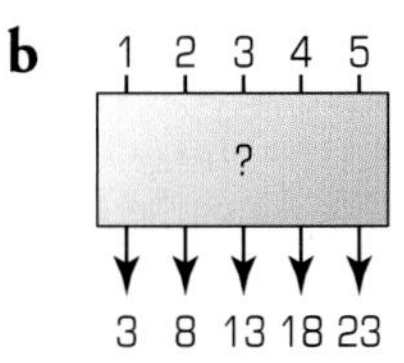

c

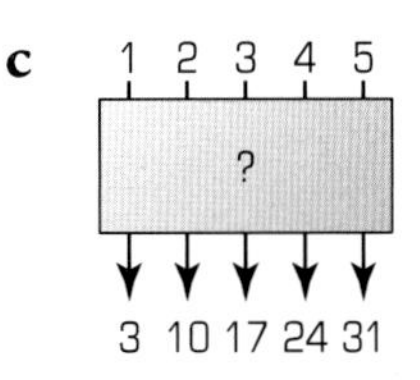

d

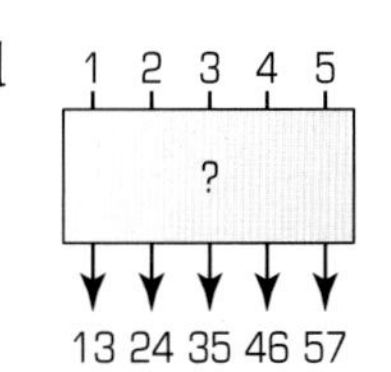

7 Challenge

a Write out the first five terms of the sequence given by the function:

$$\frac{5n-1}{6n+3}$$

b Find a formula for $T(n)$, the general term of this sequence, in terms of n, the term number:

$\frac{4}{8}, \frac{11}{10}, \frac{18}{12}, \frac{25}{14}, \frac{32}{16}, \ldots$

8 Find a formula for the general term $T(n)$, of each sequence:

a A linear sequence starting at 6 and going up in 10s.
b 2 more than the multiples of 3.

9 For each sequence, find the general term, $T(n)$, and use it to find 100th term.

a 3, 10, 17, 24, 31, ...
b 8, 12, 16, 20, 24, ...
c $\frac{5}{10}, \frac{16}{12}, \frac{27}{14}, \frac{38}{16}, \frac{49}{18}, \ldots$

10 Explain why you can't find a formula for prime numbers.

A3.3 Patterns in diagrams

This spread will show you how to:
- Generate sequences from practical contexts.
- Describe the general term of a simple sequence and justify the generalisation.

KEY WORDS
Difference
Formula
Generate
Term
Linear

You can often find sequences in diagrams.
Here is a sequence of triangles:

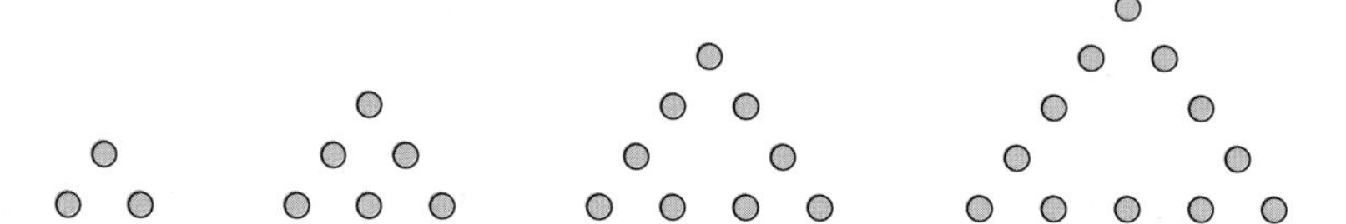

To find the number of dots in the tenth pattern you could either:

- draw the first 10 patterns, or
- find a rule.

To find a rule, start by putting the numbers into a table:

Pattern number	Number of dots
1	3
2	6
3	9
4	12

The first difference is 3, so the pattern is connected to the 3 times table.
$T(n) = 3n$, or
$D = 3n$, where D is the number of dots and n is the pattern number.

You can explain this formula easily by looking at the diagrams.

3×1

3×2

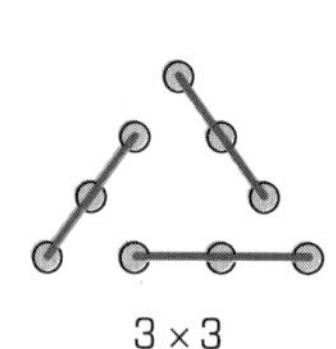

3×3

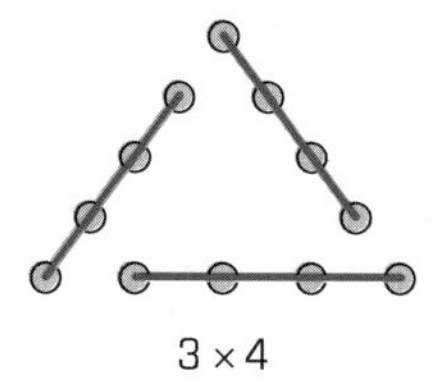

3×4

Each pattern has three lots of the pattern number.

You can use the rule to predict the number of dots in the tenth pattern.
For the tenth pattern, $n = 10$.
So $D = 3 \times 10 = 30$.
There will be 30 dots in the tenth pattern.

Exercise A3.3

1 **a** Copy and complete this table for the tile pattern:

Pattern number	1	2	3	4
Tiles				

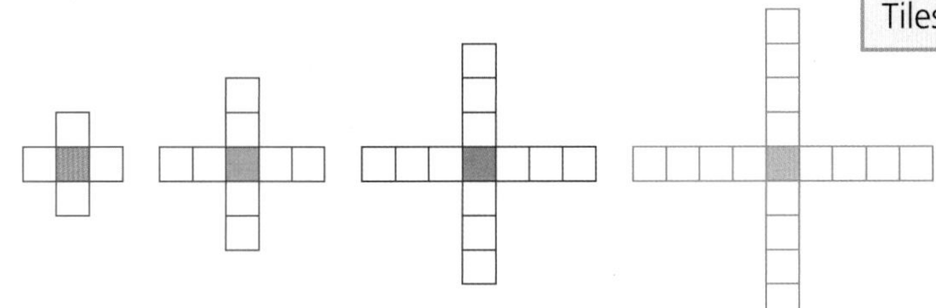

b Using differences, find a formula to connect pattern number and number of tiles.

c Explain why the formula works.

2 In chemistry, hydrocarbons are made by joining hydrogen (H) and carbon (C) together in an ordered way, using bonds (—) like this:

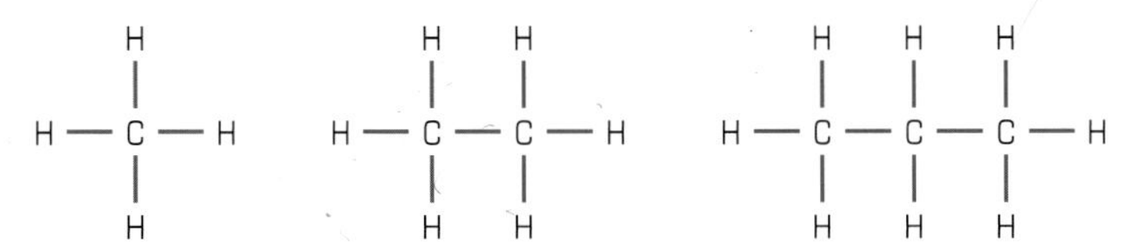

a Copy and complete these tables:

Number of carbons, C	1	2	3	4
Number of hydrogens, H			8	

Number of carbons, C	1	2	3	4
Number of bonds, B			10	

b Using differences, find a formula to connect:

i the number of carbons (C) with the number of hydrogens (H).

ii the number of carbons (C) with the number of bonds (B).

c Explain why each formula works.

3 Explain why each of the given formulae works:

a 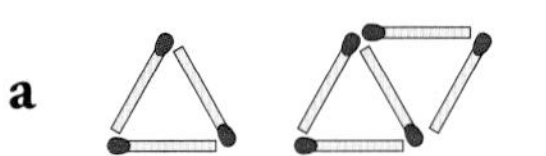 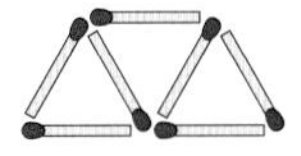

Number of matches = 2 × number of triangles + 1

$$m = 2T + 1$$

b Number of handshakes made at a party = $\dfrac{\text{(Number of people)} \times \text{(Number of people} - 1)}{2}$

$$H = \frac{n(n-1)}{2}$$

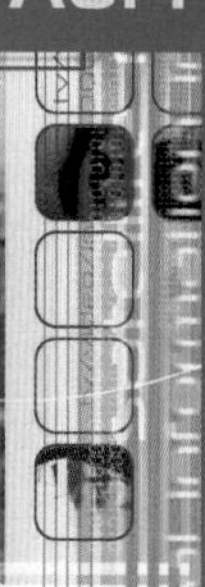

A3.4 Functions and rules

This spread will show you how to:

- Express simple functions in words and symbols.
- Represent mappings expressed algebraically.

KEY WORDS

Equation	Mapping
Function	Output
Input	Sequence
Linear	Squared

You have already learned about function machines. Remember:

- In a function machine:
  - The **input** value is the value you put in to the machine.
  - The machine performs the **function.**
  - The **output** value is the result that the machine puts out.

input → function → output

In this machine, the inputs to the machine are the counting numbers: 1, 2, 3, 4, ...

the output numbers form a sequence.

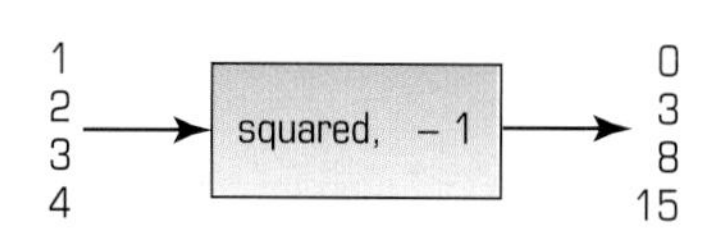

You can put the same information in a table,

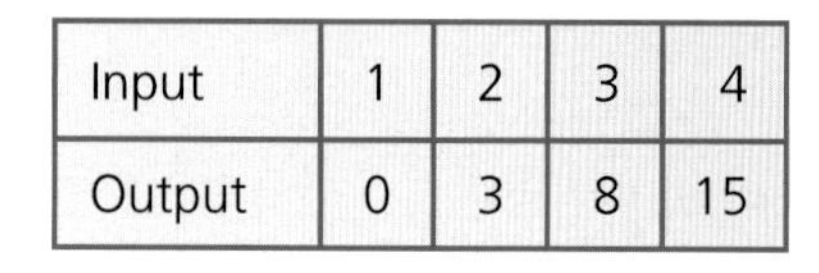

Input	1	2	3	4
Output	0	3	8	15

or use a mapping diagram.

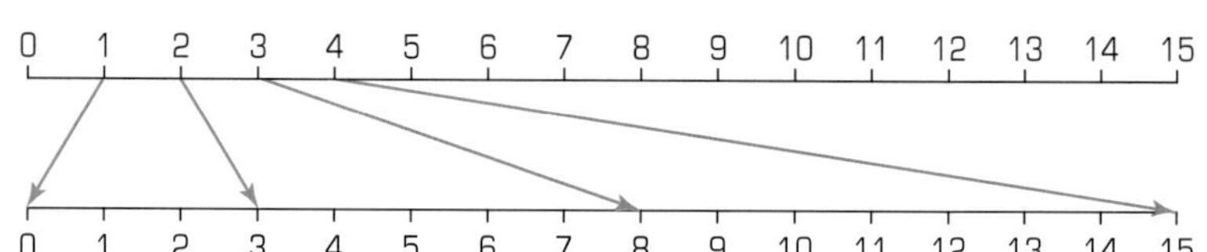

A mapping diagram uses number lines connected with arrows to show how inputs and outputs relate to each other.

You have already looked at the general term of a linear sequence.
The sequence 0, 3, 8, 15, ... is not linear because the differences change.

Using algebra: as a **mapping**, the general rule is $n \longrightarrow n^2 - 1$, where n is the input.
As an **equation**, $y = x^2 - 1$, where x is the input and y is the output.

example

Find the missing functions and express the general rule as a mapping and as an equation.

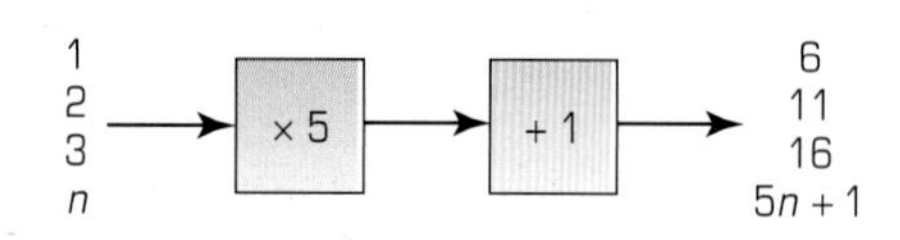

The outputs go up in 5s so the rule will include × 5
the first term is × 5 then +1
check with another term: 2 × 5 + 1 = 11

The functions are × 5 and +1
The mapping is $n \longrightarrow 5n + 1$ The equation is $y = 5x + 1$

Exercise A3.4

1 In parts **a** to **f** work out

i the function machine rule and
ii an equation for y in terms of x.

Part **a** has been done for you.

a
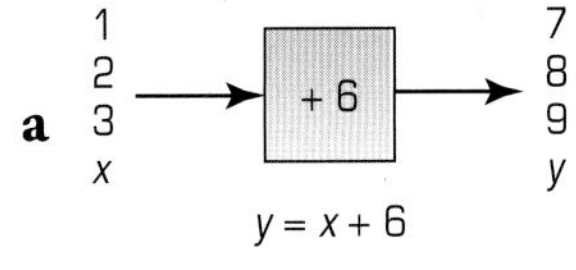

b 1, 2, 3 (x) → ? → 4, 8, 12 (y)

c
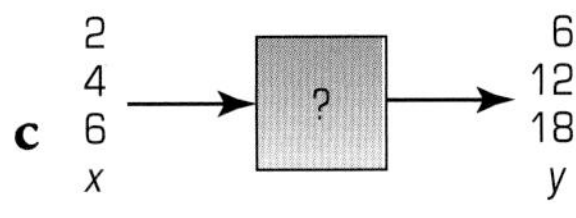

d
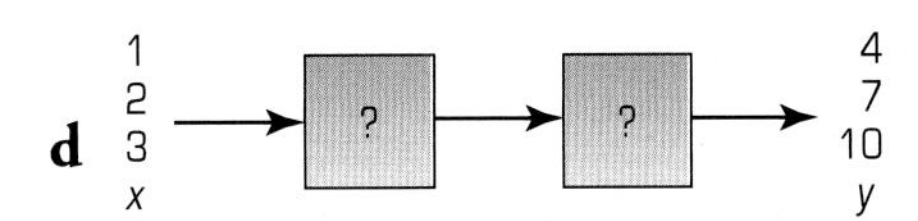

e
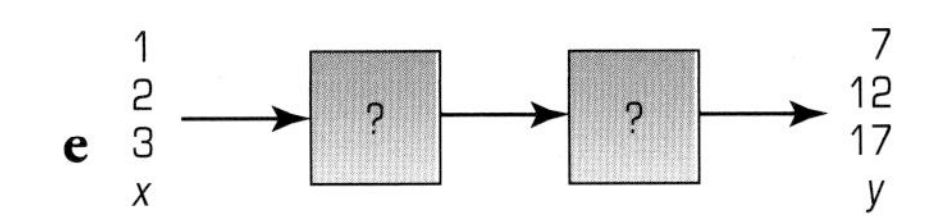

f
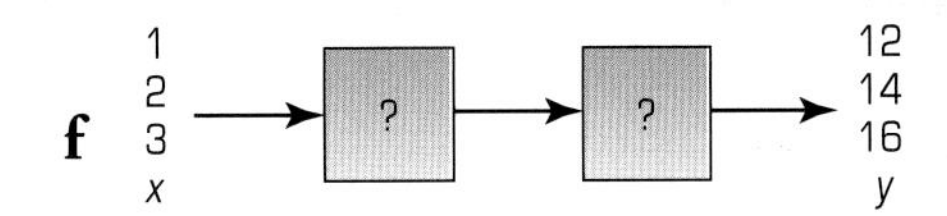

2 Match the table with the correct equation and the correct function machine.

a

x	0	1	2	3
$y =$	−1	2	5	8

b

x	0	1	2	3
$y =$	0	6	12	18

c

x	0	1	2	3
$y =$	4	7	10	13

d

x	0	1	2	3
$y =$	4	6	8	10

$y = 6x$ → × 2 → + 4 →

$y = 2x + 4$ → × 3 → + 4 →

$y = 3x - 1$ → × 6 →

$y = 3x + 4$ → × 3 → − 1 →

3 In each case an equation has been given, for y in terms of x. Make a table of values similar to the ones given in question 2, to show the input and output in each case.
a $y = 2x - 1$ **b** $y = x^2$ **c** $y = 10 - 2x$ **d** $y = 3x^2 + 1$

4 **a** The input, x of a function is 3. The output, y of the same function is 25. Find as many functions as possible that could have this mapping
b Another input-output part of the same function is $x = 4$, $y = 62$. Can you name the function?

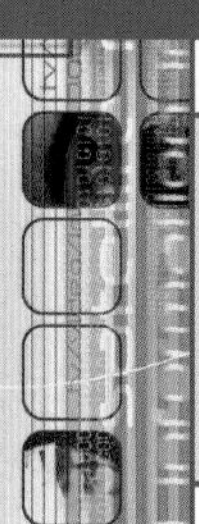

A3.5 Graphs of functions

This spread will show you how to:

- Generate points in all four quadrants and plot the graphs of linear functions.
- Begin to consider the features of graphs of simple linear functions.

KEYWORDS

Axes
Linear
Equation
Graph

In algebra you show equations using line graphs.
$x + y = 8$ is an equation.

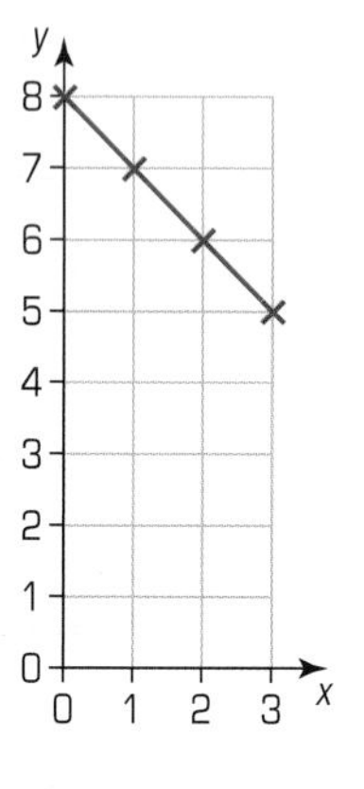

- You can choose values for x and work out the y-values.

  $x = 0, y = 8 \quad x = 1, y = 7 \quad x = 2, y = 6 \quad x = 3, y = 5$

- Now turn these pairs of values into coordinates.
  (0, 8) (1, 7) (2, 6) (3, 5)
- Plot them on a graph:
  Choose suitable axes.
  The points lie in a straight line so the relationship is **linear**.

> ▸ A linear equation forms a straight line on a coordinate grid.

You can get a more complete picture of the graph if you choose negative values of x as well. A table can help.

example

a Complete the table for the equations **i** $y = 2x + 1$ **ii** $y = {}^{-}2$

x	⁻2	⁻1	0	1	2
y					

b Using suitable axes on squared paper, plot the coordinates for each equation on a graph.

a $y = 2x + 1$

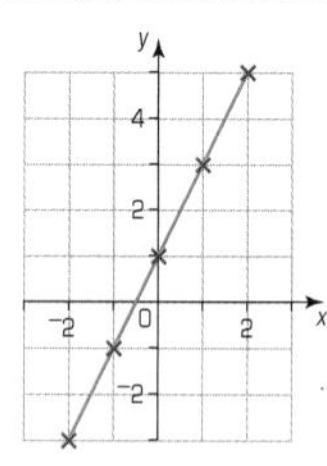

x	⁻2	⁻1	0	1	2
y	⁻3	⁻1	1	3	5

The graph is a diagonal straight line.

b $y = {}^{-}2$

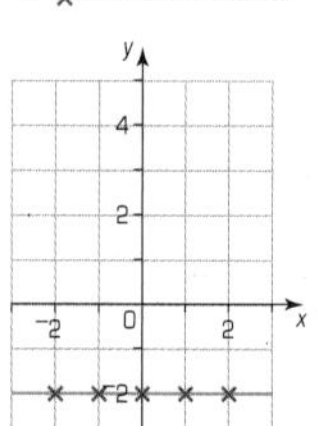

x	⁻2	⁻1	0	1	2
y	⁻2	⁻2	⁻2	⁻2	⁻2

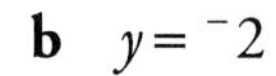

The graph is a horizontal straight line.

Exercise A3.5

You will need to work in pairs, using a grid and counters. Once you have constructed a graph in counters, make a sketch of it in your book and label it with its equation.

1 Plot the graphs given. The equation has been given and four pair of starting coordinates. Complete the final two pairs of coordinates.

a $x = 3$ (3, 1), (3, 5), (3, 4), (3, 0), (__, 2), (__, 6)

b $y = 5$ (1, 5), (6, 5), (8, 5), (2, 5), (4, __), (0, __)

c $x + y = 7$ (3, 4), (4, 3), (2, 5), (1, 6), (6, __), (0, __)

d $y = 3x$ (1, 3), (3, 9), (2, 6), (0, 0), ($\frac{1}{2}$, __), ($2\frac{1}{2}$, __)

2 List four pairs of coordinates that would lie on each of these graphs and predict if the graph will be vertical, horizontal or diagonal before you draw it. Now plot it with counters.

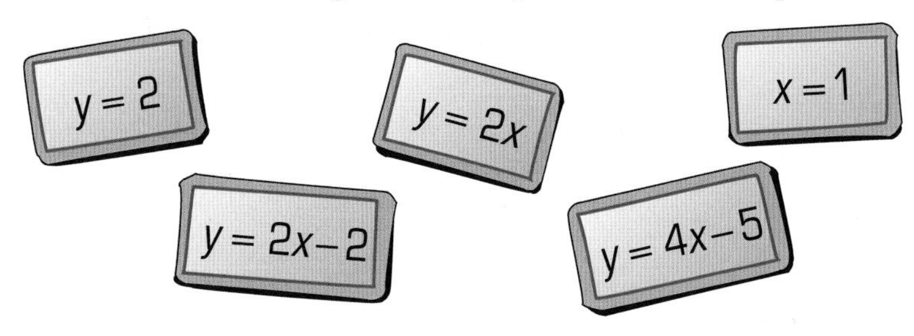

3 In each list of coordinates, there is one pair of coordinates that will not lie on the given graph. Spot this pair.

a $y = 10$ (10, 10), (10, 4), (3, 10), (6, 10)

b $y = 3x + 1$ (2, 7), (10, 31), (7, 21), (6, 19)

c $x + 2y = 15$ (1, 7), (0, $7\frac{1}{2}$), (6, $4\frac{1}{2}$), (5, 6)

d $y = 2x + 5$ ($1\frac{1}{2}$, 8), (2.3, 9.6), ($^{-}3$, $^{-}11$), ($^{-}2$, 1)

4 Draw out a set of axes from -10 to $+10$. Draw the graphs represented by the following equations.
a $y = 5$ **b** $x + y = 6$ **c** $y = 3x - 1$ **d** $x = -2$.

5 **a** Name four equations that when plotted, would give a horizonal line.
b Repeat for a vertical line.
c Repeat for a diagonal line.

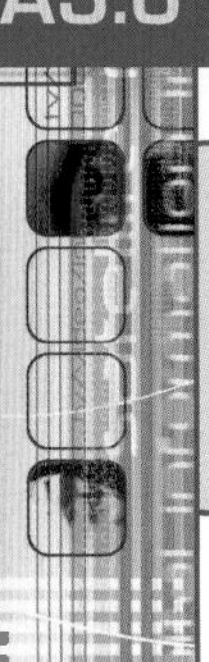

A3.6 Using a table of values

This spread will show you how to:
- Begin to consider the features of graphs of simple linear functions.
- Recognise that equations of the form $y = mx + c$ correspond to straight-line graphs.

KEYWORDS
Coordinates
Linear
Equation
Graph

It is often possible to find the equation that goes with a graph.
Here are two linear graphs:

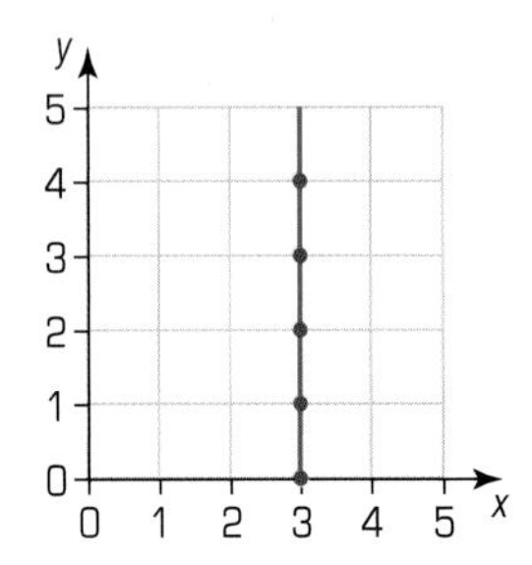

The x value is always 3.
The equation that describes the graph is $x = 3$.

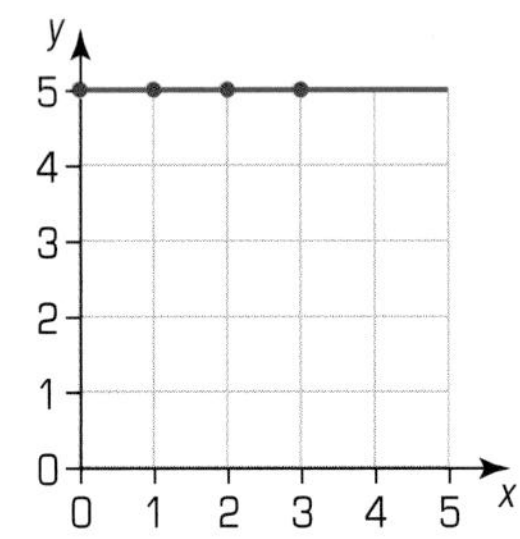

The y value is always 5.
The equation is $y = 5$.

You can read the coordinates from the graph:
(3, 4), (3, 3), (3, 2), (3, 1), (3, 0) and (0, 5), (1, 5), (2, 5), (3, 5).

- **Vertical linear graphs always have an equation $x = N$, where N is a number.**
- **Horizontal linear graphs always have an equation $y = N$, where N is a number.**

All other linear graphs are diagonal.

example

Find the equation of the graph.

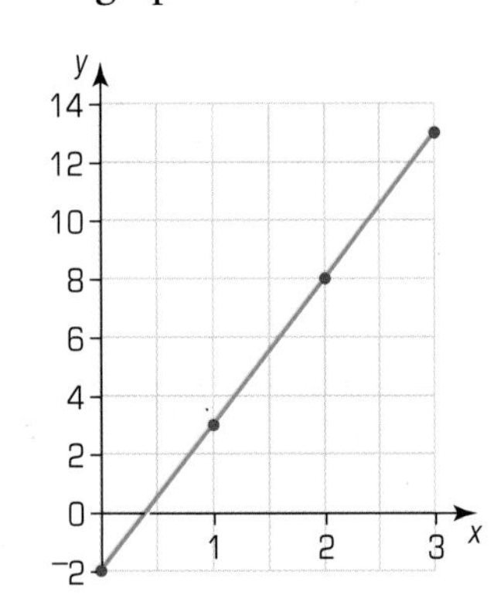

Write down some of the coordinates:

x	y
(3,	13)
(2,	8)
(1,	3)
(0,	$^-2$)

Draw a table:

x	0	1	2	3
y	$^-2$	3	8	13

y is going up in 5s, so multiply x by 5.
Try the first coordinate: $0 \times 5 = 0 \quad 0 - 2 = {}^-2$
So the rule is '×5 then $^-2$'
Check with $x = 2$:
$2 \times 5 = 10 \quad 10 - 2 = 8 \quad y = 8$, which is correct

The equation of the line is $y = 5x - 2$

Exercise A3.6

1 In each part of the question, a pair of coordinates is given.
Decide which of the graphs, given by their equations, the coordinates could lie on.

a (3, 9)

b (2, 5)

c ($^-1$, $^-6$)

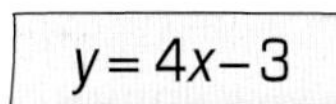

$y = 3x$

$x + y + 7 = 0$

$y = 2x + 1$

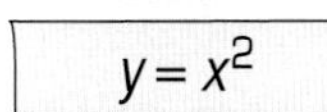

$y = x - 5$

2 For each graph, write down three pairs of coordinates and use them to find the graph's equation.

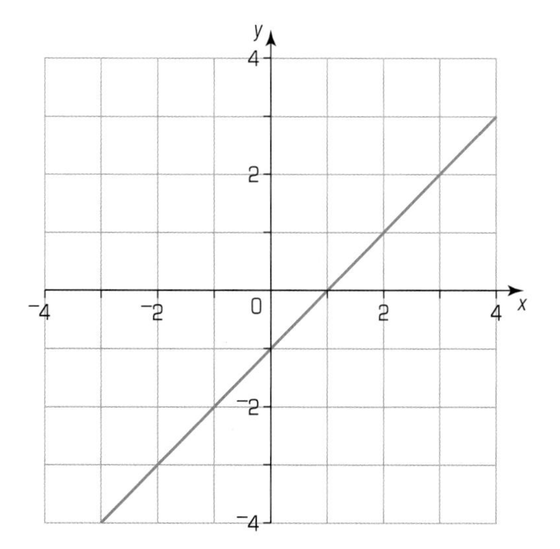

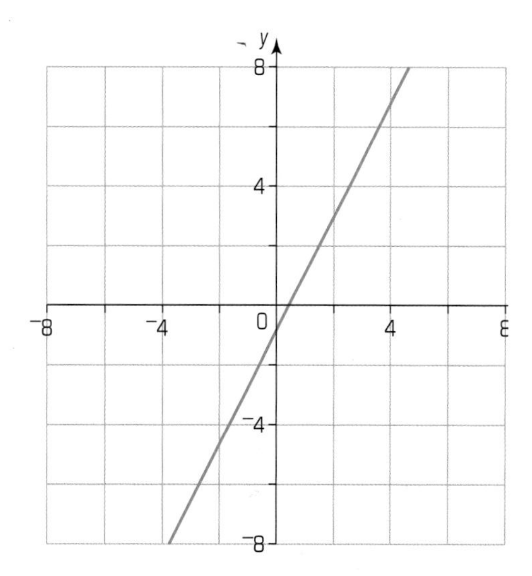

3 Use the coordinate tables to find an equation on which the coordinates would lie:

a

x	1	2	3	4
y	4	8	12	16

b

x	1	2	3	4
y	$\frac{1}{2}$	1	$1\frac{1}{2}$	2

c

x	1	2	3	4
y	2	5	8	11

d

x	1	2	3	4
y	8	15	22	29

e

x	1	2	3	4
y	9	8	7	6

f

x	1	2	3	4
y	1	4	9	16

4 Find a pair of coordinates that would lie on both of these graphs:

a $y = x$ $\quad$ $y = 4$
b $y = 2x$ $\quad$ $y = x + 3$
c $y = 3x - 1$ $\quad$ $y = x + 7$

A3 Summary

You should know how to ...

1 Recognise and use multiples, factors, primes common factor, highest common factor, lowest common multiple.

2 Begin to use linear expressions to describe the nth term of an arithmetic sequence, justifying its form.

3 Generate points in all four quadrants and plot the graphs of linear functions.

4 Recognise that equations of the form $y = mx + c$ correspond to straight-line graphs.

Check out

1 a Find the highest common factor of 24 and 60; use your answer to cancel $\frac{24}{60}$.

b True or false ... the lowest common multiple of two numbers is always the product of those two numbers. Give an example or counter example to explain your solution.

c Write 540 as a product of its prime factors.

2 a Generate an expression for the nth term of these sequences:

i 5, 8, 11, 14, 17, ...

ii 5, 16, 27, 38, 49, ...

iii 10, 8, 6, 4, 2, ...

b Generate the first five terms of these sequences:

i $T(n) = 3n + 4$

ii $T(n) = 5n - 8$

iii $T(n) = n^2 + 4$

c Explain why the following formula works:

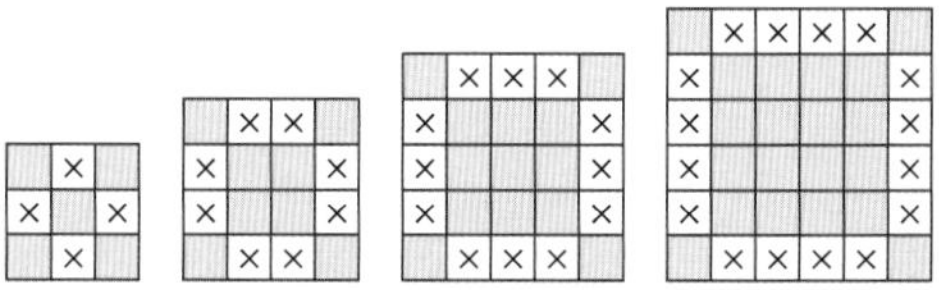

No. of crosses = 4 × (length of square − 2)

$C = 4(L - 2)$

3 Draw the graph of $y = 2x + 1$.
Use axes from $^-10$ to 10.

4 a Which of these equations correspond to a straight-line graph?

i $x = 5$ ii $y = 2x + 4$

iii $y = {}^-2$ iv $y = x^2$

Justify your answers.

b True or false: (3, 5) lies on all of these graphs. Explain.

S3 Triangles and quadrilaterals

This unit will show you how to:

- Solve geometrical problems using angle properties of triangles and special quadrilaterals.
- Use straight edge and compasses to construct:
  - a triangle
  - the mid-point and perpendicular bisector of a line segment
  - the bisector of an angle.
- Find simple loci to produce paths.
- Know and use geometric properties of cuboids and shapes made from cuboids.
- Begin to use plans and elevations.
- Solve problems and investigate in the context of shape and space.
- Represent problems and interpret solutions in geometric form.

Architects use plans and views to help them construct buildings.

Before you start

You should know how to ...

1 Calculate angles on:

a a straight line

b in a triangle

c around a point.

2 Know the names of polygons.

3 Recognise and sketch solids.

Check in

1 Calculate the unknown angles.

a

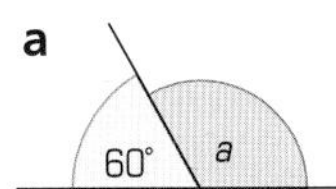

b

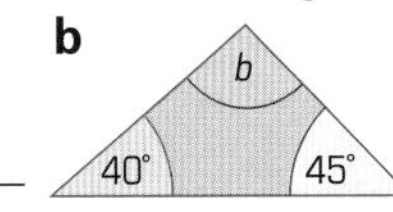

c

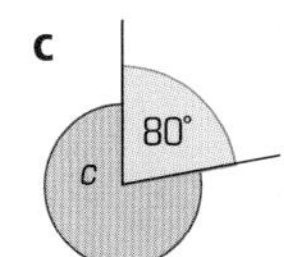

2 Name these shapes.

a

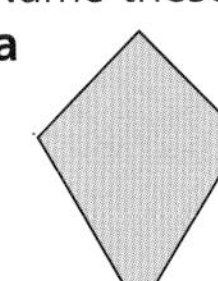

b

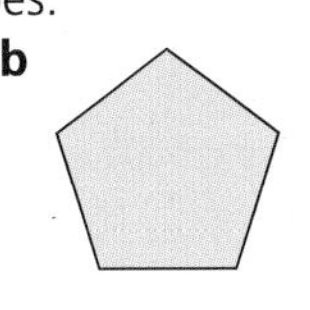

c

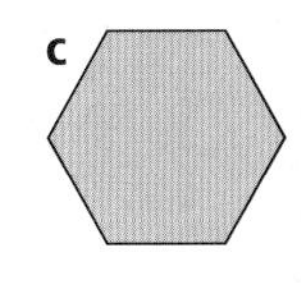

3 a Name these solids.

i

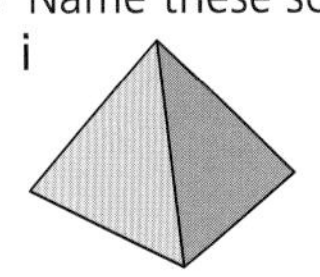

ii

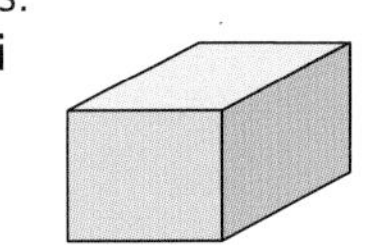

b Sketch **i** a cylinder **ii** a cone **iii** a triangular prism

S3.1 Calculating angles

This spread will show you how to:

- Calculate angles in triangles and special quadrilaterals.

KEYWORDS

Angle, Triangle, Equilateral, Quadrilateral, Axis of symmetry, Vertically opposite

Angles on a straight line add up to 180°.

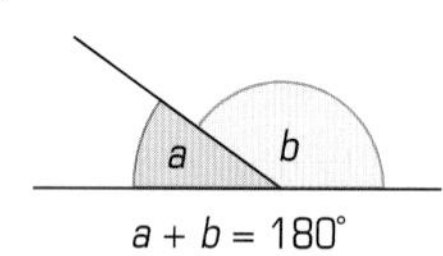

Angles in a triangle add up to 180°.

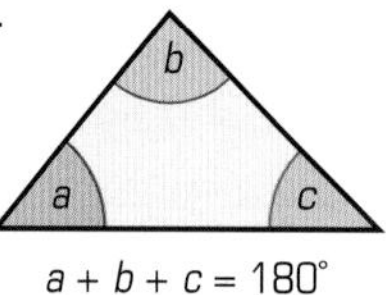

Angles in a full turn add up to 360°.

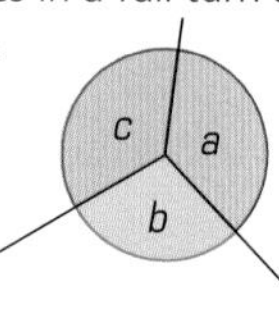

Vertically opposite angles are equal.

You can use these rules to find unknown angles.

example

Find angles a, b and c.

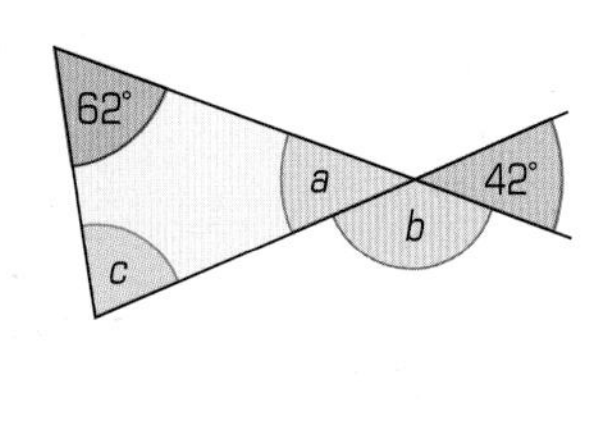

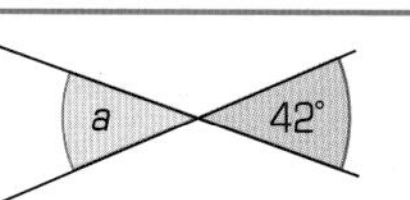

$a = 42°$
(vertically opposite angles)

$b = 180° - 42° = 138°$
(angles in a straight line)

$c = 180° - 42° - 62° = 76°$
(angles in a triangle)

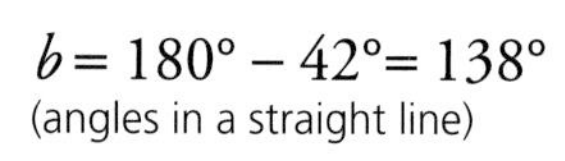

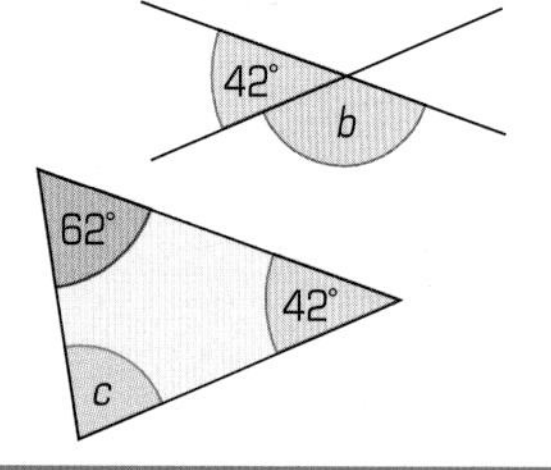

Equal lengths can often give you information about missing angles.

example

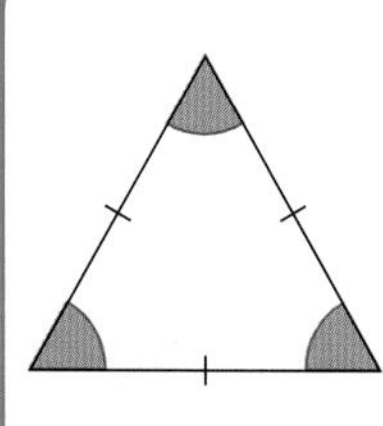

a What is the name of this triangle? **b** What is the size of each angle?

a The marks show that the lengths are equal.
It is an equilateral triangle.

b The angles of an equilateral triangle are equal.
Angles in a triangle add up to 180°.
So each angle = 180° ÷ 3 = 60°

▶ **The angles of an equilateral triangle are all 60°.**

example

Find the missing angles in these shapes.

a

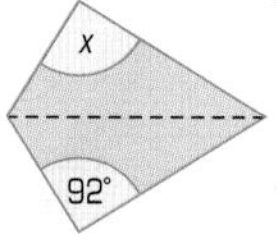

b

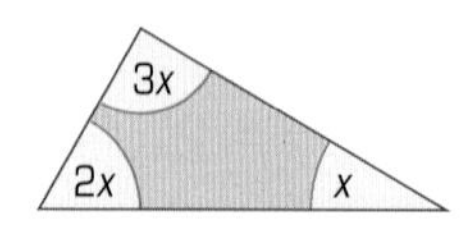

a The shape has an axis of symmetry.
This means that the lengths and angles above the line are the same as those below the line.

So $x = 92°$.

b $x + 2x + 3x = 180°$
$6x = 180°$
$x = 30°$
So the angles are 30°, 60° and 90°.

Exercise S3.1

Find the missing angles in these diagrams.

1

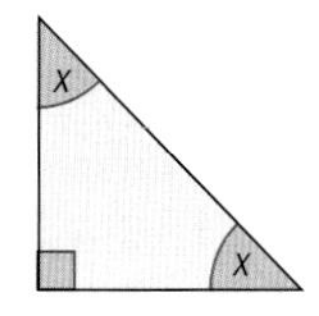

2

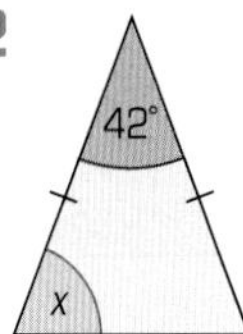

3

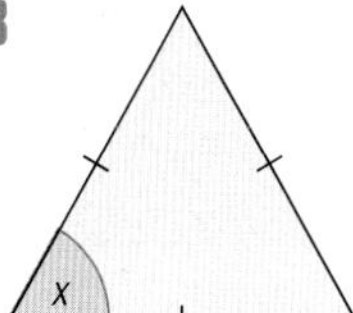

4

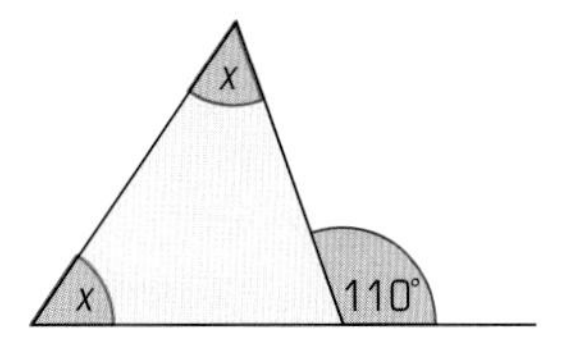

5

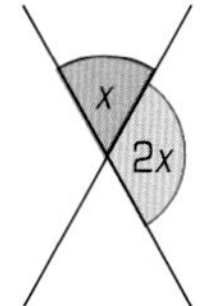

6

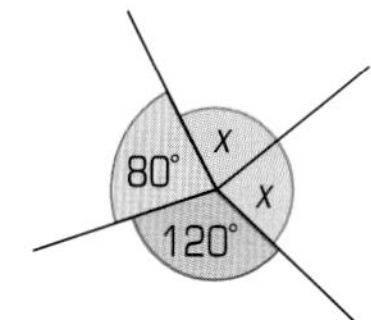

7

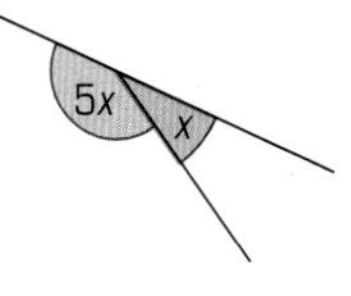

8

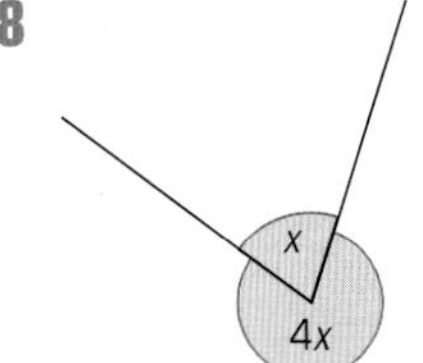

9

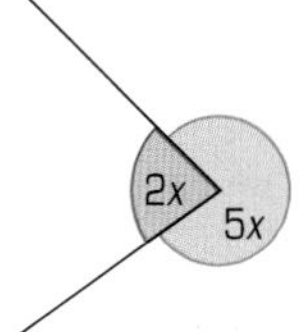

10

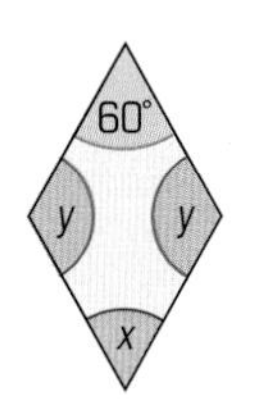

11

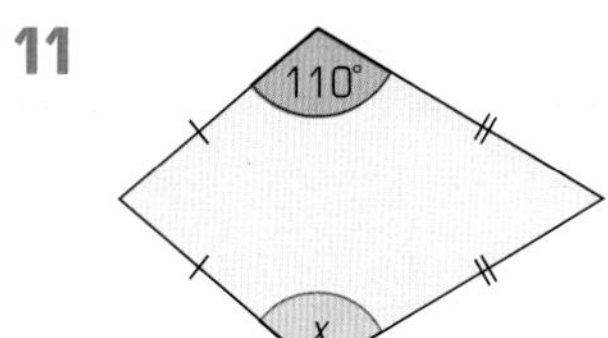

12

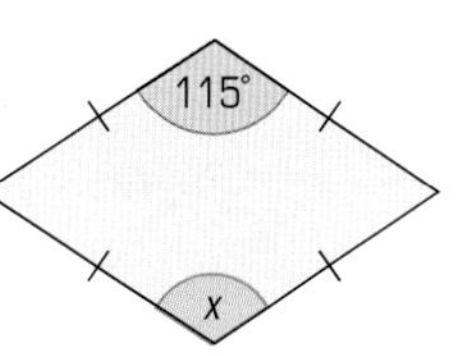

13

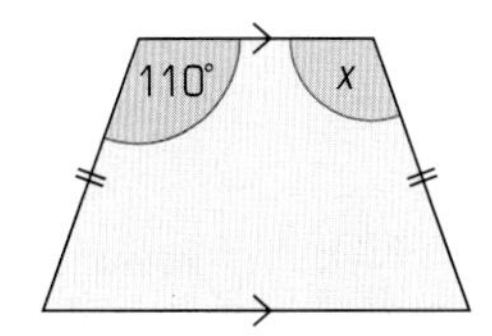

14

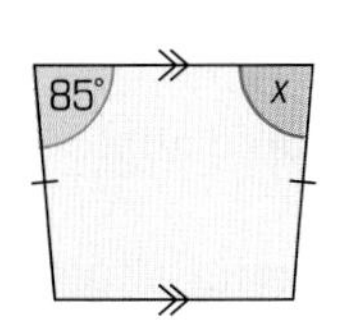

15

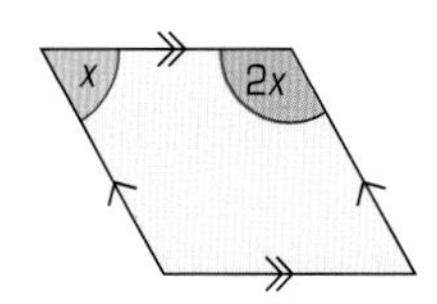

S3.2 Angles in shapes

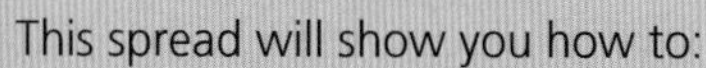

This spread will show you how to:

- Understand a proof that the sum of the angles in a quadrilateral are 360°.

KEYWORDS

Exterior, Protractor, Hexagon, Quadrilateral, Interior, Triangle

You find angles both inside and outside shapes.
Here is a hexagon:

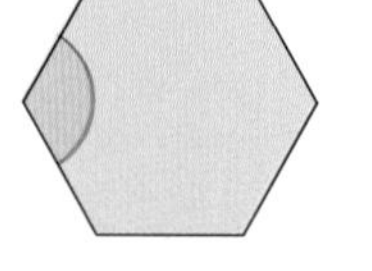

A hexagon is a shape with six sides.

The angle marked is an **interior** angle of the hexagon. It is inside the shape.

- **The angles inside a shape are called interior angles.**
- **The sum of the interior angles of a triangle is 180°.**

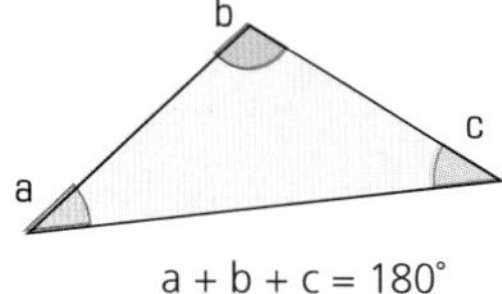

$a + b + c = 180°$

example

a Estimate and then measure accurately the interior angles of this quadrilateral.
b What do the interior angles add up to?

a **Estimate:** A is between 180° and 270° but closer to 180°, say 200°.
B is between 0° and 90° but closer to 0°, say 20°.
C looks like a right angle, so about 90°.
D is about halfway between 0° and 90°, say 45°.

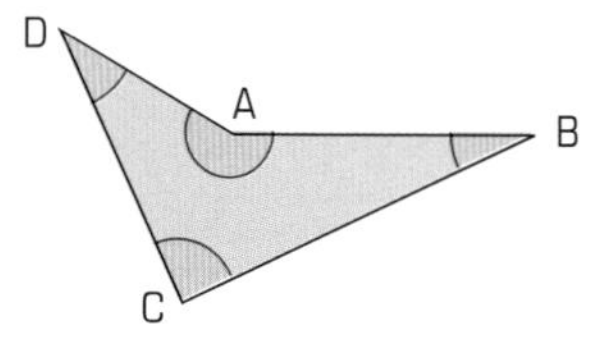

Measure: Using a protractor the angles are 210°, 25°, 90° and 35°.

b $210° + 25° + 90° + 35° = 360°$

You can identify properties of quadrilaterals from your knowledge of triangles.

Draw a quadrilateral …

… split it into triangles

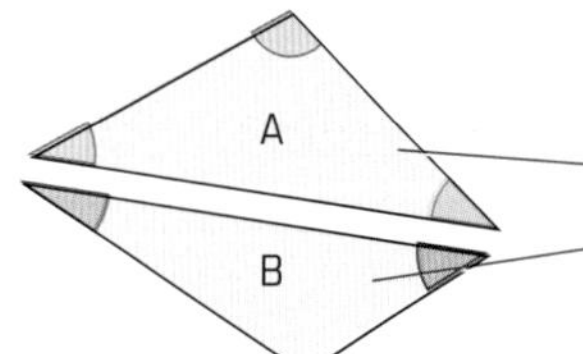

Any quadrilateral has 2 × the angle sum of a triangle.

- **The interior angles of a quadrilateral add up to 360°.**

If you extend the lines of a shape you can make angles outside the shape.

- **The angles outside a shape are called exterior angles.**
- **Exterior angle + interior angle = 180°.**

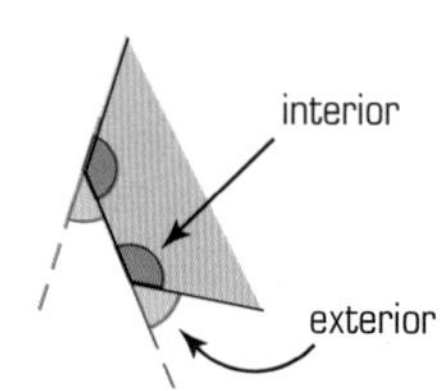

Exercise S3.2

1 Draw polygons with these numbers of sides:
a 3 **b** 4 **c** 5 **d** 6 **e** 7 **f** 8
Estimate, and then measure accurately each of their interior angles .
You can use the shapes supplied if you prefer.

a

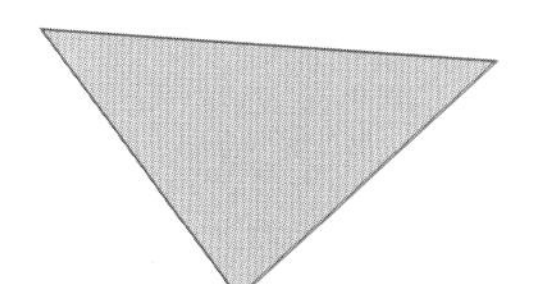

b

c

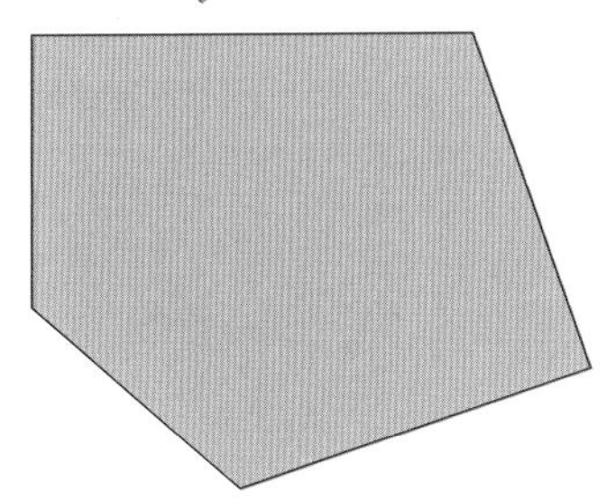

d

f

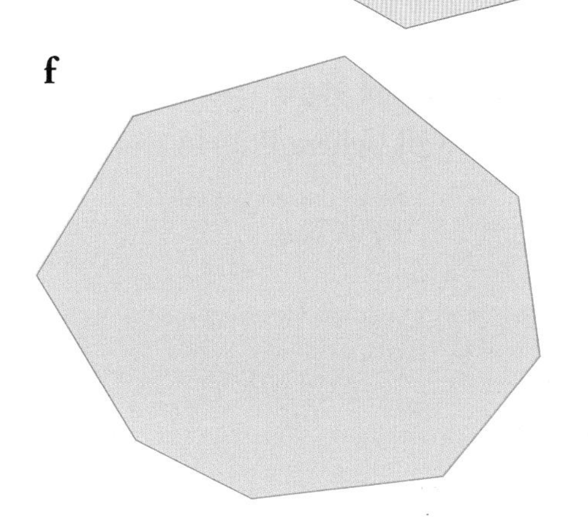

e

2 Copy and complete this table to show your results.

Number of Sides	Estimates of angles	Accurate angle sizes	Interior angle total	Exterior angle total
3				
4				
5				
6				
7				
8				

3 Extend the sides of your polygons and measure the exterior angles. Complete the last column of the table.

4 What do you notice about your results?

S3.3 Constructing triangles

This spread will show you how to:

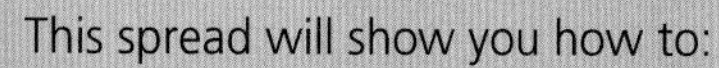

- Construct triangles using a ruler, protractor and compasses.

KEYWORDS

Arc | Construct
Compasses | Intersection

As long as you have certain information you can construct any triangle.

1. Two angles and the included side (ASA) You need a ruler and protractor.

Construct the triangle ABC where AC = 5 cm, BAC = 40° and BCA = 30°.

First sketch the triangle

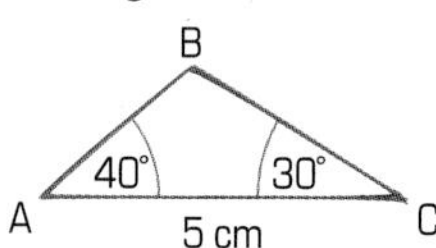

Draw the base AC with a ruler

A 5 cm C

Draw angle BAC at A

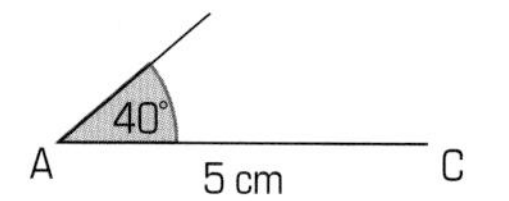

Draw angle BCA at C

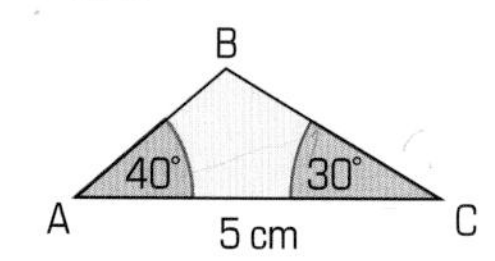

2. Two sides and the included angle (SAS) You need a ruler and protractor.

Construct the triangle PQR, where PQ = 6 cm, ∠P = 50° and PR = 4 cm.

First sketch the triangle

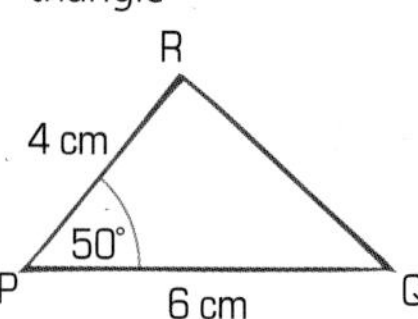

Make the longest side PQ the base

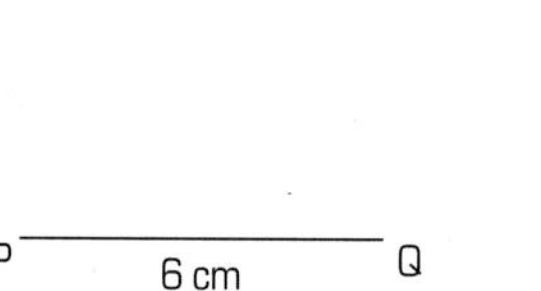

Draw an angle of 50° at P

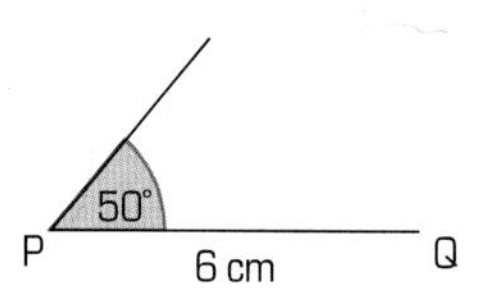

Mark R, 4 cm from P

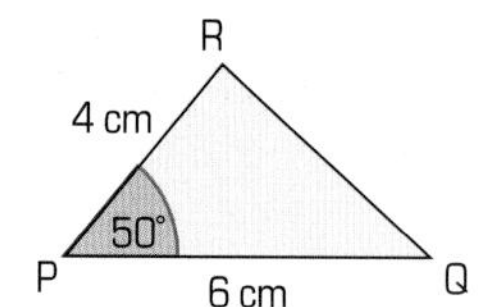

3. Three sides (SSS) You need a ruler and compasses.

Construct the triangle XYZ with lengths XZ = 3 cm, YZ = 4 cm and XY = 5 cm.

First sketch the triangle – label the vertices

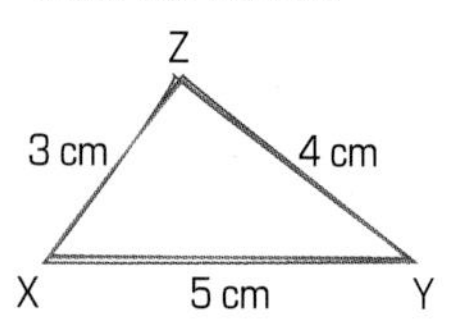

Draw a base line XY 5 cm long

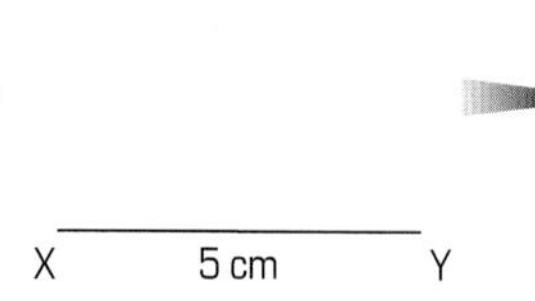

Open your compasses to 3 cm

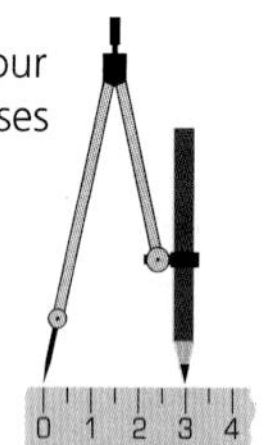

Place the point at X and draw an arc above XY

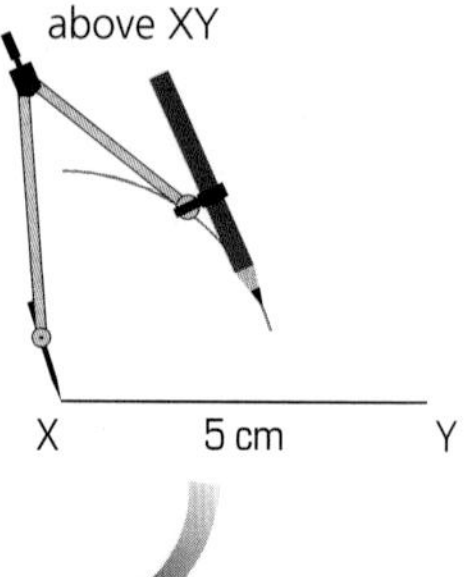

Open your compasses to 4 cm

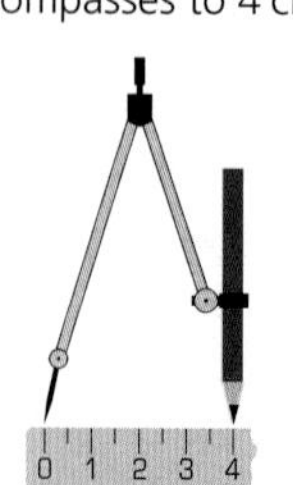

Place the point at Y and draw an arc that crosses the first one

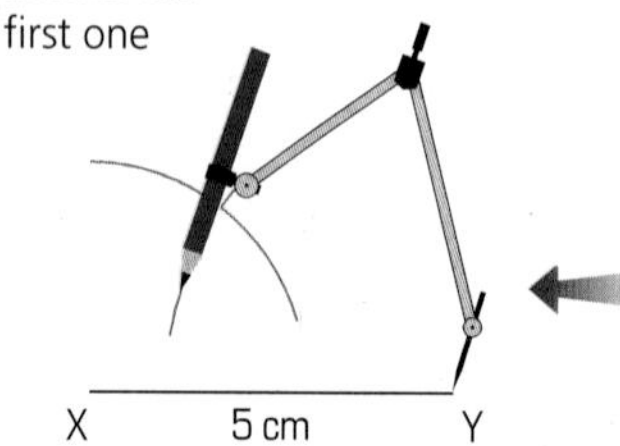

Label the intersection Z and join up the lines

Z 4 cm Y

Exercise S3.3

1 Without drawing these triangles, describe:
- ▶ What type they are (ASA, SAS or SSS)
- ▶ What special equipment you need (protractor or compasses)

a $\triangle ABC$ where AB = 4 cm, AC = 3.5 cm and $\angle CAB = 80°$.

b $\triangle PQR$ where $\angle P = 30°$, $\angle Q = 40°$ and PQ = 2 cm.

c $\triangle XYZ$ where XY = 5 cm, YZ = 3 cm and XZ = 6 cm.

2 Construct accurately the following triangle.

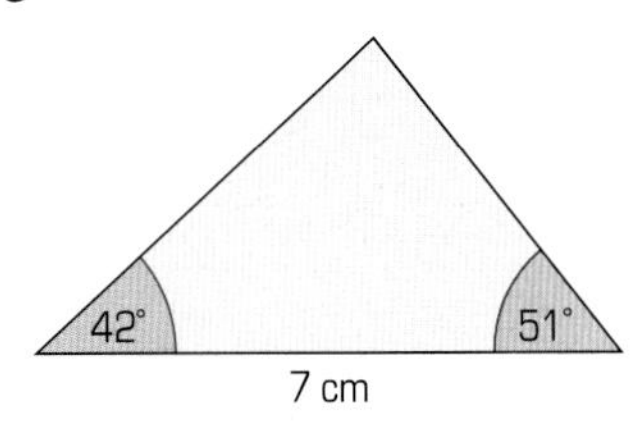

3 Construct accurately the following triangle.

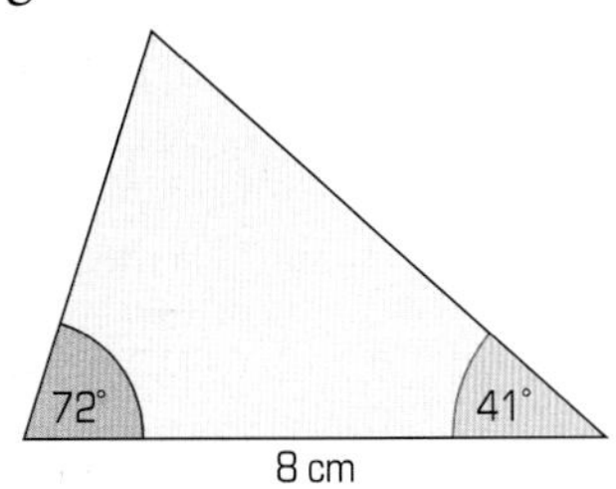

4 Construct accurately the following triangle.

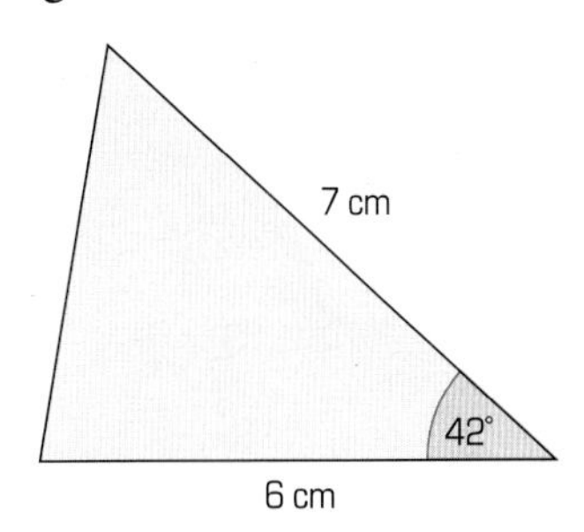

5 Construct accurately the following triangle.

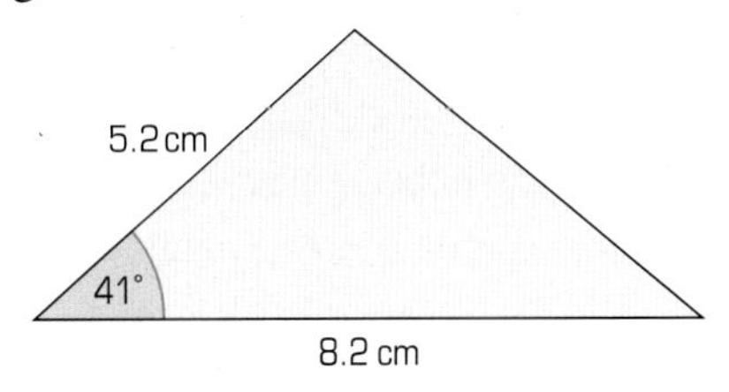

6 Construct $\triangle ABC$ where AB = 4 cm, $\angle A = 35°$, $\angle B = 42°$

7 Construct $\triangle PQR$ where PQ = 5 cm, PR = 4.5 cm and $\angle P = 35°$.

8 Construct an equilateral triangle ABC with sides 5 cm.

9 Construct an isosceles triangle EFG with EF = 3.2 cm and FG = EG = 4.7 cm.

10 Construct the quadrilateral as shown in the diagram:

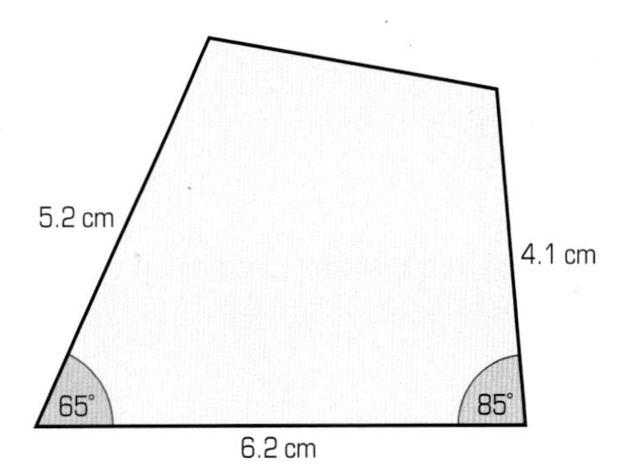

11 Construct WXYZ where WX = 4.5 cm, WZ = 3.5 cm, XY = 5.5 cm, $\angle W = 75°$ and $\angle X = 65°$. Sketch it first!

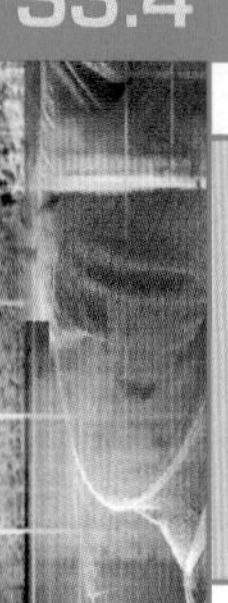

S3.4 Constructing bisectors

This spread will show you how to:
- Construct the midpoint and perpendicular bisector of a line segment.
- Construct the bisector of an angle.
- Find simple loci to produce paths.

KEYWORDS

Bisector	Midpoint
Construct	Perpendicular
Equidistant	Line segment
Locus	

▶ **A perpendicular bisector divides a straight line into two equal parts at right angles.**

Method

1 Draw a line AB.

2 Open your compasses so that they are at least half the length of AB. Place the point at A and draw an arc.

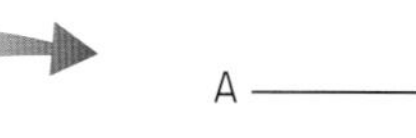

3 Keeping your compasses exactly the same, place the point at B and draw another arc.

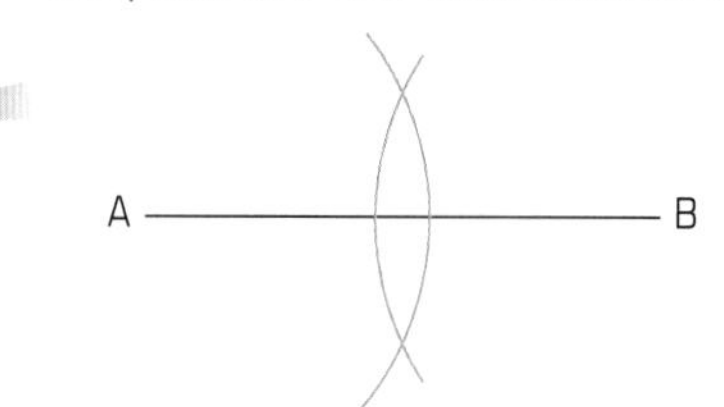

4 Join X to Y to make the perpendicular bisector of AB.

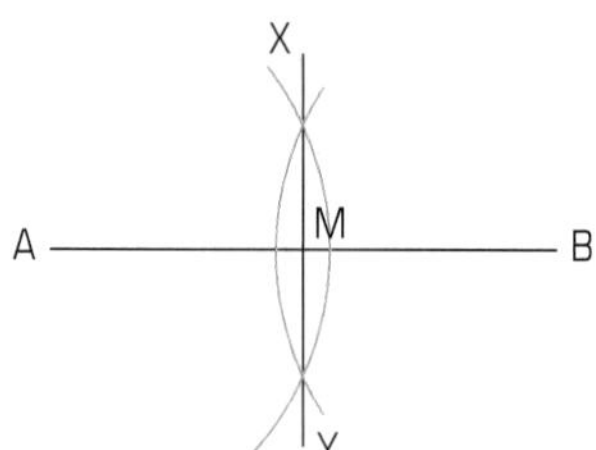

All the points on XY are **equidistant** from A and B.
M is the **midpoint** of AB.

▶ **An angle bisector divides an angle into two equal parts.**

Method

1 Draw an angle XAY.

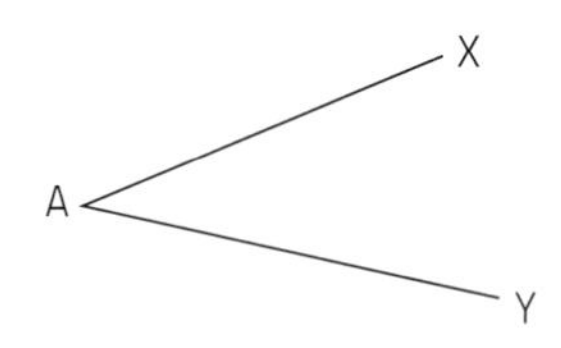

2 Open your compasses, put the point at A and draw an arc that cuts both lines.

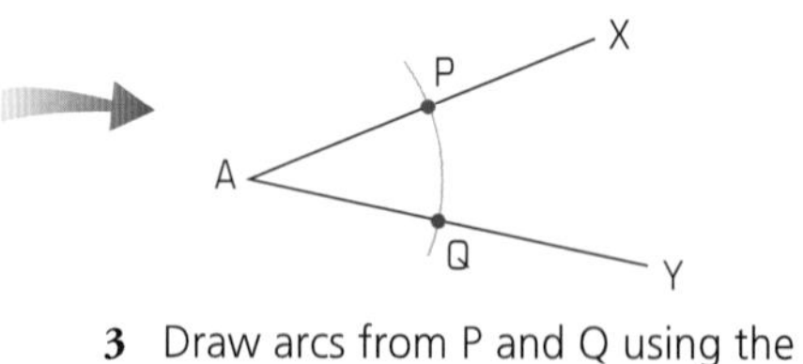

Do not remove any construction lines. These show evidence of your working.

3 Draw arcs from P and Q using the same compass settings for each one.

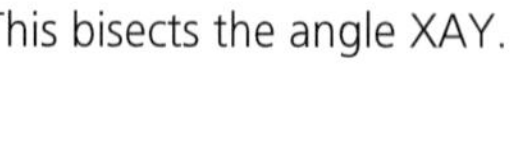

4 Join AB. This bisects the angle XAY.

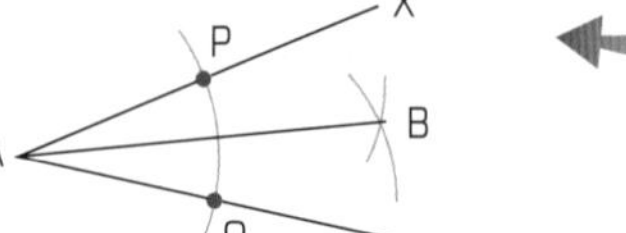

All the points on AB are equidistant from AX and AY.

Exercise S3.4

1 **a** Draw a line AB = 10 cm. Construct the perpendicular bisector of AB.
b Label the midpoint of the line X
c Construct an angle of 45° at X.

2 Using a protractor draw an angle of 60°.
a Bisect this angle to make 30°.
b Bisect one of these angles to make 15°.

3 The diagram below shows a lighthouse at A and a buoy at B.

The path traced by a point, satisfying certain conditions, is called a **locus**.

A boat sails so that it is always equidistant from A and B.
Copy the diagram and construct the path of the boat.

4 **a** Follow these instructions to construct a rhombus:
Construct a rhombus WXYZ where ∠W = 60° and the sides are 5 cm.

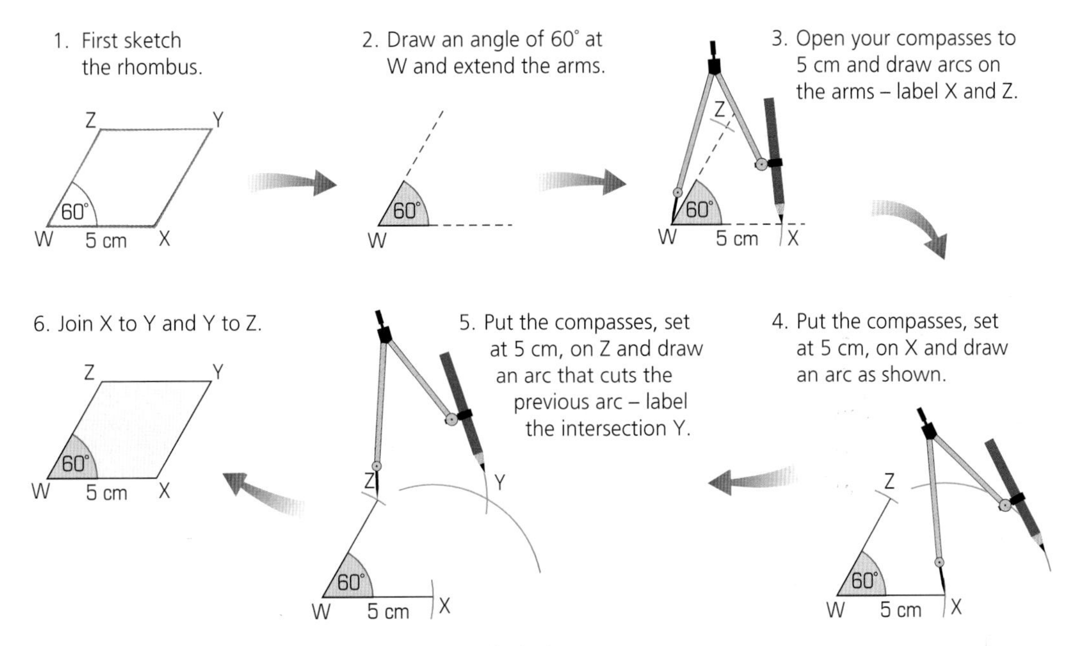

b Draw the diagonals ZX and YW. Label the intersection M.
c Measure WM and MY. What do you notice?
d Copy and complete this sentence: ZX is the ________ bisector of WY.
e Measure angle ZWY. What do you notice?
f Copy and complete this sentence: WY is the ________ bisector of ∠ZWX.

S3.5 2-D representations of 3-D shapes

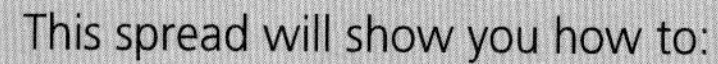

This spread will show you how to:

- Visualise and describe 3-D shapes.
- Draw plans and elevations of 3-D shapes.
- Construct shapes from their plans and elevations.

KEYWORDS

Elevation, Perpendicular, Isometric, Plan, Parallel

- You can describe a 3-D shape by its:
  - **Plan view,** as seen from the top
  - **Front elevation,** as seen from the front
  - **Side elevation,** as seen from the side

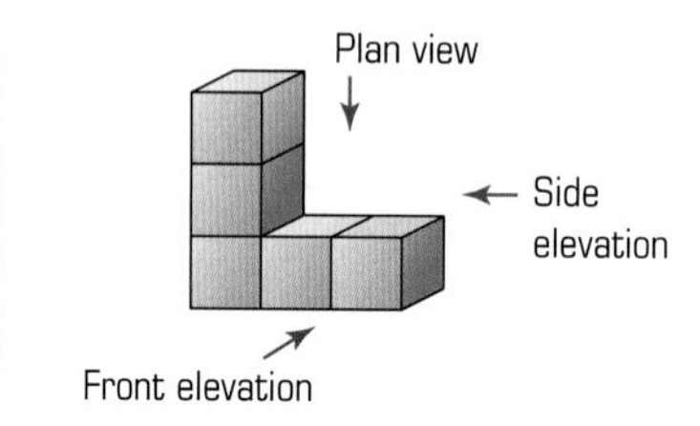

The plan view is a rectangle

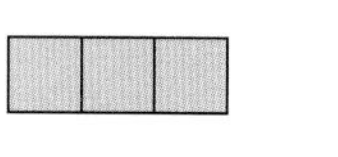

The side elevation is a rectangle

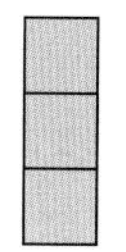

The front elevation is an L-shape

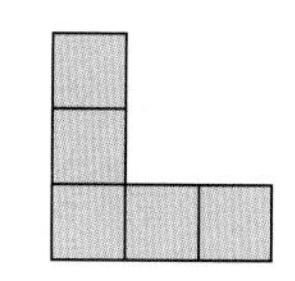

You can draw 3-D shapes using isometric paper.

example

Look at the cube ABCDEFGH. Identify:

a Two edges that meet at a vertex
b Two faces that meet at an edge
c Two lines that are neither parallel nor perpendicular

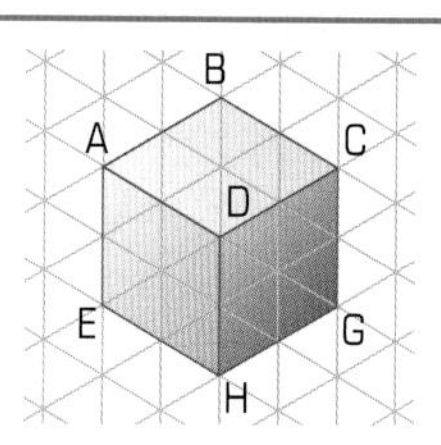

a Edges AB and BC meet at the vertex B.
b Faces ABCD and CDHG meet at the edge CD.
c Line AB is not parallel nor perpendicular to the diagonal line CH.

example

Here are the plan, front elevation and side elevation of a solid shape.
Draw the solid on isometric paper, and describe it.

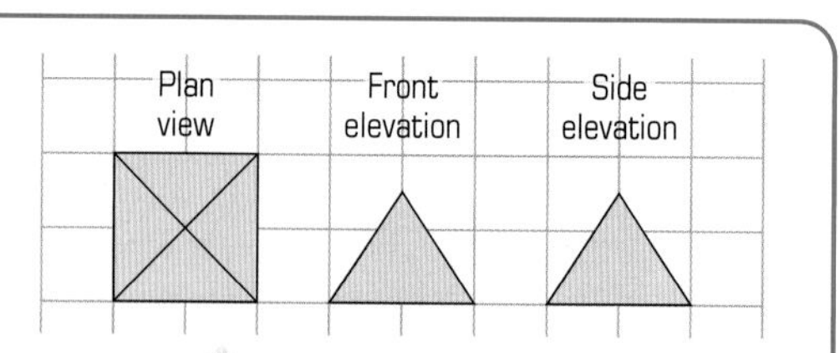

Look at the shapes and dimensions of the different views and piece the final shape together.

The shape is a square-based pyramid.

Exercise S3.5

For each of questions 1 to 6:

- ▶ name the 3-D shape
- ▶ sketch the 3-D shape on isometric paper.

Plan View	Front Elevation	Side Elevation
1		
2		
3 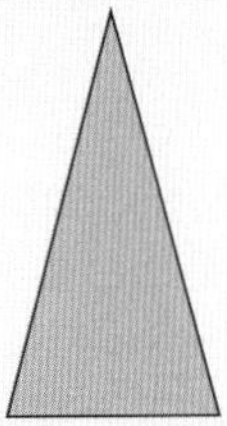		
4 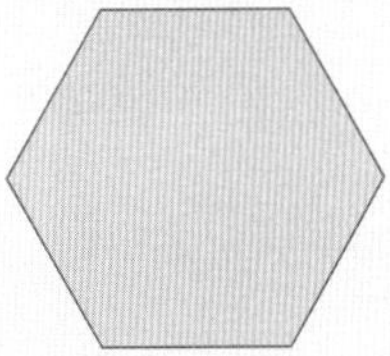		
5 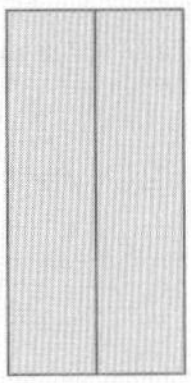	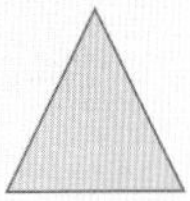	
6 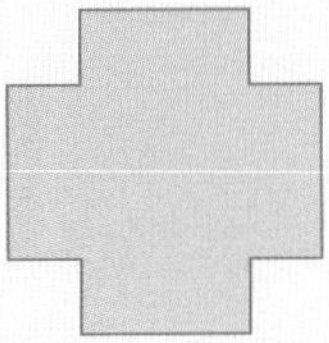	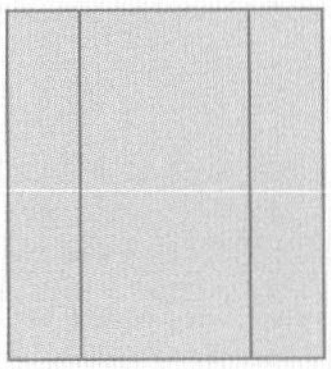	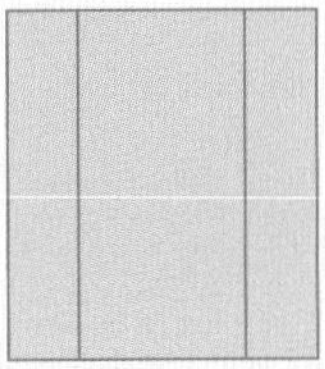

Summary

You should know how to ...

1 Solve geometrical problems using angle properties of triangles and quadrilaterals.

2 Use straight edge and compasses to construct:
- the midpoint and perpendicular bisector of a line segment.
- the bisector of an angle.

3 Construct a triangle given three sides.

4 Interpret solutions in geometric form.

5 Begin to use plans and elevations.

Check out

1 Find the missing angles:

a

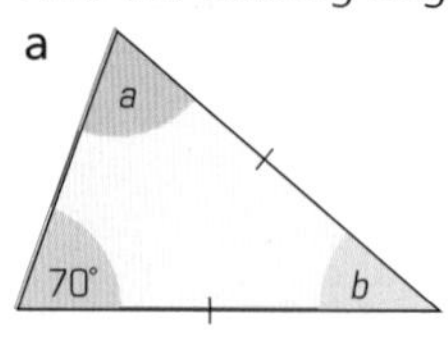

b

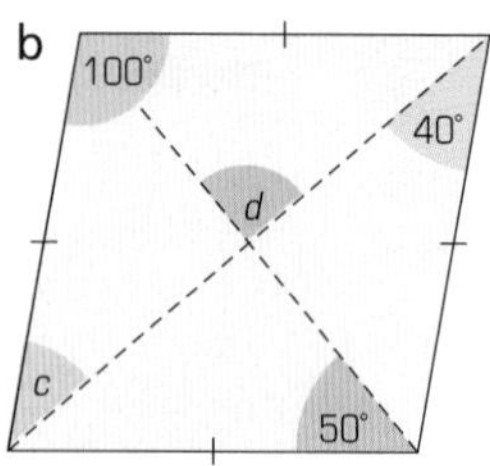

2 a Draw a line exactly 5.7 cm long
Use compasses to bisect your line.
Check your construction by measuring the mid point of the line.

b Use a ruler and a protractor to draw two lines that are 4 cm long and that meet at 80°.
Use compasses to bisect this angle.

c Bisect an angle of 52°.

3 Construct these triangles:

a ABC where AB = 10 cm, $\hat{B} = 50°$ and $\hat{C} = 60°$.

b PQR where PQ = 6 cm, QR = 7 cm, $\hat{P} = 50°$.

c EFG where EF = 3 cm, FG = 4 cm and EG = 5 cm.

4 In question 3 which of your triangles are unique?
Draw sketches to illustrate your answer.

5 Sketch this solid.

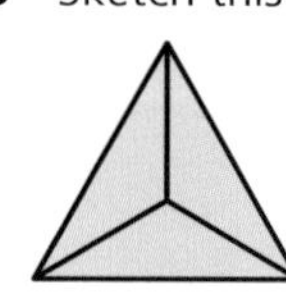

plan view

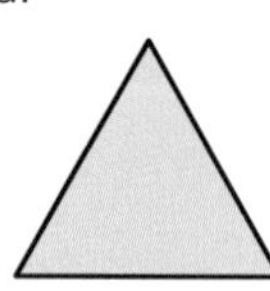

front elevation

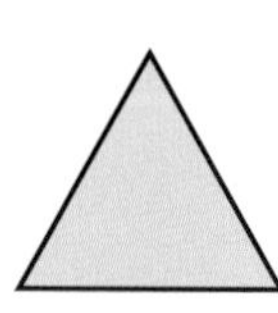

side elevation

14 Percentages, ratios and proportion

This unit will show you how to:

- Use division to convert a fraction to a decimal.
- Interpret percentage as the operator 'so many hundredths of' and express one given number as a percentage of another.
- Calculate percentages and find the outcome of a given percentage increase or decrease.
- Use the equivalence of fractions, decimals and percentages to compare proportions.
- Consolidate understanding of the relationship between ratio and proportion.
- Reduce a ratio to its simplest form.
- Divide a quantity into two or more parts in a given ratio.
- Use the unitary method to solve simple word problems involving ratio and direct proportion.
- Solve word problems and investigate in the context of number.
- Solve more complex problems by breaking them into smaller tasks.

You use ratio and proportion when you mix ingredients.

Before you start

You should know how to ...

1 Recognise simple equivalences of fractions, decimals and percentages.

2 Find simple percentages of quantities.

Check in

1 **a** Write as a decimal and as a percentage:

i $\frac{1}{4}$ **ii** $\frac{1}{5}$ **iii** $\frac{3}{10}$

b Write as a fraction in its lowest terms:

i 5% **ii** 17.5% **iii** 120%

2 Find 15% of:

a £40 **b** 640 m

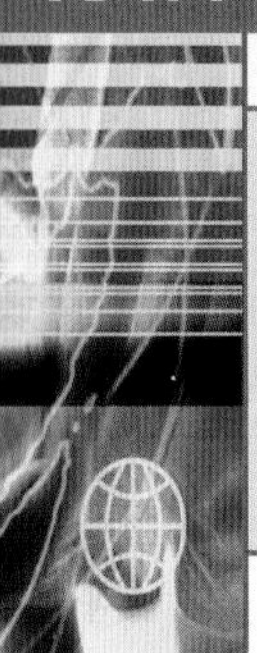

N4.1 Fraction, decimal and percentage equivalents

This spread will show you how to:

- Understand percentage as the number of parts in every 100.
- Recognise the equivalence of fractions, decimals and percentages.

KEYWORDS

Convert, Fraction, Decimal, Numerator, Denominator, Percentage, Equivalence

You can convert between decimals, fractions and percentages.

Decimals to percentages

To convert 0.23 into a percentage

- First turn it into a fraction out of 100: 0.23 $= \frac{2}{10} + \frac{3}{100}$
  $= \frac{20}{100} + \frac{3}{100}$
  $= \frac{23}{100}$
- Write this as a percentage: $\frac{23}{100}$ = 23%

Remember that a percentage is a fraction out of 100. For example, $17\% = \frac{17}{100}$

This is the same as multiplying by 100: $0.23 \times 100 = 23\%$

- **To convert a decimal to a percentage you multiply by 100.**

Fractions to percentages

To convert $\frac{4}{7}$ to a percentage

- Write it as a division: $\frac{4}{7}$ $= 4 \div 7$
- Work it out as a decimal (use a calculator): $4 \div 7$ $= 0.5312 \ldots$
  = 0.531 (to 3 decimal places)
- Multiply by 100: 0.531×100 = 53.1%

You could ×100 first and **then** divide:
$\frac{4}{7} = \frac{4}{7} \times 100\%$
$= 4 \times \frac{1}{7} \times 100\%$
$= \frac{400}{7}\%$
$= (400 \div 7)\%$
= 53.1% to 1 d.p.

- **To convert a fraction to a percentage you can either:**
  - Divide the numerator by the denominator and then multiply by 100, or
  - Multiply by 100 and then divide the numerator by the denominator.

Percentages to fractions or decimals

To convert 59% to a fraction ...

- Write it over 100: 59% $= \frac{59}{100}$

... and to a decimal

Divide by 100: 59% $= 59 \div 100 = 0.59$

This diagram might help you to convert between the three forms:

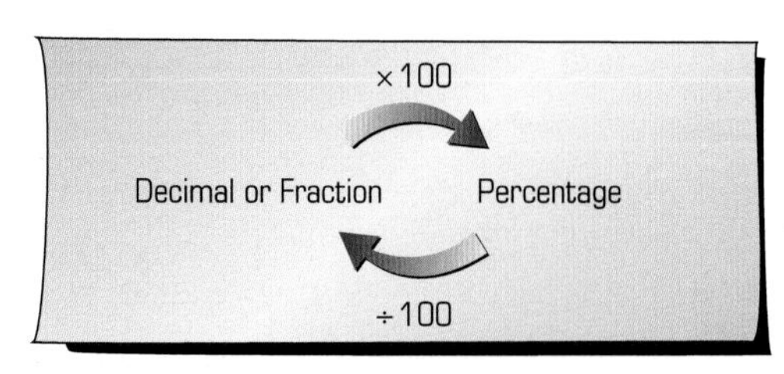

- **To convert a percentage to a fraction or decimal you divide by 100.**

Exercise N4.1

1 Complete this table:

Percentage	Decimal	Fraction in simplest form
26%		
	0.07	
		$\frac{1}{4}$
		$\frac{11}{3}$
175%		
	1.5	
	0.7	

2 Fred said that 4% is the same as 0.4 as a decimal. Explain why he is wrong.

3 Re-write each of the following fraction questions as percentage questions.
For example, $\frac{3}{5}$ of £200 can be written as 60% of £200.
Note: You do not need to work out the answer.

a $\frac{3}{8}$ of £500 **b** $0.7 \times$ £2000
c $\frac{4}{3} \times 250$ **d** 0.27×24
e $\frac{7}{8}$ of 30 sheep **f** 0.622×2000 mm

4 Work with a partner.

- Write down any four percentages.
- For each percentage describe it to your partner without using the word percentage. You can use fractions and decimal facts to describe it, for example, 23% is just less than a quarter.
- You can give up to three clues for each percentage.
- Your partner will score 3 points for guessing correctly on the first clue, 2 points on the second clue and 1 point on the third.
  Play the game until one of you has scored 20 points.

5 Calculate these showing your working and explaining your reasoning:

a Which of these is nearer to 1: $\frac{17}{19}$ or 0.93?
b Find a fraction which is halfway between $\frac{13}{19}$ and 0.65.

6 Calculate the following, in each case showing all your working out and explaining your answers.

a Axel the cat scores 43% in his Cat's Whiskas exam and $\frac{10}{23}$ in his Paw Print test. In which subject did he do best?
b In a school survey 83% of the pupils said they liked pop music. In Mr. Morrissey's class $\frac{26}{31}$ pupils said they liked pop music. How do the results of Mr. Morrissey's class compare to the rest of the school?

7 **a** What fraction of

i 123 is 74
ii 4 hours is 37 mins
iii 3 km is 1763 m
iv 18 kg is 4 kg 32 g
v 180° is 63°?

b Put your answers in order from lowest to highest.
Explain your method for doing this.
c Express each fraction as a percentage to the nearest whole number.

8 **Investigation**
Using 7 as a prime number denominator, none of the fractions $\frac{1}{7}, \frac{2}{7}, \frac{3}{7}, \frac{4}{7}, \frac{5}{7}, \frac{6}{7}$ will give an exact percentage equivalent.

$\frac{1}{7} = 0.14285 \ldots = 14.2\%$ to 1 d.p.

$\frac{2}{7} = 0.28571 \ldots = 28.6\%$ to 1 d.p.

Only $\frac{7}{7} = 100\%$ gives an exact equivalent.

Is this true for all prime number denominators? Investigate.

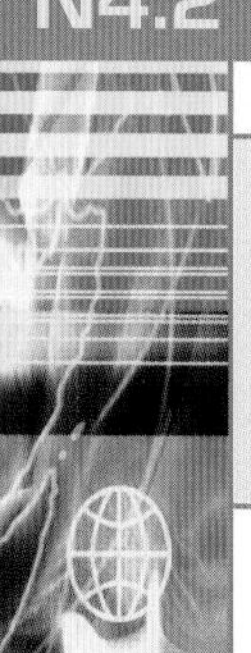

N4.2 Finding simple percentages

This spread will show you how to:

- Calculate percentages of numbers, quantities and measurements using mental methods, written methods and a calculator.
- Express one given number as a percentage of another.

KEYWORDS

Cancel
Decimal
Equivalent
Fraction
Percentage

You know that 10% is the same as $\frac{1}{10}$.
You can use this fact to find percentages of amounts.

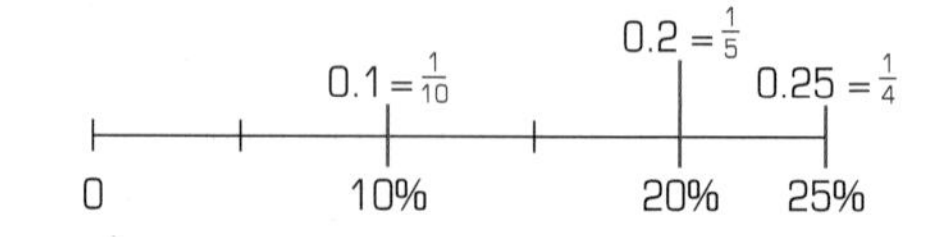

example

Find 45% of £70.

10% of £70 $= \frac{1}{10}$ of £70
$= £70 \div 10 = £7$
40% of £70 $= 4 \times 10\%$ of £70
$= 4 \times £7 = £28$
5% of £70 $= \frac{1}{2}$ of 10% of £70
$= \frac{1}{2}$ of £7 $= £3.50$
45% of £70 $= £28 + £3.50 = £31.50$

A diagram might help:
There are 4 lots of 10% ... ... and 1 lot of 5% ...

10%	10%	10%	10%	5%						
£7	£7	£7	£7	£3.50						

... making £31.50 altogether

You can also use a variety of written methods to calculate percentages of amounts.

example

Find 23% of 38 by:

a using an equivalent decimal **b** using an equivalent fraction

First approximate: 25% of 40 = 10

a 23% $= 23 \div 100 = 0.23$
23% of 38 $= 0.23 \times 38$
$= 8.74$

b 23% $= \frac{23}{100}$
23% of 38 $= \frac{23}{100} \times 38$
$= 23 \times \frac{38}{100}$
$= \frac{874}{100}$
$= 8.74$

```
  0.23
  38 ×
  0874
  0690 +
   184
   874
```

Decimal point goes 2 places from the left: 8.74

You can write an amount as a percentage of another amount.

example

What percentage of 50 cm is 28 cm?

First write it as a fraction: 28 out of 50 is $\frac{28}{50}$.
Now multiply by 100 to turn it into a percentage: $\frac{28}{50} \times 100\% = 28 \times \frac{100}{50}\%$
Cancel the fraction: $= 28 \times \frac{2}{1}\%$
$= 56\%$

So 28 cm is 56% of 50 cm.

Exercise N4.2

1 Find 10% of:

a 367 **b** 46 **c** 663

2 Find 5% of:

a 3450 **b** 60 **c** 59 000

3 Solve the following using a mixture of mental methods and jottings.

a 11% of 200 cm **b** 24% of 400 kg
c 12% of £800 **d** 5% of £40

4 Calculate the following using mental, written or calculator methods as appropriate. (Give your answers to the nearest whole unit)

a 14% of £43 **b** 8% of 89 m
c 16% of 726 m **d** 4.5% of 0.94 cm
e 72% of 56 kg **f** 7% of 36 kg
g 47% of 205 **h** 3.1% of £79
i 3% of $44 **j** 5.2% of 45 kg

5 In a football crowd, 23% of the people wore red socks and $\frac{2}{5}$ of the people didn't wear any socks. Answer the following questions:

a What percentage didn't wear socks?
b What percentage wore socks but didn't wear red socks?
c What is your answer to **b** as a decimal?
d What fraction (in its lowest terms) wore red socks?

6 There are 280 people in a school. 35% of them come to school on foot. $\frac{3}{10}$ come by car and the rest catch the bus.
How many

a come on foot **b** come by car
c catch the bus?
d What is $\frac{3}{10}$ as a percentage?
e 50% of those who walk live very close to school. How many pupils is this?

7 Make up five different percentage questions with an answer of £3.26.

8 What percentage of

a 168 pence is 49 pence?
b 3 hours is 73 mins?
c £3.20 is 49 pence?
d 7 kg is 1200 g?

9 **Game** – Target percentage

Box A – Amounts	240	370	420	390	280	325	**430**
Box B – Targets	135	156	228	359	175	288	**160**

The first player selects an amount from Box A and then a target from Box B.

Example *Amount* = *430* *Target* = *160*

The first player and the second player write down their estimates of what percentage the target is of the amount.

Example *Player 1 estimates 35%*
Player 2 estimates 38%

Each player calculates their own percentage of the amount.

Example *Player 1 calculates*
35% of 430 $\simeq$ *151*
Player 2 calculates
38% of 430 $\simeq$ *163*

The player who is nearest to the target amount wins the round. Continue until five rounds have been completed.

10 Brian invests £2400 in a savings account. Each year the total value of the money in the account increases by 6.4%.

a How much money is there by the end of the first year?
b How much money is there after 2 years?
c How much money is there after 5 years?
d When will Brian have more than £5000 in his account?

11 Calculate $\frac{2}{5}$ of 42% of £4000.

N4.3 Percentage change

This spread will show you how to:

- Find the outcome of a given percentage increase or decrease.

KEYWORDS

Decrease, Percentage, Increase

People use percentages to show a change in a quantity.

If something increases by 100% you now have 200% of it. It has doubled!

If something decreases by 100% you would have nothing left!

There are different ways you can calculate a percentage decrease.

example

In a sale a skateboard is reduced in price by 30%.
The original price was £25. What is the new price?

- First calculate the reduction:

30% of £25 = 3 × 10% of £25
= 3 × £2.50
= £7.50

- Now calculate the new price:

£25 − £7.50 = £17.50

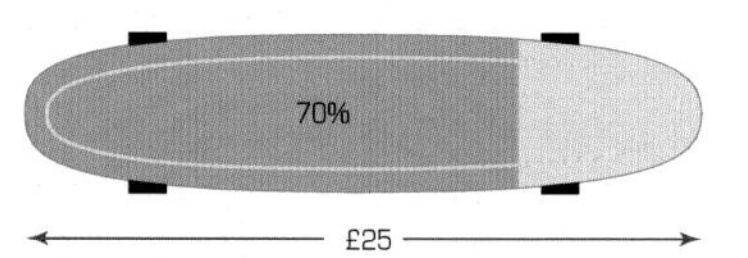

You could do the calculation without subtracting:

New price = 70% of old price
= 70% of £25
= 7 × 10% of £25
= 7 × £2.50 = £17.50

You can calculate a percentage increase in the same ways.

example

A footballer earns £2130 per week.
His wage is increased by 8%. Calculate his new weekly wage.

- First calculate the increase:

8% of £2130 $= \frac{8}{100} \times £2130$
$= £\ 8 \times \frac{2130}{100}$
$= \frac{17\,040}{100}$
= £170.40

- Now calculate the new wage:

£2130 + £170.40 = £2300.40

You could do the calculation without adding:

New wage = 108% of old wage
= 108% of £2130
$= \frac{108}{100} \times £2130$
$= £108 \times \frac{2130}{100}$
$= £\frac{230\,040}{100}$
= £2300.40

- **You can calculate a percentage increase or decrease in an amount or a measurement.**

Exercise N4.3

1 Calculate the following using mental, written or calculator methods as appropriate.

a 17% of £2400
b 85% of 240 kg
c 45% of 31.6 km
d 115% of 600 sheep
e 35% of 234 cm
f 71% of 4770 litres
g 90% of 650 people
h 13.5% of €4320

2 What percentage of:

a 300 is 6?
b 25 cm is 35 mm?
c 12 kg is 3400 g?
d 1 day is 3 hours and 20 mins?

3 Solve these problems:

a Jack invests $730 in his bank account. During the next 2 years his money has increased by 20%. How much money does he have now in his bank account?
b Sebastian wants to buy a pair of carving skis for £380. In a sale the skis are reduced by 15%. Calculate the new cost of the skis.
c David and Victoria buy a house for £1 200 000. Over the next 12 months the house increases in value by 12.4%. Calculate the new value of their house.
d A pot of pink paint contains red and white paint. 15% of the paint is red.
  i What percentage of the pink paint is white?
  ii How much red paint would be needed to make 3.5 litres of pink paint?
  iii How much white paint would be needed to make 5.2 litres of pink paint?

4 **a** In a particular show 42% of the cast are singers, 30% are dancers and the rest are 'multi-talented'. If there are 450 in the show calculate:
  i the number of dancers
  ii the number that are not singers
b A fabric consist of these materials:
30% cotton
48% nylon
22% satin
If the total area of the fabric is 250 m^2 what area is not satin?

5 **Investigation.** A car salesman increases the cost of a car costing £7800 by 12%. Later he decides to reduce the new price by 12% to bring the price back to £7800. Will this work? Explain your answer. Investigate for other percentage increases and decreases.

6 **a** A DVD costs £180. In a sale it is reduced in price by £30. What is the percentage reduction?
b Simon invested £160. After one year his interest was £9.60.
Joanne invested £140. After one year her interest was £8.96.
Who had the better rate of interest, Simon or Joanne? Explain your answer.

7 VAT is currently 17.5%. Use your calculator to find an efficient way of calculating the price of an item after VAT has been added.

8 A worker's wage increases by 7.2% each year. If his wage is now £20 000 per year,

a Calculate the wage after 1 year
b Calculate the wage after 3 years
c After how many years will his wage be more than £30 000 per year?

Proportion

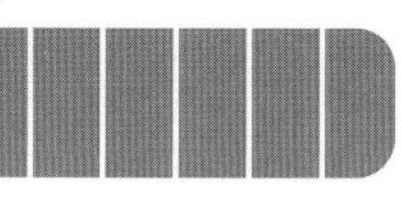

This spread will show you how to:

- Solve simple problems involving direct proportion.
- Plot a conversion graph.

KEYWORDS

Decimal, Quantity, Fraction, Scale, Proportion, Unitary

- You use a proportion to compare the size of a part to the whole.
- You can express a proportion as a fraction, a decimal or a percentage.

example

In a bag there are 20 marbles, of which 14 are green and 6 are blue. What is the proportion of green marbles?

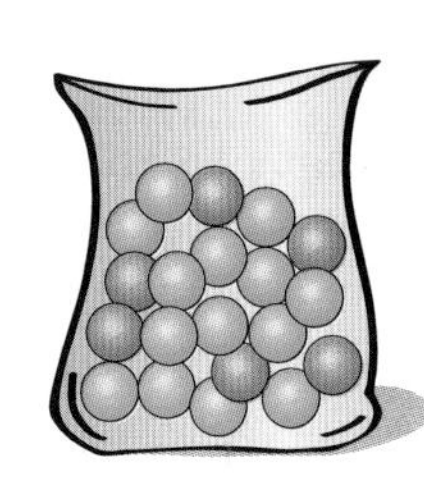

There are 14 out of 20 green marbles. You can write the proportion ...

- As a fraction: $\frac{14}{20} = \frac{7}{10}$
- As a decimal: $\frac{7}{10} = 7 \div 10 = 0.7$
- As a percentage: $0.7 \times 100 = 70\%$

Often the size of one quantity depends on the size of another quantity.

Conversion graph for sterling to euros

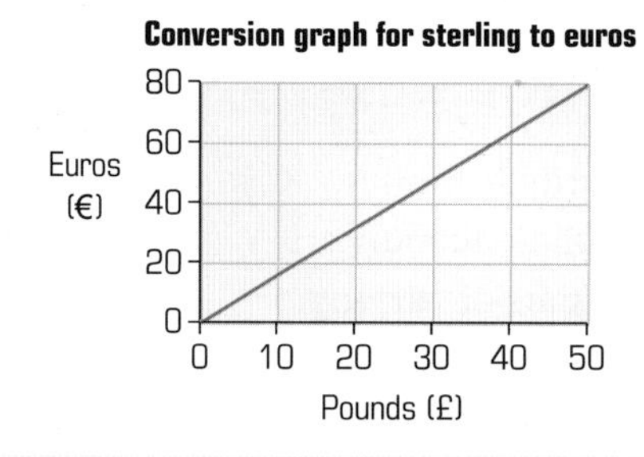

- Two quantities are in direct proportion if:
  - As one quantity increases, the other quantity increases in the same proportion
  - As one quantity decreases, the other quantity decreases in the same proportion

example

a £4 is worth 72 Austrian schillings. What is the value of £1?

Quarter the pounds ➡ quarter the schillings

£4 = 72 schillings.

÷4   ÷4

£1 = 18 schillings

b 3 kg of 'Green Slime' costs £1.20. What is the cost of 7 kg of 'Green Slime'?

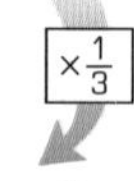

$\times\frac{1}{3}$ 3 kg of slime = £1.20 $\times\frac{1}{3}$

1 kg of slime = £0.40

×7   7 kg of slime = £2.80   ×7

Part **b** of the example uses the unitary method.
You find the value of a single unit and then scale it up or down.

Exercise N4.4

1 **a** 4 pizzas cost \$7.96.
Calculate the cost of 8 pizzas.
b 3 packets of cornflakes cost £4.20.
Calculate the cost of 1 packet of cornflakes.
c £1 is worth 1.61 Euros.
How many Euros will I get for £2.40?
d 200 g of nuts contains 97 g of fat.
How many grams of fat are there in 300 g of nuts?

2 Copy and complete this ready reckoner for the price of trees at Nahida's Nursery.

Type of Tree	1	2	3	5	10
Fir tree					£25
Birch tree		£14.40			
Rowan tree				£9.25	
Oak tree			£16.29		

3 Oak trees cost £16.29

3 Copy and complete this diagram

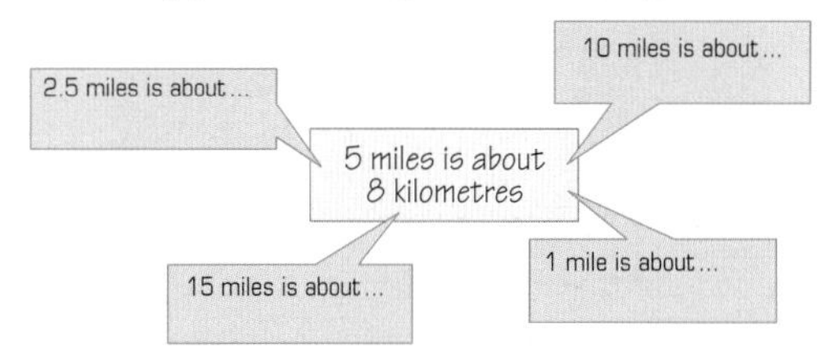

a Make up some more statements of your own and add them to the diagram.
b Work out how many miles are equivalent to 37 kilometres.

4 100 cm is about 40 inches.
a How many inches make 30 cm?
b How many cm make 36 inches?
c On graph paper draw a conversion graph to show how to change inches into centimetres.
d Use your graph to calculate how many cm are equivalent to 35 inches.

5 **a** In Anytown School the proportion of boys is 48%. The school has 1600 pupils.
How many girls attend the school?
b Asif has to pay $\frac{3}{7}$ of his salary in tax. He earns £41 400 a year. How much money does he pay in tax?
c A consumer magazine suggests that 12.7% of washing machines break down each year. There are 75 000 washing machines in Blackburn. How many washing machines will break down in Blackburn each year?

6 Work out the following
a 12 pears cost £2.88. What is the cost of 15 pears?
b A distance of 8 km is represented by 15.2 cm on my map. How many cm will represent 10 km?
c 7 books weigh 1064 grams. What is the weight of 6 books?
d With 8 gallons of petrol my car travels 248 miles. How far will it travel with 11 gallons of petrol?
e It takes 6 minutes to hard boil 3 eggs. How long will it take to hard boil 5 eggs?

7 In Great Britain there are 58 million people.
Here are the proportions of people who die each year from various causes:

Smoking	1 in 200
Car accident	1 in 4000
Struck by lighting	0.00001% (nearest whole number)
Heart disease	1 in 29
Alcohol related	1.14%

Calculate the number of people who die from each cause (to the nearest 1000).

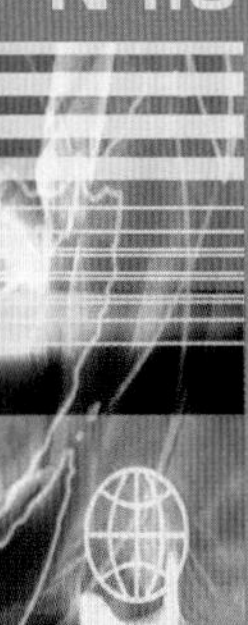

N4.5 Introducing ratio

This spread will show you how to:

- Understand the relationship between ratio and proportion.
- Simplify a ratio by cancelling.
- Divide a quantity into two or more parts in a given ratio.
- Solve simple problems using a unitary method.

KEYWORDS

Cancel	Quantity
Divide	Ratio
Proportion	Unitary

You use **proportion** to compare the size of a part to the size of the whole.
You use **ratio** to compare the size of different parts to each other.

example

In a class there are 18 boys and 12 girls.

a Find the proportion of girls as a fraction. **b** Find the ratio of boys to girls.

a The proportion of girls is $\frac{12}{30} = \frac{6}{15}$

b The ratio of boys to girls is 18 to 12.

Write this as 18 : 12
Simplify the ratio 18 : 12 = 3 : 2

- **You use the symbol : to compare quantities in a ratio.**
- **You can simplify ratios by cancelling them down.**

You cancel ratios down in the same way as fractions.

You can use ratio to divide a quantity into unequal amounts.

example

Debbie, Dionne and Dominic want to share £300 in the ratio 8 : 7 : 5.
How much will each person receive?

Debbie will get £8 for every £7 that Dionne gets and for every £5 that Dominic gets.

- First you need to divide the amount into parts.
  Work out how many parts there are: 8 + 7 + 5 = 20
- Then work out the value of one part: £300 ÷ 20 = £15
- Finally allocate parts to each person.
  Debbie will get 8 parts: 8 × £15 = £120
  Dionne will get 7 parts: 7 × £15 = £105
  Dominic will get 5 parts: 5 × £15 = £75

£15	£15	£15	£15	£15	£15	£15	£15	£15	£15	£15	£15	£15	£15	£15	£15	£15	£15	£15	£15

Debbie Dionne Dominic

- **You can use the unitary method to divide an amount in a given ratio.**
  In the unitary method you find the value of a single part.

There is an example of the unitary method in N4.4.

Exercise N4.5

1 Simplify each of the following ratios:

a 10 : 50 **b** 23 : 69
c 14 : 154 **d** 6 : 30
e 12 : 4 **f** 64 : 20
g 6 : 26 **h** 10 : 200

2 In a particular class there are 30 pupils. 12 are boys and 18 are girls. Expressed as a ratio this is 12 : 18 boys to girls respectively.

a Explain what the word 'respectively' means when dealing with ratio.
b Write the ratio 12 : 18 in its simplest form.
c What fraction, in its simplest form, of the class are boys?
d What percentage of the class are boys?

3 Simplify each of the following ratios:

a 4 : 12 : 48 **b** 12 : 60 : 120
c 40 : 16 : 4 **d** 30 : 6 : 18
e 20 : 12 : 8 **f** 80 : 34 : 20
g 5 : 60 : 25 **h** 7 : 21 : 48

4 Three chemicals A, B and C are mixed in the ratio 1 : 2 : 5

a What fraction of the mixture is chemical A?
b What fraction of the mixture is chemical B?
c Describe how you could work out the fraction that is part C using your answers to **a** and **b**.
d Write your answers to **a**, **b** and **c** as percentages.
e What should the percentages in **d** add up to?
f 24 000 litres of the mixture were made. Calculate the volume of:
i chemical A
ii chemical B
iii chemical C

5 **a** The scale of a particular map of the Jam Butty mines in Knotty Ash is 1 : 10 000.
i If the 'kiln' is shown on the map as being 1.25 cm long, what is its actual length?
ii In real life the mine shaft is 100 m long. How long will this be on the map?
b In the 'school weekly draw' you are told that the chance of winning is 1 in 15. If 450 tickets are sold how many are:
i likely to win?
ii likely to lose?

6 **a** The angles in a pentagon add to 540°. If the ratio of the angles is 2 : 5 : 2 : 1 : 1, what is the size of each angle?
b Three students volunteer for a litter collecting competion. They share their prize money of £44 in the ratio of litter they collected.
Amy collected four times as much as Beatrice and Colin collected half as much as Beatrice.
Write the ratio using : notation and work out how much each student receives.

7 **Investigation**
Find 2 numbers whose ratio is 4 : 7 and whose difference is 39.
Make up another problem for your partner to solve.

8 Simplify these ratios.

a 4 m : 30 cm
b 2 km : 450 m
c 400 g : 3 kg
d 420 mm : 30 cm : 0.7 m

N4 Summary

You should know how to ...

1 Use the equivalence of fractions, decimals and percentages to compare proportions.

Check out

1 A cement mix uses 35% sand for every kg of cement.

a How much sand is contained in 4.5 kg of cement?

b If you had 2 kg of sand, how much cement could you make (assuming you had plenty of all the other ingredients)?

2 Calculate percentages and find the outcome of a given percentage increase or decrease.

2 a A sunflower grows by 10% from 12 cm. What is its new height?

b A dress costing £45 is reduced by 20% in a sale. What is its new price?

3 Divide a quantity into two or more parts in a given ratio.

3 Divide £160 between three people in the ratio 3 : 15 : 2.

4 Use the unitary method to solve simple word problems involving ratio and direct proportion.

4 Work out the following using mental or written methods.

a 4 apples cost 92 pence. How much would 13 apples cost?

b 24 eggs cost £3.36. How much would 7 eggs cost?

c 7 kg of sugar cost £4. How much would 10 kg of sugar cost?

d 51 calculators cost £229.50. How much would 20 calculators cost?

4 Linear equations

This unit will show you how to:

- Construct and solve linear equations with integer coefficients.
- Multiply a single term over a bracket.
- Know that algebraic operations follow the same conventions and order as arithmetic operations.
- Solve word problems and investigate in the context of algebra.
- Identify the necessary information to solve a problem.

The speed and performance of a car can be described by equations.

Before you start

You should know how to ...

1 Use algebraic conventions, including brackets.

2 Use negative number in calculations.

3 Convert improper fractions to mixed numbers.

Check in

1 Write these sentences using algebra:

a A number, multiplied by 6 and 7 added.

b A number, subtract 4 and divide by 2.

c A number, multiplied by itself, add 9 and multiply by 10.

2 Evaluate:

a $^-36 \div 9$ **b** $30 \div {^-5}$ **c** $^-15 \div {^-3}$

d $\frac{-42}{7}$ **e** $\frac{100}{^-4}$ **f** $\frac{^-121}{^-11}$

3 **a** Convert these improper fractions to mixed numbers:

i $\frac{36}{7}$ **ii** $\frac{19}{4}$ **iii** $\frac{-20}{3}$ **iv** $\frac{-37}{5}$

b Put these numbers in ascending order:

$\frac{43}{7}$ $5\frac{7}{9}$ $\frac{71}{8}$ $8\frac{2}{3}$

A4.1 Solving algebraic equations

This spread will show you how to:

- Understand and use inverse operations.
- Construct and solve simple linear equations.

KEYWORDS
Equation, Solve, Expression, Solution

$6x - 8$ is an **expression**.
If you make it equal to a value, it becomes an **equation**: $6x - 8 = 16$
To find x, you must solve the equation.

You can think of a pair of scales:

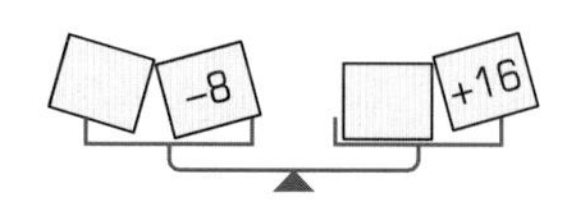

$6x - 8$ is on one side and 16 is on the other.

Add 8 to the left hand side:

To stop the scales unbalancing, add 8 to the right as well:

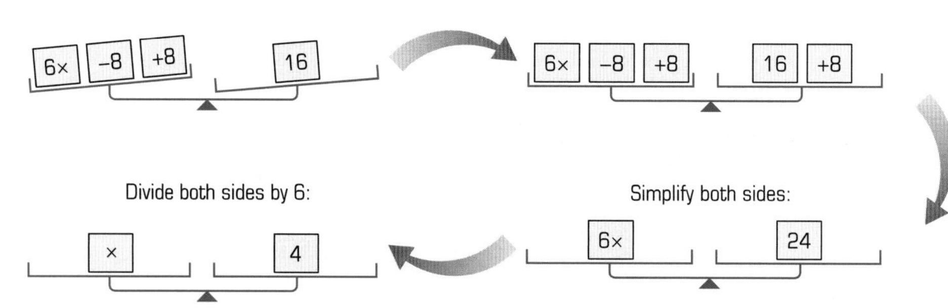

So $x = 4$ Check: $6 \times 4 - 8 = 16$.

You do not need to use a pair of scales to solve equations each time.
You just need to know about inverse operations.

- Subtraction is the inverse operation to addition.
- Division is the inverse operation to multiplication.

example

Solve $\dfrac{3(x-7)}{2} = 6$

Read the equation:
Think of a number, x, subtract 7, multiply by 3 then divide by 2. The answer is 6.
Now do the reverse:
Multiply by 2: $3(x - 7) = 12$
Divide by 3: $x - 7 = 4$
Add 7: $x = 11$
Check the solution is $x = 11$ by substituting:
$\dfrac{3(11-7)}{2} = \dfrac{3 \times 4}{2} = 6$ ✓ correct.

Exercise A4.1

1 Solve these equations:

a $x + 9 = 17$

b $x - 3 = 10$

c $10 - x = 2$

d $8 - y = 12$

e $2x - 1 = 19$

f $8 - 3x = 2$

g $\frac{x}{4} + 5 = 7$

h $11 - \frac{8}{x} = 7$

2 Solve these equations.

a $5(x + 3) = 60$

b $\dfrac{x - 4}{2} = 10$

c $\dfrac{6x + 14}{4} = 20$

d $2x^2 = 98$

e $\dfrac{2(x + 5) - 14}{2} = 8$

f $8x - 13 = 17$

g $\dfrac{2x + 18}{3} = 2$

h $\dfrac{3 - y}{2} + 4 = 2$

3 Solve this equation for x in the same way as in question 2.

$$\frac{w(px + q)}{k} = m$$

4 In a number pyramid the numbers next to each other are added to finally get the top total.
What are the missing numbers in this pyramid? Use equations to help you find them.

Hint: let the missing number in the bottom row be x

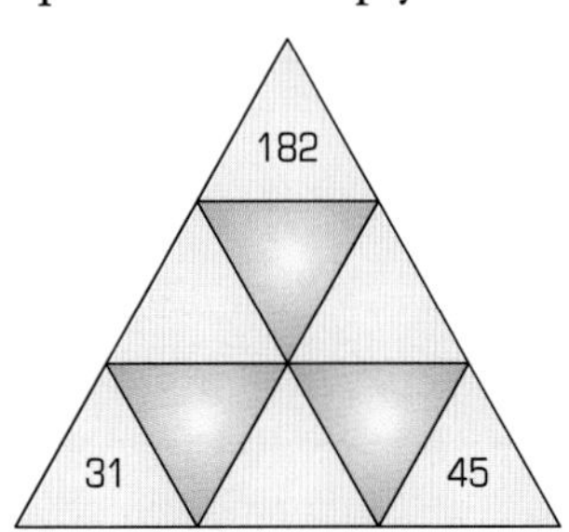

5 The angles in a triangle are $a°$, $(a + 10)°$ and $(a + 20)°$. Find the value of each angle.

6 There are 376 stones in 3 piles. The second pile has 24 more stones than the first pile. The 3rd pile has twice as many stones as the 2nd. How many are in each pile?

A4.2 Solving harder equations

This spread will show you how to:
- Understand and use inverse operations.
- Consolidate forming and solving linear equations.

KEYWORDS
Equation
Term
Negative
Solve

Sometimes the unknown values x can appear more than once in an equation.
You can solve equations by balancing one side of the = with the other.

example

Solve $6x - 2 = 2x + 10$

$6x - 2 = 2x + 10$

$4x - 2 = 10$ Subtract $2x$ from both sides.

Now think of a number, ×4 then ⁻2 to give you 10.

$4x = 12$ Add 2 to both sides

$x = 3$ Divide both sides by 4.

Check: $6 \times 3 - 2 = 16$ $2 \times 3 + 10 = 16$

You need to be careful with negative algebra terms.

An equation is like a mirror: added terms on one side become subtracted on the other.

example

Solve $18 - 3x = 5x - 6$

$18 - 3x = 5x - 6$

$18 = 5x + 3x - 6$ **Add** $3x$ to both sides

$18 = 8x - 6$ Think of a number: ×8, ⁻6 to give you 18.

$24 = 8x$ **Add** 6 to both sides

$3 = x$ Divide both sides by 8

So $x = 3$ Check: $18 - 3 \times 3 = 9$ $5 \times 3 - 6 = 9$

▶ **When x appears on both sides, you should remove the smallest algebraic term from both sides first.**

Exercise A4.2

1 Solve these equations:

a $16 - 5x = 4x + 2$ **b** $8x - 35 = 25 - 2x$

c $13 - 4x = 6x - 2$ **d** $5x - 4 = 3x + 12$

e $4x - 16 = 2x + 2$ **f** $4x + 27 = 9x - 8$

2 Solve these equations; each answer is a negative integer or a fraction.

a $11 + 5x = 10 - 3x$ **b** $6 - 8x = 15 + 3x$

c $9 - 4x = 15 - 12x$ **d** $6x - 8 = 3x + 11$

e $4x + 9 = 8x - 12$ **f** $10x + 15 = 5x - 20$

3 In each of the following, write your own equation to describe the problem, then solve your equation.

a I think of a number, multiply it by 4 and subtract 13 then I get the same answer as when I multiply the number by 3 and *subtract it* from 29.

b I think of a number, multiply it by 7 and subtract 12 and I get the same answer as when I multiply it by 2 and add 8.

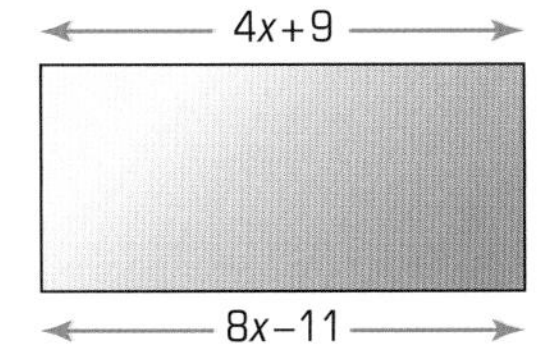

d

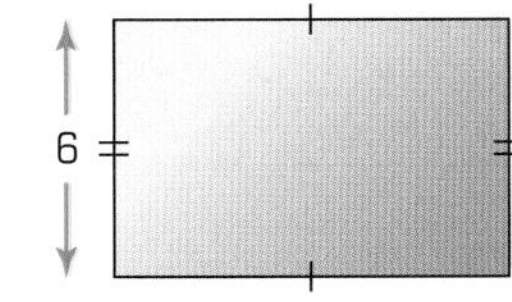

$x+2$

The perimeters of the rectangle and the square are equal.

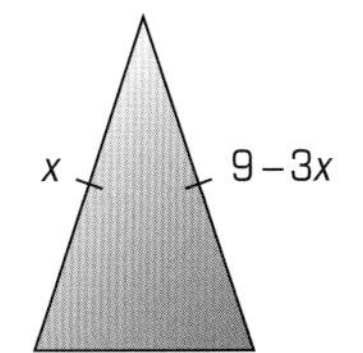

f

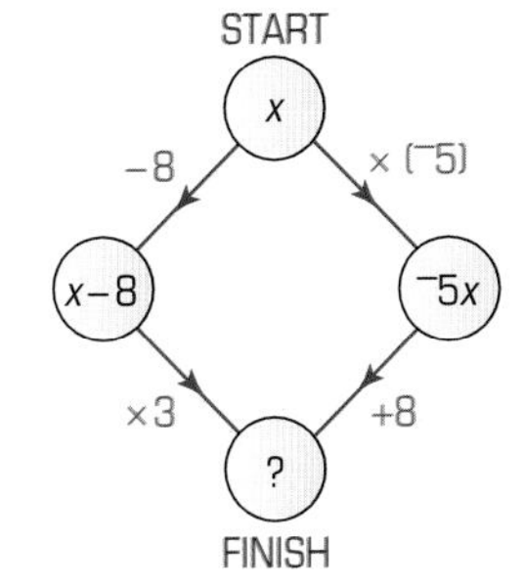

To get the same solution using either route, what number should you begin with?

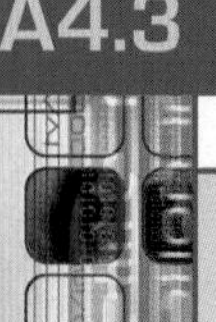

A4.3 Expanding brackets

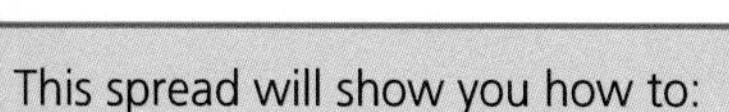

This spread will show you how to:

- Simplify or transform linear expressions by collecting like terms.
- Multiply a single term over a bracket.

KEYWORDS

Collect, Simplify, Expand, Term, Expression

You often use brackets when you want to multiply a whole expression.

Here is an expression: $2x + 7$

To multiply it by 4, you can write: $4(2x + 7)$

This means $2x + 7$ **all** multiplied by 4.

Often you need to remove brackets from an expression.

For example, $4(2x + 7) = 4 \times 2x + 4 \times 7$

$= 8x + 28$

Removing brackets from an expression by multiplying is called **expanding** the brackets.

- **You can expand a bracket by multiplying each term within the bracket by the term outside the bracket.**

example

Expand $p(p + 5)$

$p(p + 5) = p \times p + p \times 5$

$= p^2 + 5p$

Remember: p^2 is called 'p squared'.

Expanding brackets can help you to simplify long expressions.

example

A rectangle has length $3(x + 4)$ and width 6.

a Find an expression for its perimeter.

b Simplify your expression by expanding.

c Find an expression for the area, using a bracket.

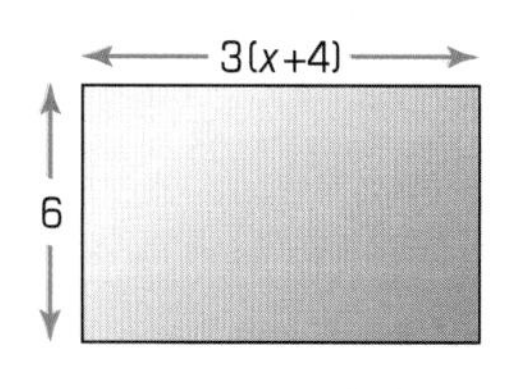

a Perimeter $P =$ length + width + length + width

$P = 3(x + 4) + 6 + 3(x + 4) + 6$

$P = 3(x + 4) + 3(x + 4) + 12$

b $3(x + 4) = 3x + 12$ expand the bracket

$P = 3x + 12 + 3x + 12 + 12$

$P = 6x + 36$ combine like terms

So $P = 6x + 36$

c Area $A =$ length $\times$ width

$A = 3(x + 4) \times 6 = 6 \times 3 \times (x + 4)$

$A = 18 \times (x + 4)$ So $A = 18(x + 4)$

Exercise A4.3

1 Expand these brackets:

a $3(x+4)$ **b** $6(y+7)$ **c** $13(a+13)$

d $4(2a+5)$ **e** $5(y-2)$ **f** $11(b-4)$

g $x(x+6)$ **h** $p(p-4)$ **i** $q(r-q)$

2 **a** Write an expression for the area, A, of this rectangle. Expand your answer.

b Write a formula for the perimeter, P, of this rectangle. Simplify your answer.

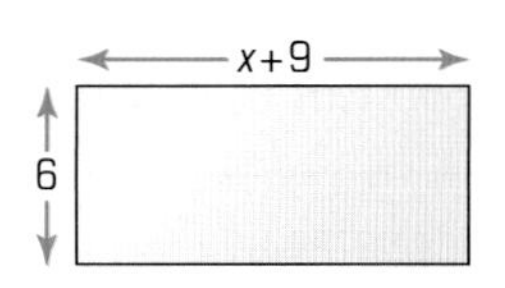

3 For each rectangle, write an expression for the shaded area using brackets. Expand your brackets.

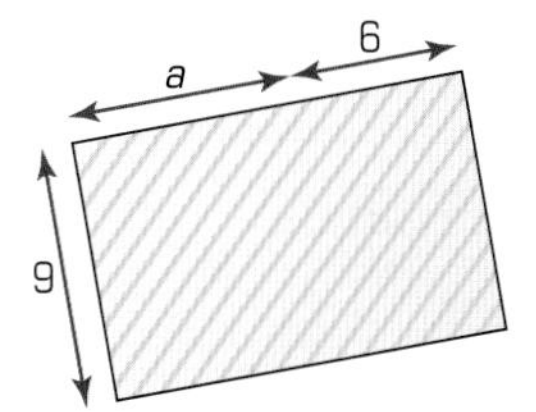

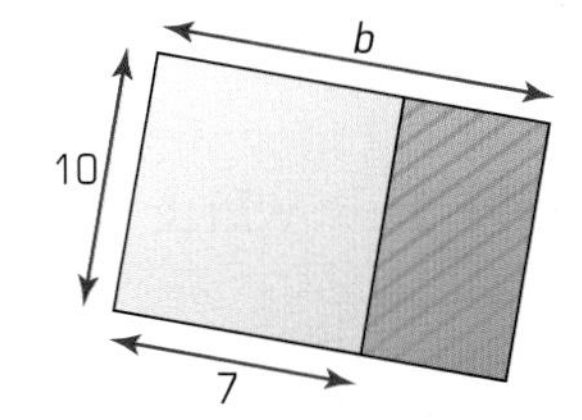

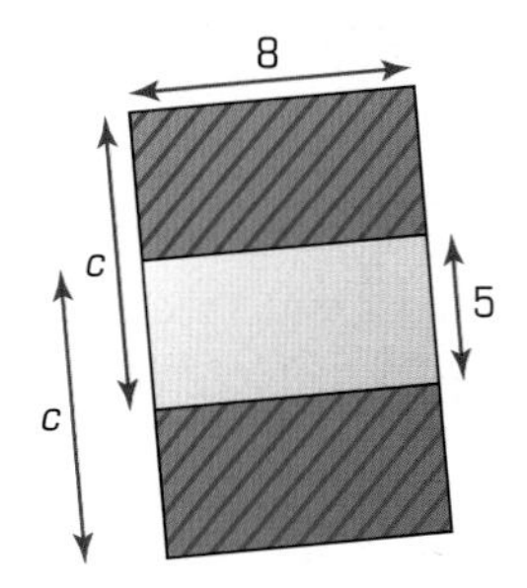

4 These expressions have already been expanded. Copy and complete the expressions to show how they looked with brackets.

a $6x + 12 = 6(\_\_ + \_\_)$

b $8y + 24 = 4(\_\_ + \_\_)$

c $x^2 + 2x = x(\_\_ + \_\_)$

d $6ab + 4a = 2a(\_\_ + \_\_)$

e $12x + 24 = 2(\_\_ + \_\_) = 4(\_\_ + \_\_) = 6(\_\_ + \_\_) = 12(\_\_ + \_\_)$

5 If $3(x+9) = 3x + 27$, then $\dfrac{3x+27}{3} = x + 9$.

Use this example to simplify these fractions.

a $\dfrac{4x+8}{4}$

b $\dfrac{12x+18}{6}$

c $\dfrac{10p+75}{5}$

A4.4 Constructing equations

This spread will show you how to:

- Form linear equations and solve them by transforming both sides in the same way.

KEYWORDS
Construct
Solve
Equation

You can use algebra as a tool for solving problems.

example

Humph, Derek and Sairha win a big prize on the lottery, receiving £11 950 between them.
Derek receives £100 less than Humph.
Sairha gets £50 more than twice what Humph receives.
How much money does each person get?

- Let the amount that Humph receives be x, where x is measured in pounds.
- Then Derek receives £100 less: $x - 100$
- Sairha receives twice Humph's amount plus £50: $2x + 50$

Write an equation:
The total prize money is £11 950, so: $x + (x - 100) + (2x + 50) = 11\,950$

Solve the equation:

$$\begin{aligned} x + x + 2x - 100 + 50 &= 11\,950 \\ 4x - 50 &= 11\,950 \\ 4x &= 11\,950 + 50 \\ 4x &= 12\,000 \\ x &= 12\,000 \div 4 \\ x &= 3000 \end{aligned}$$

So Humph receives £3000
Derek receives £3000 − £100 = £2900
Sairha receives 2 × £3000 + 50 = £6050 Check: 3000 + 2900 + 6050 = 11 950

Many problems can be worked out by constructing and solving equations.

example

The area of this rectangle is 80 cm^2.
Calculate x and find the dimensions of the rectangle.

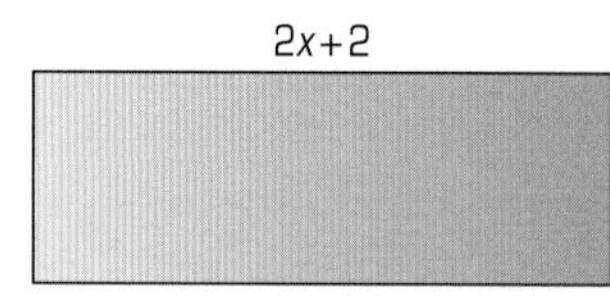

The lengths are equal, so:

$$\begin{aligned} 2x + 2 &= 4x - 12 \\ 2 &= 2x - 12 \\ 14 &= 2x \\ 7 &= x \end{aligned}$$

Find the length by substituting $x = 7$ into one of the expressions: $2 \times 7 + 2 = 16$
(Check: $4 \times 7 - 12 = 16$)
Length × width = area, so:
16 × width = 80 Width = 80 ÷ 16 = 5

The length is 16 cm and the width is 5 cm.

Exercise A4.4

1 Form an equation to find x in each diagram:

a

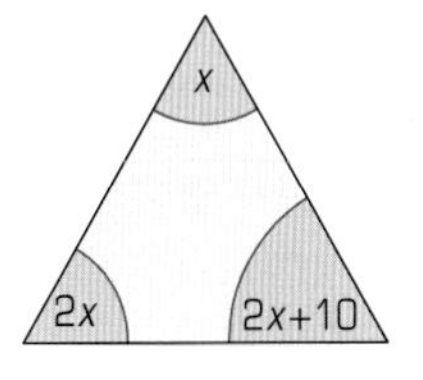

b

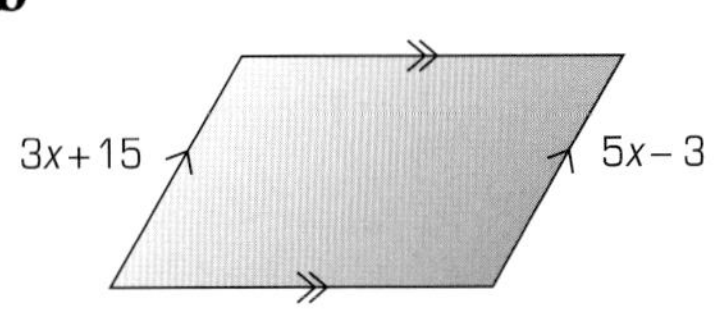

c

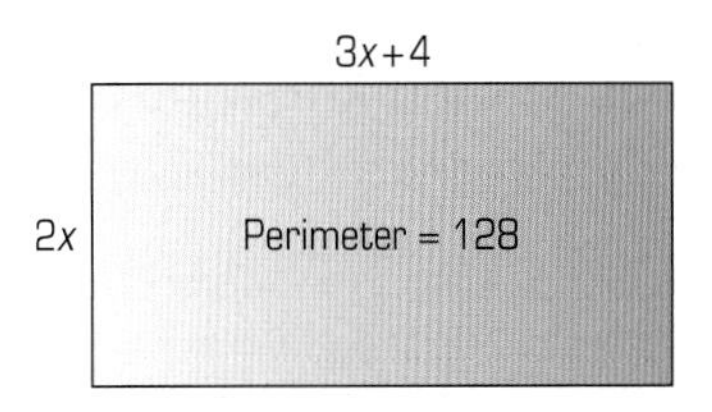

2 In a number pyramid, two neighbouring numbers are added to get the one above:

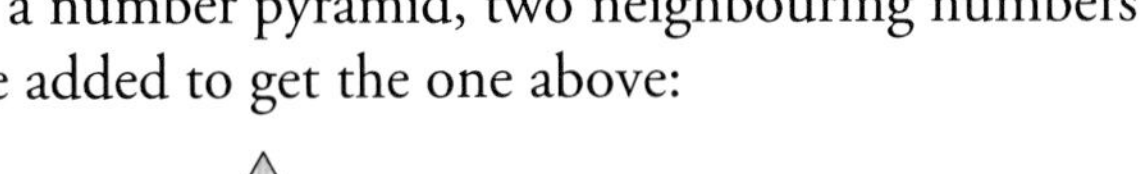

Find the value of x in this pyramid:

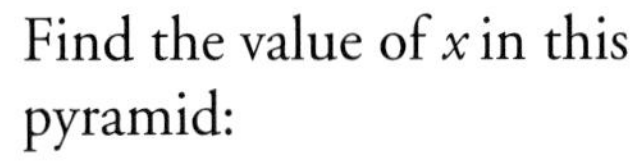

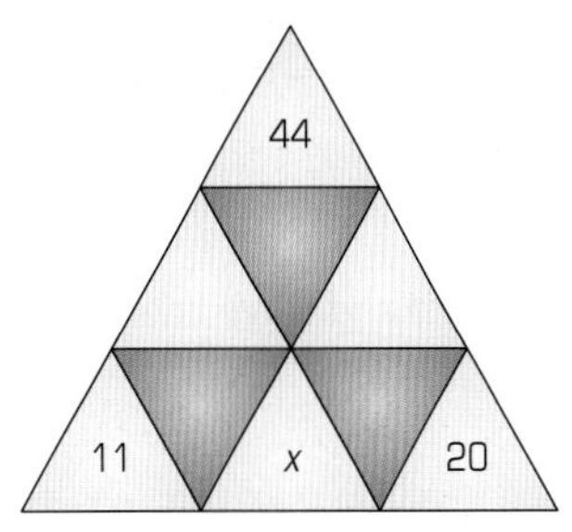

3 **a** If I take a certain number, multiply it by 5 and add 10, I get the same answer as when I multiply it by 9 and take 20. What is the number?

b If I take a certain number, subtract 7 and then multiply by 4, I get the same answer as when I add 3 and multiply by 2. What is the number?

c If I take a certain number, multiply it by 5 and subtract 9, I get the same as when I multiply it by 3 and take it from 12. What is the number?

4 Which value of x gives the same output for these chains?

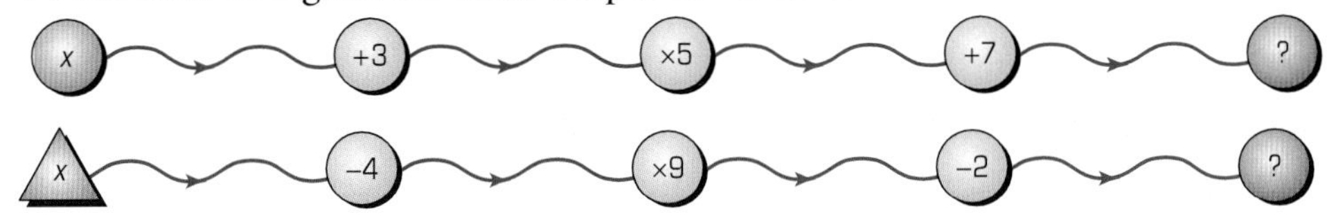

5 In a maths competition:

- A correct answer is worth 5 points
- An incorrect answer is worth $^{-}3$ points

The team answers three times as many questions correct as incorrect.
They score 60 points.
How many questions were there?

A4 Summary

You should know how to ...

1 Simplify or transform linear expression by collecting like terms.

Check out

1 Simplify:

a $3a + b - 2a - 3b$

b $a \times b - 3a + 2b + 2ab$

c $m + p - q + m + 2p - 3q - 2m + 2q$

d $x \times x \times y \times y \times z \times 3 \times x$

2 Multiply a single term over a bracket.

2 a Expand:

i $3(x + 5)$ ii $6(2x + 4)$ iii $10(3.6x + 9)$

iv $p(p + 8)$ v $z(w + z)$ vi $6b(2b + 3a)$

b Simplify:

i $\frac{10x+20}{5}$ ii $\frac{33x+99}{11}$ iii $\frac{169x+52}{13}$

3 Construct and solve linear equations, with integer coefficients.

3 a Solve these equations:

i $3x + 4 = 5x - 10$

ii $4(x - 2) = 2(x + 6)$

iii $24 - 9x = 11x + 4$

b Solve these equations:

i $8x - 14 = 3x + 17$

ii $3(2x + 9) = 4(5x + 11)$

iii $13 - 8x = 5x + 21$

c Construct an equation in n and solve it to find the original number:

i I think of a number, multiply it by 7, subtract 11 and get 59.

ii I think of a number, double it and add eight.
I get the same answer as when I multiply it by five and subtract 10.

iii I think of a number, multiply it by 4 and subtract it from 29. I get the same answer as when I treble it and add 11.

4 Solve word problems and investigate in the context of algebra

4 Use algebra to solve the following problems:

a Five years ago, Clare was 20 years younger than her father. This year she will be half her father's age. How old is Clare?

b Find the angles in this triangle if the two equal angles are 5° more than the other angle.

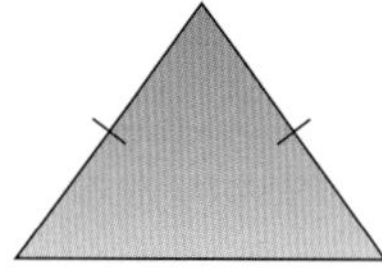

4 Transformations

This unit will show you how to:

- Transform 2-D shapes by rotations, reflections and translations and by simple combinations.
- Identify all the symmetries of 2-D shapes.
- Understand the language and notation associated with enlargement.
- Enlarge 2-D shapes, given a centre of enlargement and a positive whole-number scale factor.
- Solve problems and investigate in the context of transformations.

You can find symmetry in nature.

Before you start

You should know how to ...

1 Know symmetry properties of 2-D shapes.

Check in

1 How many lines of symmetry has:

a a square

b a parallelogram

c an equilateral triangle

d a circle?

You should know how to ...

2 Plot coordinates in all four quadrants.

Check in

2 Plot the points (1, 1) ($^-2$, 1) ($^-3$, $^-2$) (0, $^-2$) and join them in order.
What shape have you drawn?

Reflection

This spread will show you how to:

- Understand reflection in two dimensions.
- Reflect a 2-D shape in a mirror line.

KEYWORDS

Coordinate | Reflection
Image | Mirror line

When you look in a mirror you see a reflection of yourself.

The reflection is called an **image**.

2-D shapes are reflected using a **mirror line**.

You can describe a mirror line drawn on a coordinate grid.

example

The rectangle R has vertices A(1, 2), B(1, 4), C(4, 4), and D(4, 2).

a Draw the line $x = {}^{-}1$ and reflect the rectangle in it. Label the image R′.

b Reflect R in the line $y = 1$ and label the image R″.
You can use tracing paper to help.

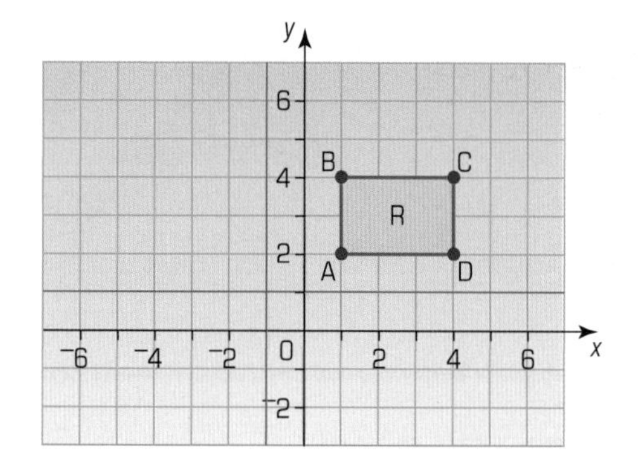

1. Draw the line $x = {}^{-}1$.

2. Reflect R in the line $x = {}^{-}1$, and label all the vertices. This is R′.

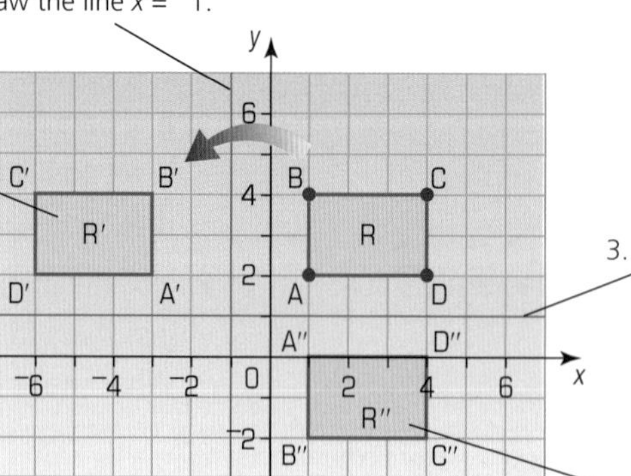

3. Draw the line $y = 1$.

4. Reflect R in the line $y = 1$. This is R″.

Equations of lines parallel to the x- and y-axes are explained in A3.6.

Exercise S4.1

1 Copy the axes drawn on the grid into your exercise book. Draw the kite and label it R.

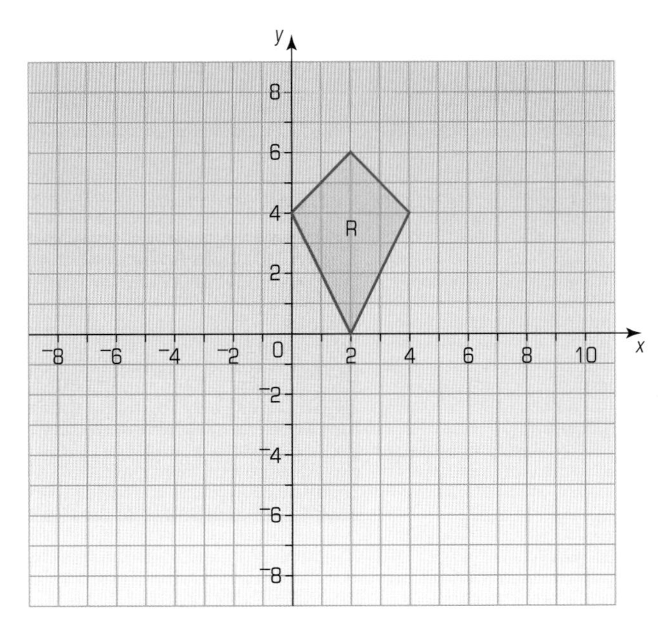

a Reflect R in the line $x = 0$. Label the new shape R′.
b Reflect R in the line $y = 0$. Label the new shape R″.
c Reflect R′ in the line $y = 0$. Label your new shape R‴.

2 Draw axes from $^{-}7$ to $+7$. On your axes draw the shape A with coordinate vertices (3, 1), (3, 7), (5, 7), (5, 4)
a Reflect A in the line $x = 4$. Label the new shape A′.
b Reflect A in the line $y = 2$. Label the new shape A″.
c Reflect A′ in the line $y = 2$. Label the new shape A‴.

3 Draw axes from $^{-}7$ to $+7$. On your axes draw the shape P with coordinate vertices (0, 1), (0, 7), (5, 7), (5, $^{-}1$)
a Reflect P in the line $x = 2$. Label the new shape P′.
b Reflect P in the line $y = 0$. Label the new shape P″.
c Reflect P″ in the line $y = 0$. Where does your shape end up?

4 Draw axes from $^{-}7$ to $+7$. On your axes draw the shape B with coordinate vertices (2, 2), (2, 4), (1, 4), (1, 6), (5, 6), (5, 3), (4, 3), (4, 2)
a Reflect B in the line $y = 2$. Label the new shape B′.
b Reflect B in the line $x = 2$. Label the new shape B″.
c Reflect B′ in the line $y = 2$. Where does your shape end up?
d Reflect B″ in the line $x = 2$. Where does the shape end up?

5 Draw axes from $^{-}7$ to $+7$. Draw your own shape on the grid and move it through three reflections to return it to its original position. Describe your reflections.

Enlargement

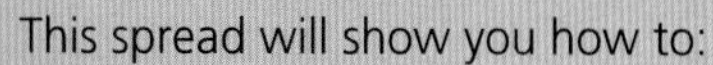

This spread will show you how to:

- Understand enlargement.
- Enlarge 2-D shapes by a positive scale factor.

KEYWORDS

Enlargement, Object, Image, Scale factor

When you enlarge a photograph you multiply all its lengths by a certain amount.

The photograph on the left has been enlarged by 100%. All the lengths have become twice the size.

You can enlarge shapes on a grid in a similar way.

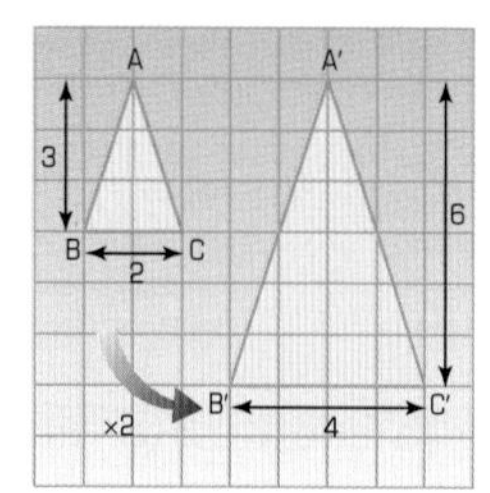

A′B′C′ is an enlargement of ABC by a **scale factor** 2.

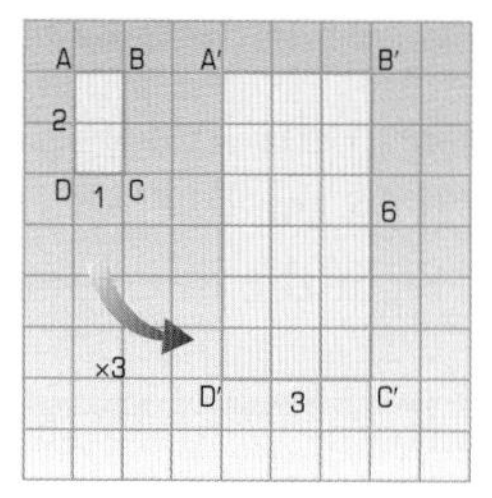

A′B′C′D′ is an enlargement of ABCD by a **scale factor** 3.

- **The scale factor of an enlargement is the number by which you multiply a length in the object to get the corresponding length in the image.**

To fix the position of the image you need to specify the **centre of enlargement**.

In the diagram triangle A′B′C′ is the image of ABC after an enlargement of scale factor 4 where X is the centre of enlargement.

Compare the lengths in the table:

Original	Image
XA = 1	XA′ = 4
XB = 3	XB′ = 12

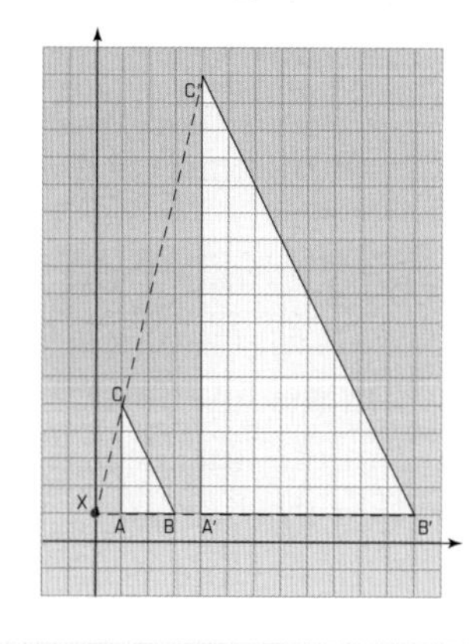

- **If you extend the lines between corresponding points on the image and the object, they intersect at the centre of enlargement.**

Exercise S4.2

1 Copy the grid and enlarge the shape ABCDEF by a scale factor 2.

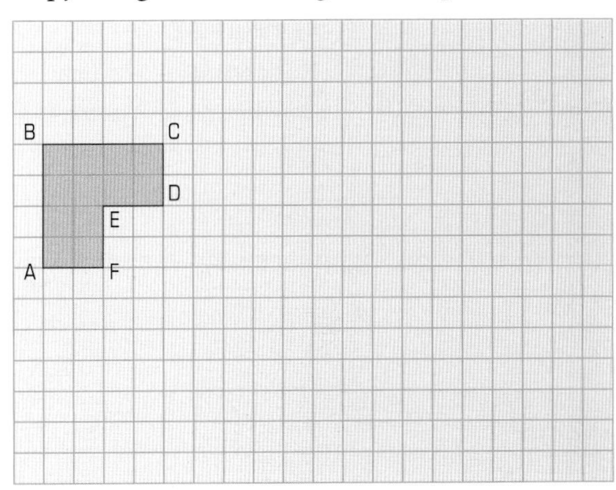

2 **a** Copy and complete the table for the shapes on the grid.

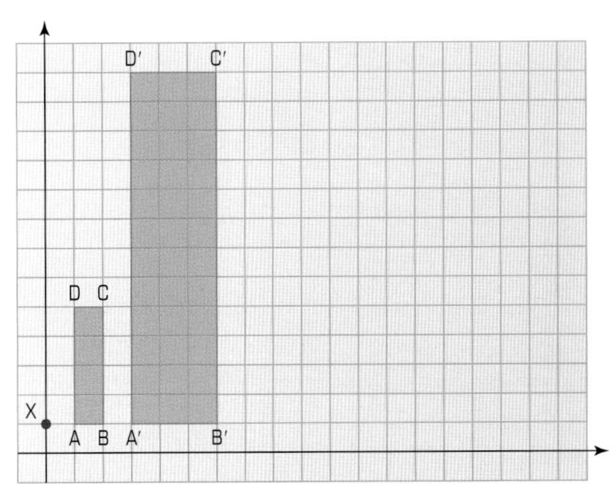

Original	Image
XA =	XA′ =
XB =	XB′ =
XC =	XC′ =
XD =	XD′ =

b Describe the enlargement fully.

3 Draw a pair of axes from 0 to 20. Plot the points A (1, 2), B (1, 5), C (4, 2).
Draw the enlargement of ABC using a scale factor of 3 and centre of enlargement (0, 1).

4 Draw a pair of axes from 0 to 25. Plot the points P (1, 1), Q (1, 4), R (5, 1).
Draw the enlargement of PQR using scale factor 4 and centre of enlargement (1, 0).

5 Draw axes from 0 to 10.
Plot the points E(3, 4), F(5, 4), G(5, 6), H(3, 6).
Draw the enlargement of EFGH using a scale factor of 2 and centre of enlargement (3, 4).

S4.3 Translation

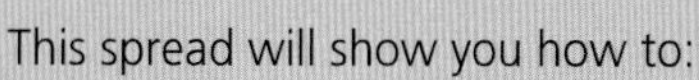

This spread will show you how to:

- Understand translation.
- Translate a shape using a vector.

KEYWORDS

Horizontal, Vertical, Translation, Vector

In a reflection the image is reversed.

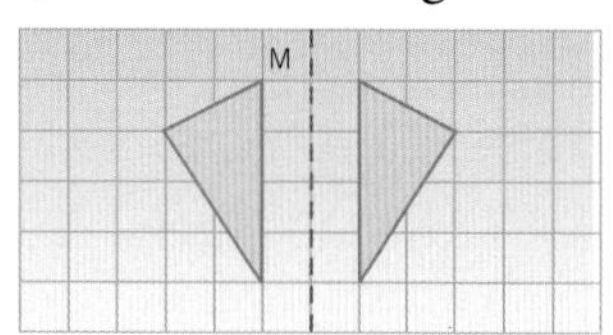

The triangle is reflected in the mirror line M.

In a **translation**, the image stays the same way up.

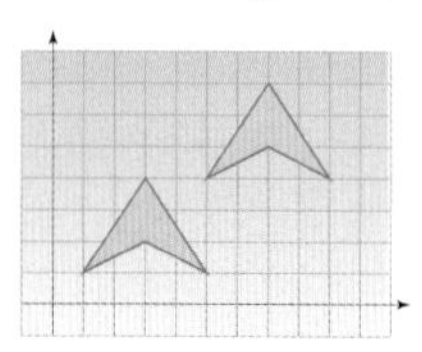

The arrowhead is moved, or **translated**, four squares to the right and three squares up.

- **In a translation, you move an object horizontally and vertically.**

You can write a translation using shorthand, as a **vector**.
Triangle ABC has been moved five squares to the right and two squares up.

You can write this as the vector $\begin{pmatrix} 5 \\ 2 \end{pmatrix}$.

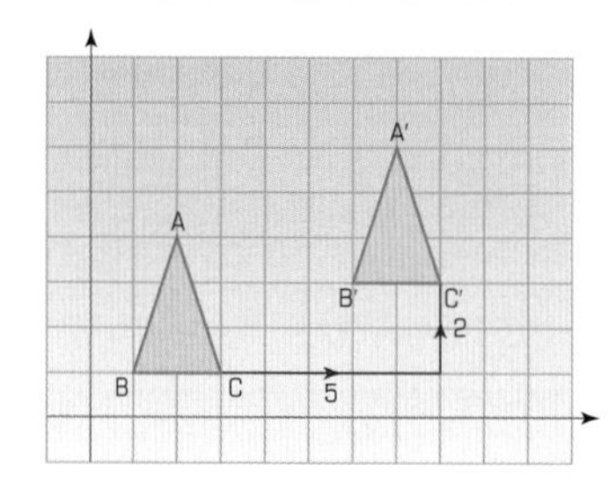

- **The numbers in a vector represent:** $\begin{pmatrix} \text{number of units horizontally} \\ \text{number of units vertically} \end{pmatrix}$

Vectors work rather like coordinates: to the right and up are positive, to the left and down are negative.

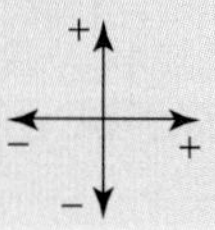

example

Describe the translation that sends square P to:

a Q **b** R

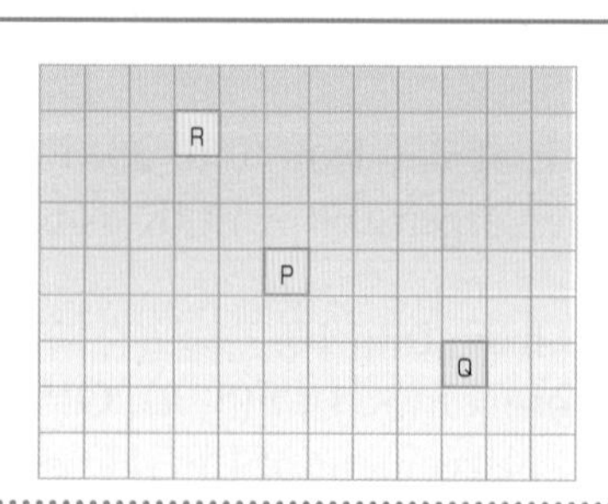

a Q is 4 to the right and 2 down.

The translation from P to Q is $\begin{pmatrix} 4 \\ ^-2 \end{pmatrix}$.

b R is 2 to the left and 3 up.

The translation from P to R is $\begin{pmatrix} ^-2 \\ 3 \end{pmatrix}$.

Exercise S4.3

1 Draw the rectangle and label it R on a copy of these axes.

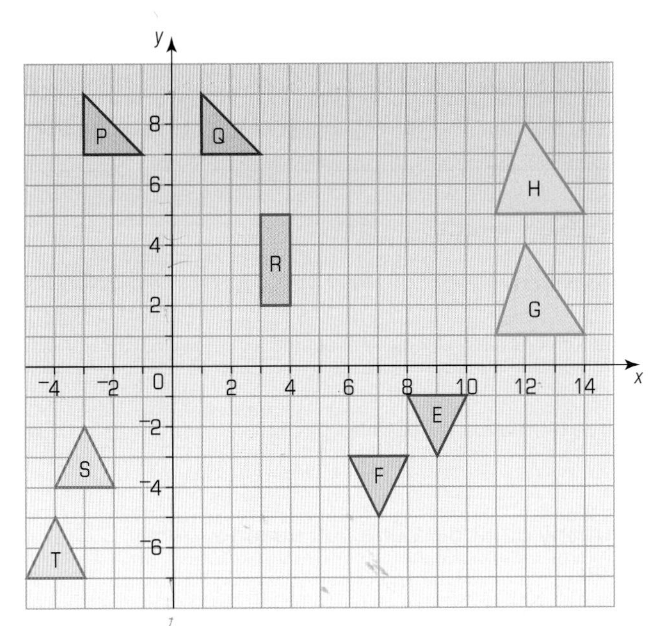

a Translate R 4 squares to the left and 3 squares down. Label the new shape R′.
b Translate R 2 squares to the left and 6 squares down. Label the new shape R″.
c Translate R 0 squares to the left and 7 squares down. Label the new shape R‴.

2 On the grid in question 1 what vectors send
a P to Q **b** S to T **c** G to H **d** E to F?

3 Draw axes from $^{-}10$ to +10. On your axes draw the shape **A** with coordinate vertices (2, 2), (2, 7), (5, 7), (5, 4).
a Translate A four squares to the left and eight squares down. Label the new shape A′.
b Translate A five squares to the right and three squares up. Label the new shape A″.
c Translate A two squares to the right and 0 squares up. Label the new shape A‴.

4 Draw axes from $^{-}10$ to +10. On your axes draw the shape B with coordinate vertices (0, 2), (0, 7), (5, 7), (5, 4).
a Translate B using the vector $\begin{pmatrix} 0 \\ 1 \end{pmatrix}$ and label the new shape B′.

b Translate B using the vector $\begin{pmatrix} ^{-}1 \\ 2 \end{pmatrix}$ and label the new shape B″.

c Translate B using the vector $\begin{pmatrix} 0 \\ ^{-}3 \end{pmatrix}$ and label the new shape B‴.

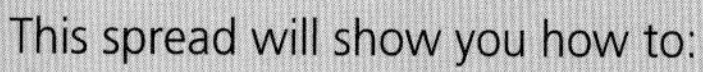

S4.4 Rotation

This spread will show you how to:
- Understand rotation.
- Rotate 2-D shapes about a given point.

KEYWORDS
Anti-clockwise
Image
Object
Rotation

In a **rotation**, you turn an object around a point.

The grid shows a triangle labelled T.

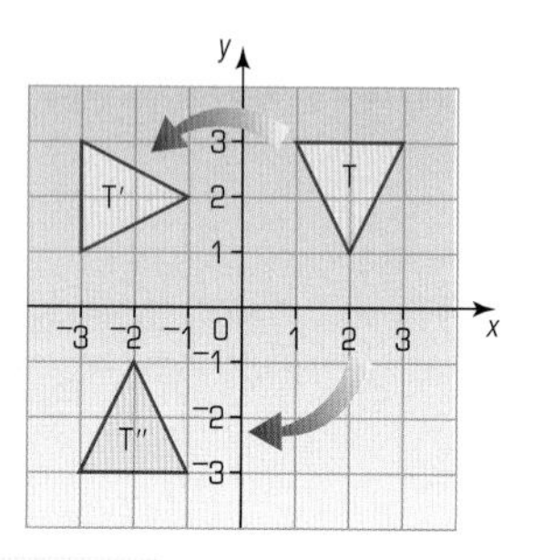

You can rotate T through a quarter turn (90°) anti-clockwise about the origin to give the shape T′.

You can rotate T through a half turn (180°) about the origin to give the shape T″.

- To rotate a shape about any point on a grid you need to specify:
  - The centre of rotation
  - The angle in which you are turning the shape

Anticlockwise angles are positive; clockwise angles are negative.

Sometimes the centre of rotation is inside the shape.

example

a Rotate shape D through 90° clockwise about the point ($^-1$, 1). Label the new shape E.

b What rotation will send D to F?

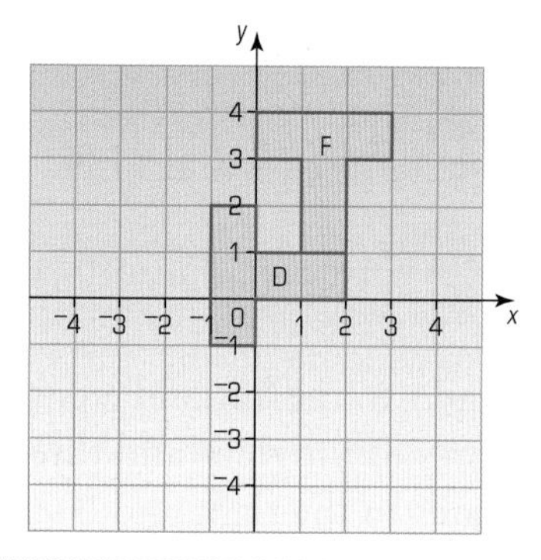

a Use tracing paper to help you rotate the shape:

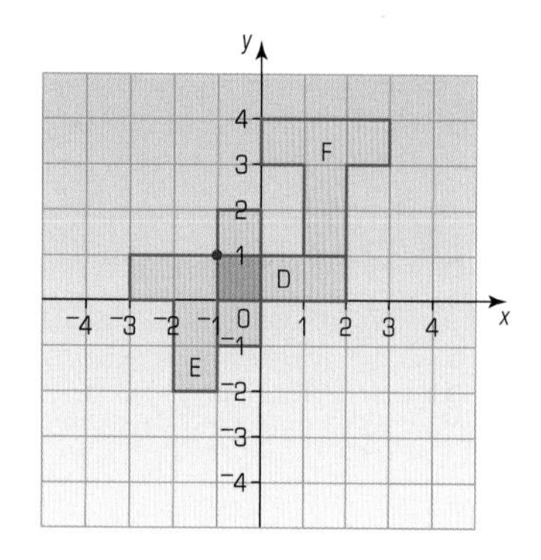

b Use tracing paper and try different centres of rotation until you find the right one:

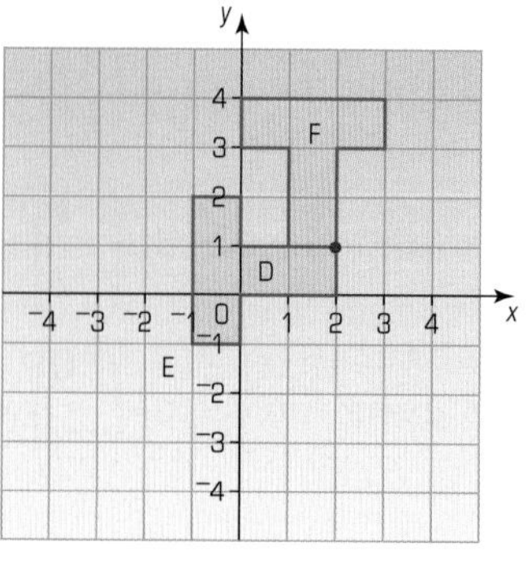

The rotation is 90° clockwise about (2, 1).

Exercise S4.4

1 Copy these axes. Draw the rectangle and label it R.

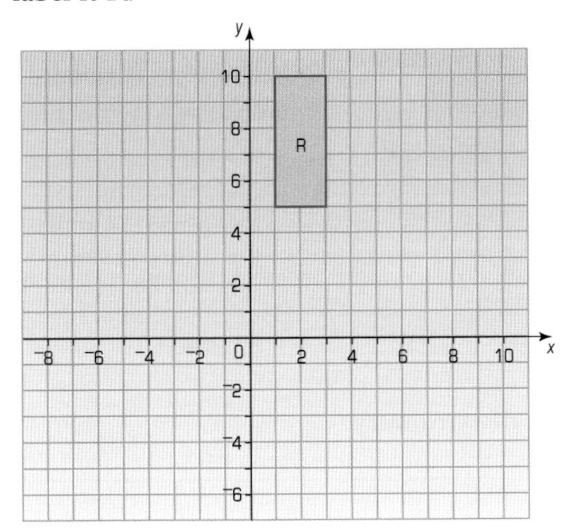

a Rotate R through 90° clockwise about the origin and label the image R′.
b Rotate R through 180° about the origin and label the image R″.
c Rotate R through 270° clockwise about the origin and label the image R‴.
d What rotation will send R‴ to R?
e What rotation will send R″ to R?

2 Draw axes from ⁻7 to +7. On your axes draw the shape A with coordinate vertices (1, 2), (1, 7), (5, 7), (5, 2).
a Rotate A through 90° clockwise about the point (0, 1) and label the image A′.
b Rotate A through 180° about the point (0, 1) and label the image A″.
c Rotate A through 270° clockwise about the point (0, 1) and label the image A‴.
d What rotation will send A‴ to A?
e What rotation will send A″ to A?

3 Draw axes from ⁻7 to +7. On your axes draw the shape B with coordinate vertices (2, 2), (2, 6), (5, 6), (5, 3), (4, 3), (4, 2).
a Rotate B about the point (2, 2) for 90° in a clockwise direction. Label the new shape B′.
b Rotate B about the point (2, 2) for 90° in an **anti**-clockwise direction. Label the new shape B″.
c What rotation sends B″ to B?
d What rotation sends B′ to B?

4 Draw axes from ⁻7 to +7. On your axes draw the shape Q with coordinate vertices (1, 2), (1, 7), (5, 7), (5, 2).
a Rotate Q through 90° clockwise about the point (2, 4) and label the image Q′.
b Rotate Q through 180° about the point (2, 4) and label the image Q″.
c Rotate Q through 270° clockwise about the point (2, 4)and label the image Q‴.
d What rotation will send Q‴ to Q′?
e What rotation will send Q″ to Q‴?

5 Shape A has been rotated:

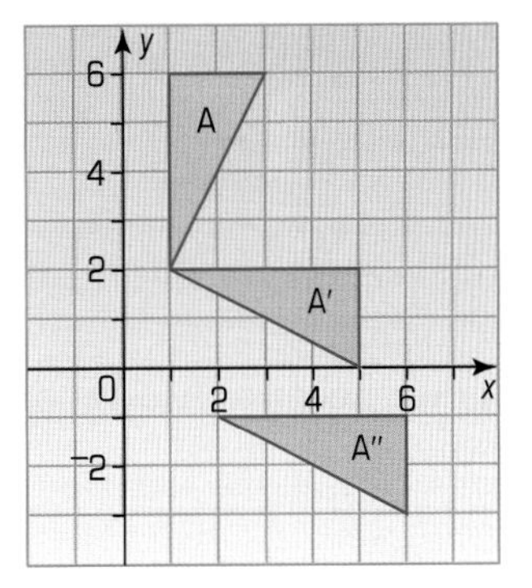

Describe the rotation that sends:
a A to A′
b A to A″
c A′ to A

S4.5 Symmetry

This spread will show you how to:

- Recognise and visualise symmetry of 2-D shapes.

KEYWORDS

Pentagon, Self-inverse, Reflection, Symmetry, Rotation

You can place a mirror line inside a shape.
This isosceles triangle has **reflection symmetry**.
The two halves are identical mirror images of one another.

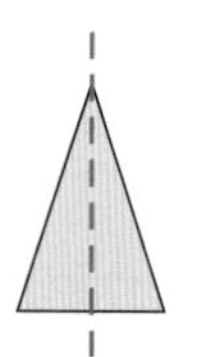

Reflection is **self-inverse**: the mirror line reflects the object to the image, and also reflects the image to the object.

- A shape has reflection symmetry if it has a **line of symmetry**.

example

Reflect triangle ABC in the mirror line AC.
Describe the new shape.

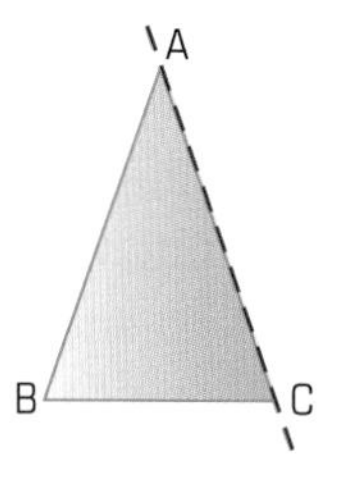

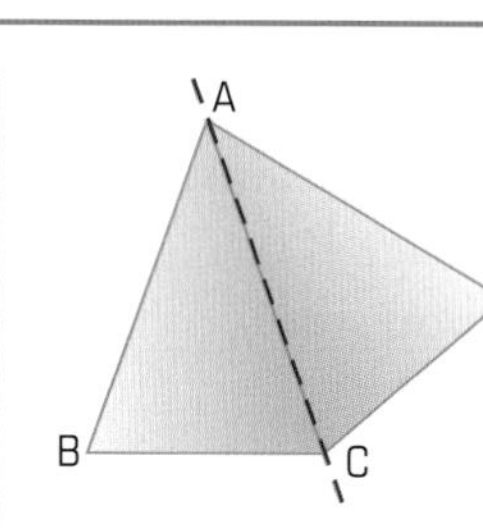

The new shape is a kite.
It has reflection symmetry through the line AC.

You have learned that you can rotate a shape about a centre of rotation.
The centre of rotation can be either ...

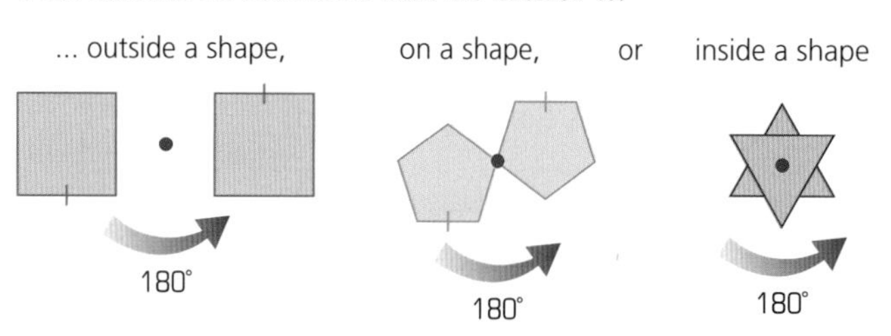

- **The inverse of a rotation is** an equal rotation about the same point in the opposite direction.

90° clockwise

Inverse is 90° anticlockwise.

A shape has rotational symmetry if it fits onto itself more than once in a 360° turn.
The number of times it fits onto itself is the order of rotational symmetry.

example

What is the order of rotational symmetry of a regular pentagon?

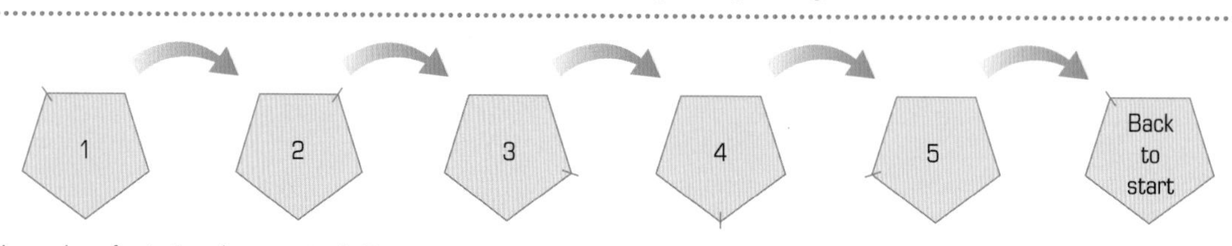

The order of rotational symmetry is 5.

Exercise S4.5

Copy these shapes accurately.
Rotate each shape through 180° about the point marked X.
Draw the shape again and reflect it in the red mirror line.
For each shape that you create, describe it as fully as you can.
Use the terms 'reflective symmetry' and 'order of rotational symmetry'.

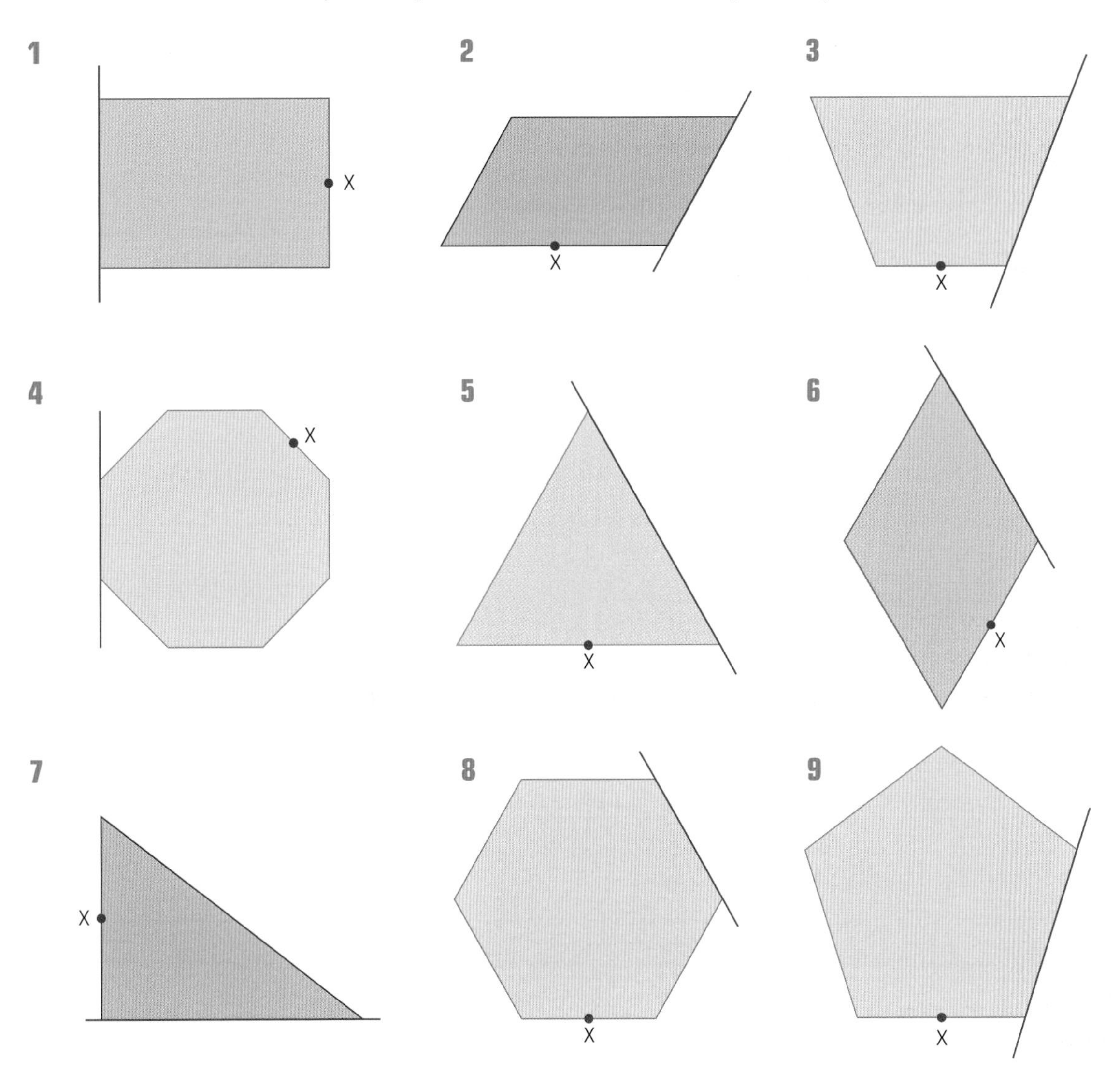

10 **a** How many lines of symmetry does the *original* shape in question 4 have?
b What is the order of rotational symmetry of this shape?
c Name the shape.
d Repeat a, b and c for question 8.

S4.6 Transformations

This spread will show you how to:
- Transform 2-D shapes by repeated, reflections, rotations or translations.
- Explore the effect of combining transformations.

KEYWORDS
Congruent, Rotation, Transformation, Translation, Reflection, Enlargement

You can use transformations to move a shape around a grid.

▶ A transformation is a change applied to a shape.

Reflection, rotation, translation and enlargement are all transformations.

example

Describe the transformations that have been applied to the shape A to achieve:
a shape A′ **b** shape A″ **c** shape A‴.

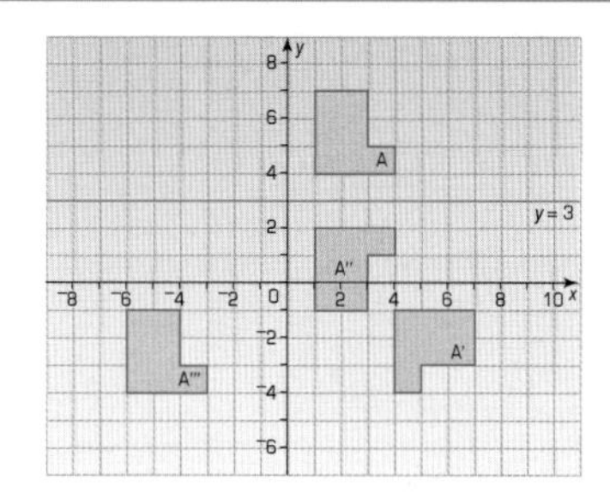

a A moves to A′ by a rotation through 90° clockwise about (0, 0).
b A moves to A″ by a reflection about $y = 3$.
c A moves to A‴ by a translation $\begin{pmatrix} ^{-}7 \\ ^{-}8 \end{pmatrix}$.

▶ A **congruent** transformation leaves the size and shape of an object unchanged.
Reflection, rotation and translation are all congruent transformations.

You can combine transformations on a grid.

example

Describe the transformations that move:
a P to Q
b Q to R

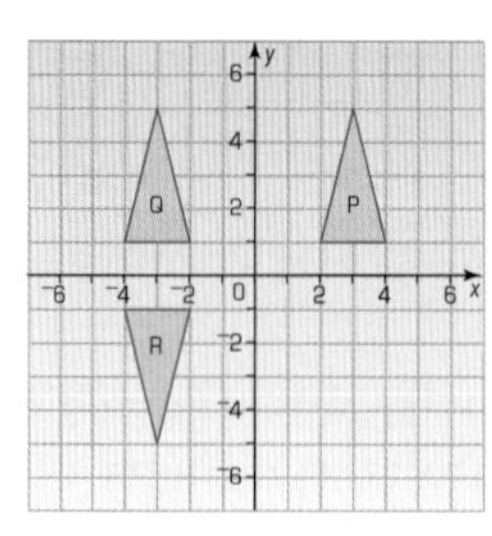

c What single transformation would move P to R?

a Reflection in the y-axis.
b Reflection in the x-axis.
c Rotation about (0, 0) through 180°.

Exercise S4.6

1 Copy the grid and shape A.

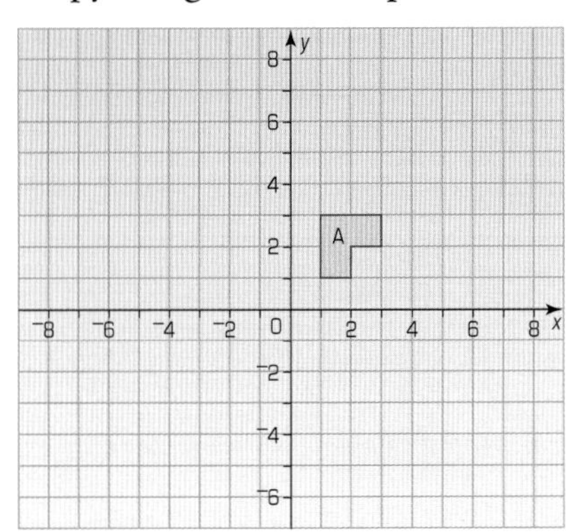

Transform A by:

a enlarging it by scale factor 3 centre (0, 0). Label it A′.

b reflecting it in the *x*-axis. Label it A″.

c rotating it through 180° about (0, 0). Label it A‴.

d translating it by the vector $\begin{bmatrix} ^-10 \\ ^-8 \end{bmatrix}$. Label it A⁗.

e Describe two repeated transformations that would be equivalent to the rotation in part **c**.

2 Look at the shapes on this grid.

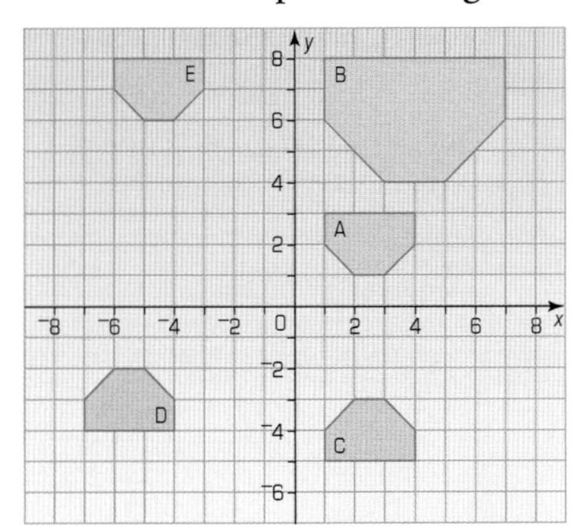

Describe the transformations which send

a A to B	**b** A to C	**c** A to D	**d** A to E
e B to A	**f** C to A	**g** D to A	**h** E to A

3 Describe an easy way to tell whether a transformation is:

a an enlargement

b a reflection

c a rotation

d a translation.

Use words like congruent, similar, bigger, smaller etc.

Summary

You should know how to ...

1 Transform 2-D shapes by simple combinations of rotations, reflections and translations.

2 Identify all the symmetries of 2-D shapes.

3 Enlarge 2-D shapes given a centre of enlargement and a positive whole-number scale factor.

Check out

1 a

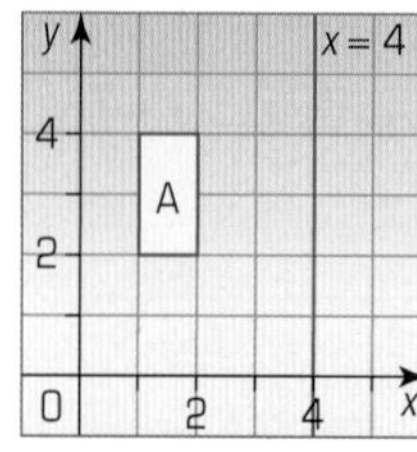

Reflect A in line $x = 4$

b

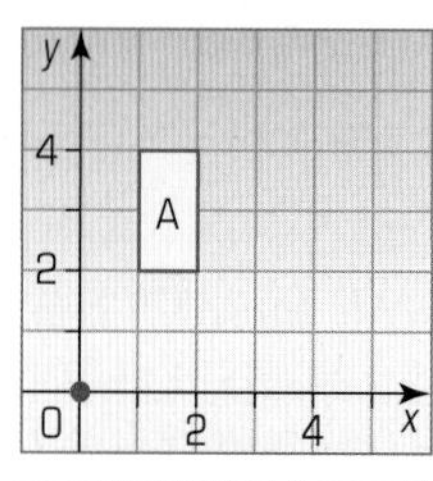

Rotate A about the origin through 90° clockwise

c

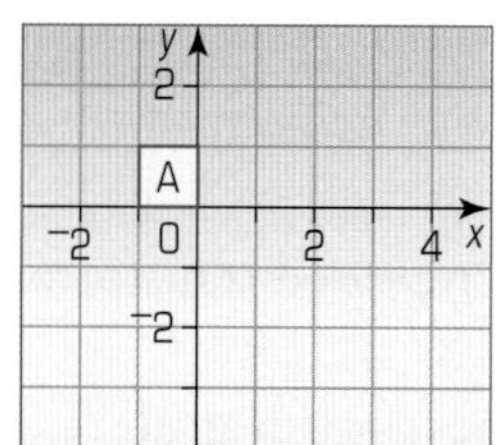

Translate A by the vector $\begin{pmatrix} 2 \\ -3 \end{pmatrix}$

2 What is the order of rotational symmetry of a regular octagon?

3 a Enlarge A by a scale factor of 2 using centre (0, 2)

b Find the centre of enlargement that maps A onto A′

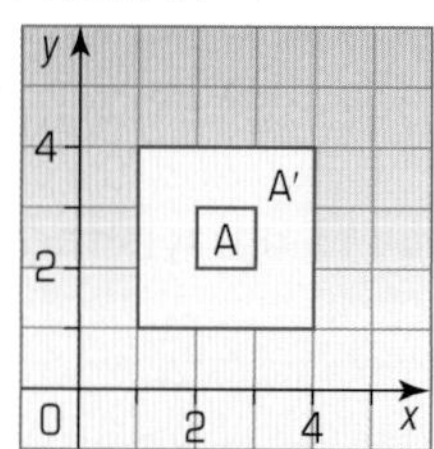

15 More number calculations

This unit will show you how to:

- Know that a recurring decimal is a fraction.
- Round positive whole numbers to any given power of 10.
- Round decimals to the nearest whole number or to one or two decimal places.
- Recognise and use multiples, factors, common factor, HCF, LCM and primes.
- Use known facts to derive unknown facts.
- Consolidate and extend mental methods.
- Use standard column procedures for multiplication and division.
- Understand where to position the decimal point by considering equivalent calculations.
- Order, add and subtract fractions by writing them with a common denominator.
- Recall known facts, including fraction to decimal conversions.
- Calculate fractions of quantities.
- Use the equivalence of fractions, decimals and percentages to compare proportions.
- Calculate percentages and find the outcome of a given percentage increase or decrease.
- Solve word problems and investigate in the context of number.
- Suggest extensions to problems, conjecture and generalise.

Many values make more sense when they are rounded.

Before you start

You should know how to ...

1 Round decimal to the nearest whole number.

2 Convert between simple fractions, decimal and percentages.

3 Find a fraction of an amount.

Check in

1 Round to the nearest whole number:

a 2.6 **b** 3.15 **c** 20.02

2 Convert:

a $\frac{1}{8}$ to a decimal

b 0.35 to a fraction

c 18% to a fraction

3 Calculate:

a $\frac{3}{4}$ of 12.8 km **b** $\frac{2}{3}$ of £22.80

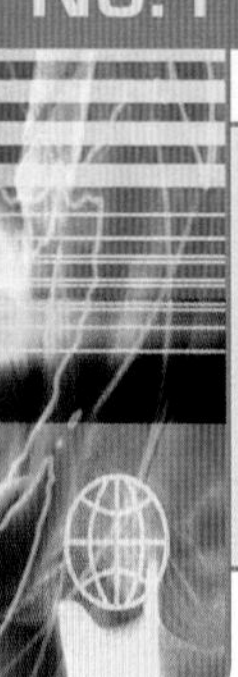

N5.1 Rounding

This spread will show you how to:
- Round positive whole numbers to a given power of 10.
- Round decimals to the nearest whole number or to one or two decimal places.
- Recognise recurring decimals.

KEYWORDS

Decimal	Recurring
Digit	Rounding
Fraction	Integer

In real life it often helps if you round numbers off.

There were 11 373 at the match

About 10 000 were at the match

The lottery jackpot is £12 103 567.64

The jackpot is about £12 000 000

- **You can round whole numbers to a given power of 10.**

For example, 2837 = 3000 to the nearest 1000
= 2800 to the nearest 100
= 2840 to the nearest 10

You can round decimals to the nearest integer or to a certain number of decimal places.

example

Round 2.5478 to:

a 2 decimal places **b** 1 decimal place **c** the nearest whole number

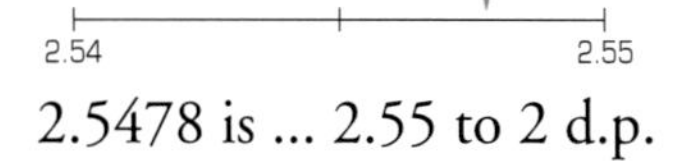

2.5478 is ... 2.55 to 2 d.p.

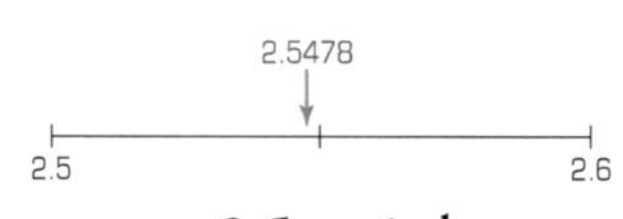

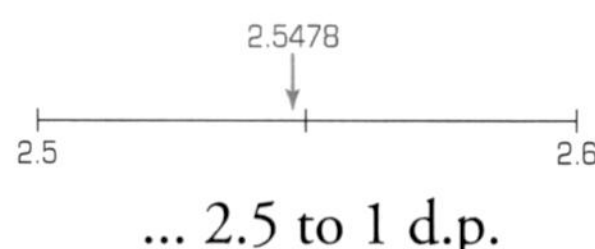

... 2.5 to 1 d.p.

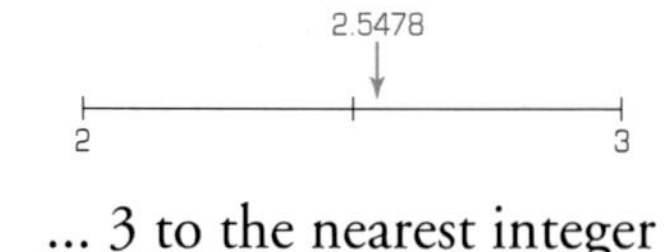

... 3 to the nearest integer

- **When you round to n decimal places you need to look at the $(n + 1)$th digit.**
  If it is halfway or more, you round the number up.

0.33333333

When you convert some fractions into decimals the digits seem to go on forever.

$\frac{1}{3} = 1 \div 3 = 0.33333 \ldots$ $\frac{1}{6} = 1 \div 6 = 0.16666 \ldots$ $\frac{5}{33} = 5 \div 33 = 0.151515 \ldots$

These are called recurring decimals and you can write them using a dot notation.

$\frac{1}{3} = 0.3333 \ldots = 0.\dot{3}$ $\frac{1}{6} = 0.16666 \ldots = 0.1\dot{6}$ $\frac{5}{33} = 0.151515 \ldots = 0.\dot{1}\dot{5}$

Exercise N5.1

1 A rope is measured by Charlie and found to be 86.4 cm long. What is its length

a to the nearest cm

b to the nearest 10 cm?

2 **a** The weight a of prize onion was recorded as 640 g, to the nearest 10 g. List all the weights, to the nearest g, that it could have been.

b Is it possible to weigh the mass of an onion 'exactly'? Explain your answer.

3 Round each of the following numbers to

a the nearest 1000

i 23 467 **ii** 87 664 **iii** 736 547

b the nearest 1000 000

i 23 987 876 **ii** 39 765 346

iii 456 765 543

4 In a moment of madness Charlie Chalk counts the number of grains of sand in the sand-pit. The teacher asks for the answer to the nearest 100 000. Charlie loses count and doesn't know whether the answer is 345 527 324 or 345 478 564. Does it matter? Explain your answer.

5 **a** Write the number 765 456 456 003 in words.

b 10^7 has 7 zeros: 10 000 000.
Round the above number to the nearest:

i Billion (10^9)

ii One hundred million (10^8)

iii Ten million (10^7)

iv One million (10^6)

v One hundred thousand (10^5)

vi Ten thousand (10^4)

vii One thousand (10^3)

viii One hundred (10^2)

ix Ten (10^1)

6 **a** Round each of these decimal numbers to 2 decimal places:

i 45.467 **ii** 35.4589

iii 3.46784 **iv** 4.9205

b Round the numbers to 1 decimal place.

7 **a** Write these fractions using 'recurring' notation: $\frac{9}{11}$, $\frac{2}{3}$, $\frac{41}{133}$

b Round each decimal to 2 d.p.

c Why do you think using 'recurring' notation can sometimes be better than rounding off?

8 **a** Copy and complete this table

Number	Nearest whole number	to 1 d.p.	to 2 d.p.
11.374			
3.599			
18.9672			
7.999			

b Write down what you notice.

9 Sean measures the length of his desk to the nearest whole number as 142 cm.
Ian measures the length of his desk to 2 d.p. as 142.00 cm.
Are the two measurements the same?
Explain your answer.

10 **Investigation**

a Use your calculator to express these fractions as decimals using recurring notation. The repeating block contains 6 digits.

i $\frac{1}{7}$ **ii** $\frac{2}{7}$ **iii** $\frac{3}{7}$ **iv** $\frac{4}{7}$ **v** $\frac{5}{7}$ **vi** $\frac{6}{7}$

b Investigate with other recurring decimals.

c How would you express $\frac{3}{17}$ as a recurring decimal?

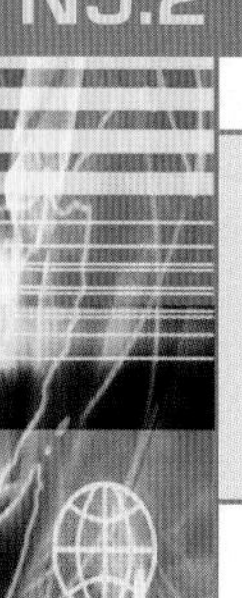

N5.2 Factors, multiples and primes

This spread will show you how to:

- Find the prime factor decomposition of a number.
- Use prime factors to find the highest common factor and lowest common multiple of a set of numbers.

KEYWORDS

Factor, Prime, Multiple, Product, Power

A **prime factor** is a factor that is also a prime number.

For example, the prime factors of 10 are 2 and 5 because:

- they are both factors of 10
- they are both prime numbers

$10 = 2 \times 5$

- **You can express any whole number as a product of its prime factors.**

example

Express 45 as a product of its prime factors.

First draw a diagram:

5 and 9 are factors of 45 $(5 \times 9 = 45)$

3 and 3 are factors of 9 $(3 \times 3 = 9)$

$$45 = 5 \times 9 = 5 \times 3 \times 3 = 5 \times 3^2$$

You could find the prime factors of 45 by repeatedly dividing by prime numbers:
45 ÷ 2 won't go exactly ...
... so try 3: 3)45
Try 3 again: 3)15
Now try 5: 5)5
1

$45 = 3 \times 3 \times 5 = 3^2 \times 5$

You can use prime factors to find the highest common factor or the lowest common multiple.

example

Find: **a** the highest common factor (HCF) of 100 and 80.
b the lowest common multiple (LCM) of 100 and 80.

a $100 = 2^2 \times 5^2 = 2 \times 2 \times 5 \times 5$
$180 = 2^2 \times 3^2 \times 5 = 2 \times 2 \times 5 \times 3 \times 3$
$2 \times 2 \times 5$ are common to both products.
The HCF of 100 and 80 is $2 \times 2 \times 5 = 20$.

b The LCM should contain the highest power of each prime number:
$100 = 2^2 \times 5^2$ $180 = 2^2 \times 3^2 \times 5$

- Together, these contain the prime numbers 2, 3 and 5.
- The highest power of each is 2^2, 3^2 and 5^2.
- The LCM is $2^2 \times 3^2 \times 5^2 = 900$.

You could draw a Venn diagram of the prime factors:

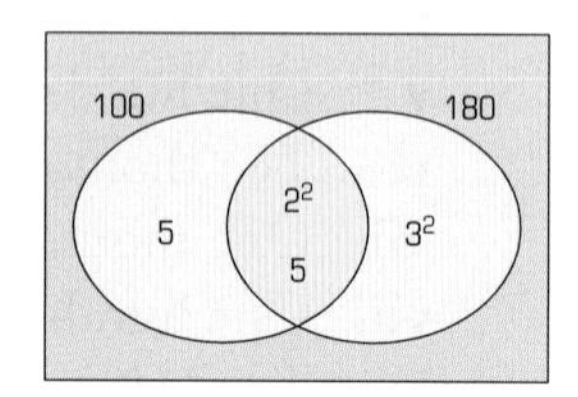

HCF = $2^2 \times 5$
LCM = $2^2 \times 3^2 \times 5^2$

Exercise N5.2

1 Find all the factors of
- a 10
- b 20
- c 130
- d 40
- e 56
- f 150
- g 72
- h 142
- i 224

2 **Investigation**
- a Find 4 numbers less than 100 with exactly 12 factors.
- b Find a number with exactly 15 factors.

3 Express the following numbers as products of their prime factors
- a 12
- b 22
- c 32
- d 44
- e 150
- f 72
- g 130
- h 224
- i 142

4 Find the HCF and LCM of:
- a 44 and 72
- b 32 and 44
- c 22 and 12
- d 12 and 130

5 **Investigation**
- a Find the HCF of 130 and 150.
- b Find the LCM of 130 and 224.
- c Express 168 and 120 as products of their prime factors.
- d Describe how to find the lowest common multiple of 2 numbers using prime factors.

6 **Investigation**
- a Which numbers less than 100 have an odd number of factors?
- b What do the numbers in part **a** have in common?

7
- a Two hands move around a dial. The faster hand moves around in 24 seconds and the slower 34 seconds. If the 2 hands start together at the top of the dial, how many seconds does it take before they are next at the top?
- b A floor is to be covered in square tiles. The floor measures 420 cm by 810 cm. What is the largest size of *square* tile which can be used to tile the floor, without breaking any of the tiles?
- c A garden measures 31.5 m by 29.4 m. It is to be fenced in using posts which must be an equal distance apart. What is the greatest distance apart the posts can be?
- d
  - i Four bells ring at intervals of 9 s, 15 s, 20 s and 12 s. If they all start together, how long will it be, before they all ring at the same time again?
  - ii What is the shortest time before 2 of the bells ring together?
  - iii What is the shortest time before 3 of the bells ring together?

8 **Investigation**

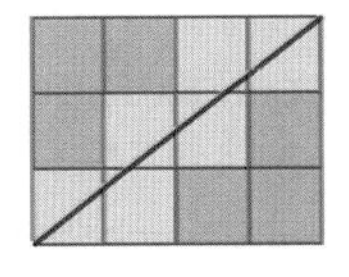

This diagonal passes through 6 squares.
- a How many squares will the diagonal of a 6×8 rectangle pass through?
- b Investigate different sized rectangles.

9 **Investigation**

Which of the odd numbers between 10 and 30 have a prime factor in common? Are they related in any other way?

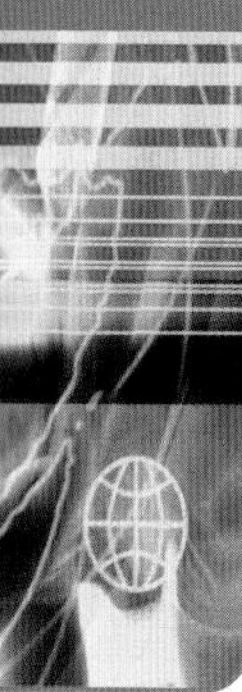

N5.3 Multiplying and dividing mentally

This spread will show you how to:
- Extend strategies for mental multiplication and division.
- Consolidate the rapid recall of number facts and use known facts to derive unknown facts.
- Use tests of divisibility.

KEYWORDS

Digit, Divisibility, Factor, Partitioning

If you can calculate in your head then you can work out problems wherever you are. Here are two useful methods for multiplying in your head.

Using factors to simplify numbers

Break down the numbers into factors you can multiply with easily:

$0.42 \times 30 = 0.42 \times 10 \times 3$

$0.42 \times 10 = 4.2 \quad 4.2 \times 3 = 12.6$

so $0.42 \times 30 = 12.6$

$42 \times 0.03 = 42 \times 0.01 \times 3$

$42 \times 0.01 = 0.42 \quad 0.42 \times 3 = 1.26$

so $42 \times 0.03 = 1.26$

Using partitioning

Remember: $7 \times 8 = 56$ so $56 \div 8 = 7$; $7 \times 4 \times 2 = 56$ so $56 \div 4 \div 2 = 7$

Break down the numbers by partitioning to make them easier to multiply:

$21 = 20 + 1$

$$^{-}6.3 \times 21 = {}^{-}6.3 \times 20 + (-6.3 \times 1) = {}^{-}126 + {}^{-}6.3 = {}^{-}132.3$$

$9 = 10 - 1$

$$0.57 \times 9 = (0.57 \times 10) - (0.57 \times 1) = 5.7 - 0.57 = 5.13$$

You can use factors to divide in your head:

$$645 \div 15 = 645 \div 5 \div 3 = 129 \div 3 = 43$$

You can check whether numbers will divide exactly by using a divisibility test:

- $\div 12$ the number divides by 4 **and** it divides by 3
- $\div 15$ the number divides by 5 **and** it divides by 3

example

Does 2220 divide by 15?

- Check $2220 \div 5$: 2220 ends in zero so it divides by 5
- Check $2220 \div 3$: $2 + 2 + 2 + 0 = 6 \Rightarrow$ the sum of the digits is divisible by 3 so 2220 is divisible by 3.

Therefore 2220 does divide by 15.

Check: $2220 \div 15 = 148$

Exercise N5.3

1 Calculate the following mentally.
You may need to make some jottings.

a $4.2 \times 30 =$
b $3.1 \times 60 =$
c $5.2 \times 9 =$
d $1.8 \times 7 =$
e $132 \div 6 =$
f $1192 \div 8 =$
g $276 \div 12 =$
h $306 \div 9 =$

2 This is a game for two players and a dice.
Here is an example:

- First choose a decimal Target Number, say 5.6.
- Player 1 throws a 6 on the dice.
- Player 2 throws a 5.
  Player 2 chooses a place value of 0.5 and subtraction: $6 - 0.5 = \mathbf{5.5}$
- Player 1 throws a 1.
  Player 1 chooses a place value of 0.1 and addition: $5.5 + 0.1 = \mathbf{5.6}$, which is the Target Number.
- Player 1 wins.

You can use any place value and any mathematical operation.

3 Use the fact:

$$45 \times 36 = 1620$$

to answer the following questions.

a $1620 \div 36 =$
b $360 \times 4.5 =$
c $18 \times 90 =$
d $810 \div 45 =$
e $0.162 \div 0.45 =$
f $180 \times 18 =$
g $135 \times 36 =$
h $1620 \div 72 =$

4 Copy and complete this diagram:

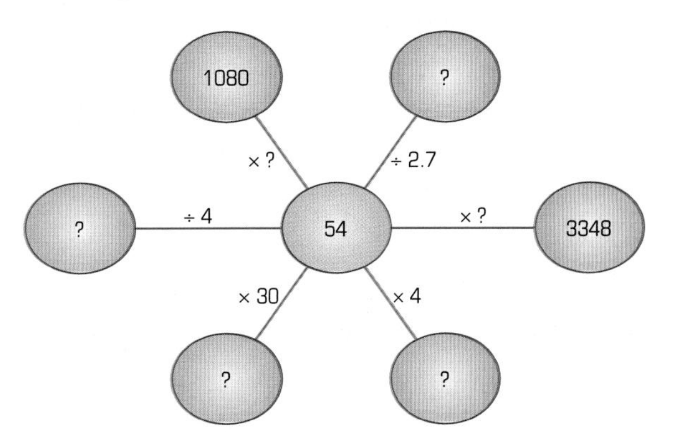

5 Calculate mentally:

a $555 \div 15$ **b** 34×0.05
c $558 \div 18$ **d** $^{-}2.5 \div 0.1$
e $0.48 \div 0.01$ **f** 4.7×0.2
g $^{-}14 \times 2.5$ **h** 3.8×3.5
i 7.8×31 **j** 19×2.6
k 1.9×13 **l** 26×3.1
m 72×19 **n** $0.8 \times {}^{-}21$

6 Solve these arithmagons:

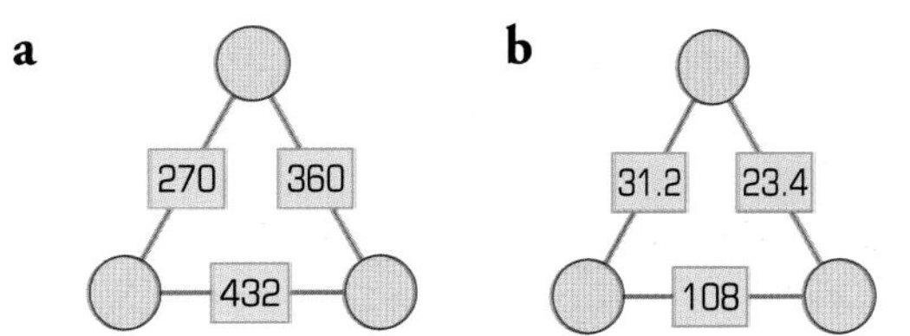

7 **Investigation**

a Does multiplying always make a number bigger?
b Does dividing always make a number smaller?

8 **a** Write you own set of divisibility rules for numbers from 11 to 30.
b For which numbers have you not been able to find a rule?
Which numbers have more than one rule?

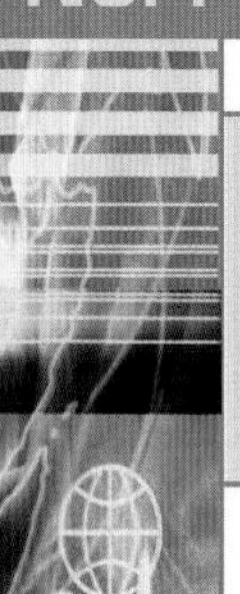

N5.4 Standard written multiplication

This spread will show you how to:

- Refine written methods of multiplication and division of whole numbers to ensure efficiency, and extend to decimals.
- Make and justify estimates and approximations.

KEYWORDS

Approximate Equivalent
Convert Multiply
Decimal

You can use the grid method for multiplying decimals up to 2 decimal places.

32.4×6.23

First approximate: $30 \times 6 = 180$

	30	2	0.4
6	180	12	2.4
0.2	6	0.4	0.08
0.03	0.9	0.06	0.012

Add up the numbers:

$$\begin{array}{r} 180 \\ 12 \\ 6 \\ 2.4 \\ 0.4 \\ 0.9 \\ 0.08 \\ 0.06 \\ +\ 0.012 \\ \hline 201.852 \\ \hline \end{array}$$

$32.4 \times 6.23 = 201.852$

The grid method becomes awkward when you are dealing with this many decimal places. It is easier to use the standard method.

example

Calculate 32.4×6.23 using the standard method of multiplication.

First approximate: $30 \times 6 = 180$

Convert to an equivalent calculation: $32.4 \times 10 \times 6.23 \times 100 \div 1000$
$= 324 \times 623 \div 1000$

$\div$ 1000 to balance the $\times 10 \times 100$

Work out 324×623:

$$\begin{array}{r} 623 \\ \times \quad 324 \\ \hline 2\,492 \\ 12\,460 \\ +\ 186\,900 \\ \hline 201\,852 \\ \hline \end{array}$$

$324 \times 623 = 201\,852$

Divide by 1000: $32.4 \times 6.23 = 201\,852 \div 1000$
$= 201.852$

- **You can multiply with decimals by:**
  - **Converting any decimals to whole numbers**
  - **Using the standard method.**

Don't forget to convert your answer back to decimals.

Exercise N5.4

1 Work out estimated answers to each of these multiplications.

a 34×23 **b** 65×37
c 94×87 **d** 36×98
e 654×34 **f** 365×63
g 534×79 **h** 634×62

2 Work out the exact answers to question 1. Use your estimated answers to check that these exact answers are correct.

3 Fredrica is organising a trip to the theatre to watch a Christmas pantomine. Each coach will hold 53 people and there are 147 adults, 132 children and 52 O.A.P.s going. The price of a coach ticket is £3.41 for adults and £2.11 for children. Fredrica's boss wants rough answers to the following. Provide these answers for Fredrica.

a How much will be collected for the childrens' seats?
b How much will be collected for the adults' seats?
c How much will be collected for the O.A.P.s' seats?
d How much will be collected altogether? Explain how you worked this out.
e Discuss with a partner how they worked out **d**. Is there another way to get this answer?
f How many coaches will be needed?

4 **a** Work out the exact answers to the problem in question 3.
b Explain why the estimated answer to part **f** is the same as the exact answer.

5 Work out estimated answers to each of these multiplications.

a 5.4×2.1 **b** 9.5×3.5
c 10.4×8.7 **d** 8.6×98
e 45.4×44 **f** 2.67×63
g 5.34×9.9 **h** 434×6.2

6 Work out the exact answers to question 5. Use your estimated answers to check that these exact answers are correct.

7 Give the answers to question 5 correct to 3 significant figures.

8 Give estimated answers to the following:

a $23.465 + 0.045 - 3.45$
b $3.45 - 0.0004$
c $39 + 0.76 + 0.345 - 2.534$
d $3.045 - 3 - 0.047$
e $45.83 + 1\,000\,000 - 0.34$

9 Work out the exact answers to question 8.

10 Work out estimated answers to help you decide which is the heavier: a bag containing 182 stones weighing 4.2 g or one containing 143 stones that each weigh 7.6 g.

11 Work out exact answers for question 10 showing all of your working out. Do your answers agree on which is the heavier bag?

12 Explain how to do long multiplication for 0.0345×5.3.

13 Show how you would set out a written method for a question where the number you are multiplying by has more than two digits, for example 3.54×0.645.

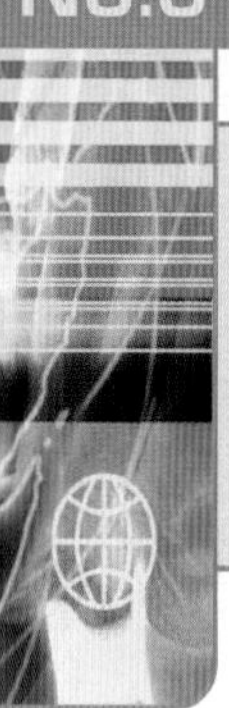

N5.5 Standard written division

This spread will show you how to:

- Refine written methods of division of whole numbers to ensure efficiency, and extend to decimals.
- Interpret the display on a calculator in different contexts.
- Use checking procedures.

KEYWORDS

Approximate, Equivalent, Decimal, Quotient, Divisor

You can divide with decimals by using an equivalent calculation.

These are equivalent calculations: $8 \div 4 = 2$ $80 \div 40 = 2$ $800 \div 400 = 2$
In the same way: $234 \div 4.2 = 2340 \div 42$

- You can divide with decimals by:
  - multiplying both divisor and dividend by the same appropriate power of 10 to remove the decimals
  - using the repeated subtraction method

example

Work out **a** $234 \div 4.2$ **b** $321.6 \div 7.8$

a First estimate: $200 \div 4 = 50$
Now convert: $234 \div 4.2 = 2340 \div 42$

```
42)2340
  -2100    42 × 50
    240
   -210    42 × 5
    30.0
   -29.4   42 × 0.7
    0.60
   -0.42   42 × 0.01
    0.18
```

$50 + 5 + 0.7 + 0.01 = 55.71$

$234 \div 4.2 = 55.7$ to 1 d.p.

b First estimate: $320 \div 8 = 40$
Now convert: $321.6 \div 7.8 = 3216 \div 78$

```
78)3216
  -3120    78 × 40
     96
    -78    78 × 1
    18.0
   -15.6   78 × 0.2
    2.40
    2.34   78 – 0.03
    0.06
```

$321.6 \div 41.2$ to 1 d.p.

The result of a division is called a **quotient**. You may need to round it.

example

Multi-packs contain 12 packets of crisps. How many multi-packs can you make from 90 packets of crisps and how many are left over?

$90 \div 12 = 7$ remainder 6 so you can make 7 multi-packs and there are 6 packets left over.

Exercise N5.5

1 Work out these divisions using any method you like, giving a remainder where there is one.

a $23 \div 4$ **b** $40 \div 6$
c $30 \div 7$ **d** $54 \div 5$
e $37 \div 8$ **f** $74 \div 8$
g $47 \div 9$ **h** $33 \div 4$
i $58 \div 3$ **j** $26 \div 15$

2 Use a 'standard method' to work out these divisions, giving a remainder where this is one.

a $123 \div 5$ **b** $410 \div 6$
c $630 \div 7$ **d** $954 \div 5$
e $937 \div 8$ **f** $724 \div 8$
g $347 \div 9$ **h** $133 \div 4$
i $158 \div 3$ **j** $326 \div 6$

3 Use a 'standard method' to work out these divisions, giving a remainder where this is one.

a $123 \div 52$ **b** $410 \div 65$
c $630 \div 71$ **d** $954 \div 52$
e $937 \div 82$ **f** $724 \div 83$
g $347 \div 95$ **h** $133 \div 42$
i $158 \div 13$ **j** $326 \div 62$

4 Work out these divisions, giving a remainder to 1 d.p. where there is one.

a $32.3 \div 5$ **b** $6.10 \div 6$
c $156.4 \div 7$ **d** $185.4 \div 5$
e $243 \div 18$ **f** $72.6 \div 14$
g $33.57 \div 19$ **h** $212.3 \div 14$
i $138.58 \div 23$ **j** $362.6 \div 26$

5 Work out these divisions, giving a remainder to 1 d.p. where there is one.

a $923 \div 3.2$ **b** $510 \div 4.5$
c $830 \div 51$ **d** $854 \div 6.2$
e $337 \div 82$ **f** $824 \div 8.3$
g $547 \div 7.5$ **h** $533 \div 3.2$
i $258 \div 2.3$ **j** $37.4 \div 6.2$

6 Calculate to 1 d.p.

a $324.6 \div 8.7$
b $216.7 \div 4.2$
c $423.9 \div 0.7$
d $28.6 \div 6.1$
e $18 \div 13$
f $12.5 \div 4$
g $216.9 \div 7.3$

7 **Investigation** By rolling two ordinary dice, how many different fractions can you make? Write them all down (for example rolling a 3 and a 4 can make two fractions $\frac{3}{4}$ and $\frac{4}{3}$).
Convert each of these fractions to a decimal and group together the equivalent fractions.

8 Mrs Laughalot teaches five classes. The number of pupils in each of her classes is as follows:

Year 7	26
Year 8	23
Year 9	25
Year 10	29
Year 11	24

At Christmas she buys five big tubs of sweets (one for each of her classes), each tub containing 2000 sweets.

a Work out how many sweets the children in each of her classes would get and state the remainder in each case.

b If you put all remainders together and gave them to Year 7, how many extra sweets would each child in this class receive?

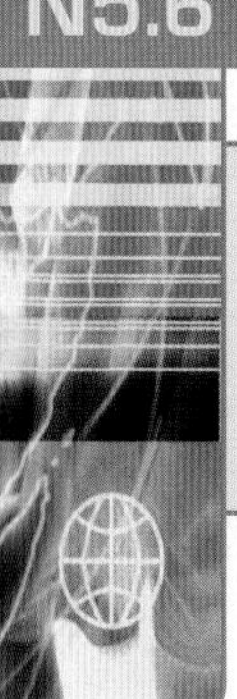

N5.6 Using equivalent fractions

This spread will show you how to:
- Order fractions.
- Add and subtract fractions.

KEYWORDS
Denominator Multiple
Factor Numerator
Fraction Prime

To compare fractions with different denominators, you need to change them to equivalent fractions.

example

Which is bigger, $\frac{4}{5}$ or $\frac{8}{11}$?

- First find the lowest common multiple (LCM) of 5 and 11.
  $5 = 5 \times 1$ $11 = 11 \times 1$
  So the LCM $= 5 \times 11 = 55$
- Now find equivalent fractions with 55 as the common denominator.
  $\frac{4}{5} = \frac{44}{55}$ $\frac{8}{11} = \frac{40}{55}$
- Now compare the fractions: $\frac{44}{55} > \frac{40}{55}$, so $\frac{4}{5} > \frac{8}{11}$.

Remember that to find the LCM you break a number down into its prime factors.

You can use this method to add or subtract fractions with different denominators.

- **To add or subtract fractions:**
  - **Find equivalent fractions with a common denominator**
  - **Add or subtract the numerators**

example

Find:

a $\frac{4}{13} - \frac{1}{8}$

b $\frac{3}{5} + \frac{7}{9}$

a $13 = 13 \times 1$
$8 = 2 \times 4$
$= 2 \times 2 \times 2 = 2^3$

The LCM of 13 and 8 is $13 \times 2^3 = 104$

$\frac{4}{13} = \frac{32}{104}$

$\frac{1}{8} = \frac{13}{104}$

$\frac{4}{13} - \frac{1}{8} = \frac{32}{104} - \frac{13}{104}$
$= \frac{19}{104}$

b $5 = 5 \times 1$ $9 = 3^2$
The LCM of 5 and 9 is $5 \times 3^2 = 45$

$\frac{3}{5} = \frac{27}{45}$

$\frac{7}{9} = \frac{35}{45}$

$\frac{3}{5} + \frac{7}{9} = \frac{27}{45} + \frac{35}{45}$
$= \frac{62}{45}$
$= 1\frac{17}{45}$

- **When you add or subtract fractions you may need to convert your answer to a mixed number.**

Exercise N5.6

1 In each case state which is the larger fraction. Show your working out.

a $\frac{1}{3}$ or $\frac{4}{11}$ **b** $\frac{3}{8}$ or $\frac{7}{25}$ **c** $\frac{4}{19}$ or $\frac{3}{18}$ **d** $\frac{5}{11}$ or $\frac{2}{7}$

2 In a particular game Jane won $\frac{1}{6}$ of the games, Harry won $\frac{2}{5}$ and Bill won the rest.

a What fraction of games did Bill win?

b What fraction of games did Harry and Jane win together?

c What is the difference, as a fraction, of the games won by Bill and Jane?

3 Draw a diagram to show why $\frac{2}{9} + 1\frac{4}{9} = 1\frac{2}{3}$.

4 In a playground game two pupils are playing 'twirls'. Anna twirls through $3\frac{1}{3}$ turns, and Carl twirls through $4\frac{1}{5}$ turns.

a What is the difference between the number of twirls they did?

b How many twirls did they do altogether?

5 John said that the answer to $\frac{2}{3} + \frac{1}{4} = \frac{3}{7}$.

a Explain, in your own words, what he has done wrong.

b Work out the correct answer, showing your working.

6 Calculate:

a $\frac{1}{4} + \frac{4}{5}$ **b** $\frac{3}{5} + \frac{2}{3}$ **c** $\frac{4}{5} + \frac{2}{7}$

d $\frac{2}{9} + \frac{3}{5}$ **e** $\frac{3}{10} + \frac{3}{4}$ **f** $2\frac{5}{6} + 1\frac{4}{5}$

g $3\frac{7}{9} - 1\frac{4}{9}$ **h** $4\frac{2}{3} - 6$ **i** $3\frac{2}{11} + \frac{1}{4}$

7 Work out:

a $\frac{1}{4} - \frac{2}{3} + \frac{3}{7}$ **b** $\frac{5}{6} + \frac{1}{3} - \frac{4}{5}$

c $\frac{6}{7} + \frac{2}{3} + \frac{1}{9}$ **d** $\frac{2}{5} - \frac{1}{3} + \frac{2}{7}$

8 Answer these in the same way as question 7. Convert the mixed numbers to improper fractions to begin:

a $3\frac{1}{4} - 2\frac{4}{5}$ **b** $4\frac{3}{7} + 3\frac{4}{9}$ **c** $2\frac{3}{4} + 6\frac{3}{7}$

d $5\frac{5}{8} - 4\frac{3}{10}$ **e** $3\frac{3}{11} + 2\frac{2}{7}$

9 Show that

$$\frac{x}{3} + \frac{y}{2} = \frac{2x + 3y}{6}$$

by choosing any two numbers for x and y.

10 Add these algebraic fractions:

a $\frac{x}{4} + \frac{x}{5}$ **b** $\frac{x}{3} + \frac{x}{8}$ **c** $\frac{x}{5} + \frac{x}{3}$ **d** $\frac{x}{3} + \frac{x}{7}$

11 Here is an example of four fractions, each with a different denominator, that total 1.

$\frac{2}{3} - \frac{17}{30} + \frac{4}{5} - \frac{9}{10}$

Make up 3 more fraction 'strings' similar to the one above that total 1.
Each fraction must have a different denominator.

12 Work out which is the greater in each part:

a $\frac{2}{5}$ of 45 or $\frac{1}{4}$ of 80

b 23% of 50 or $\frac{1}{3}$ of 33

c 0.6×86 or $\frac{1}{8}$ of 36

13 **a** Write the next two terms in this sequence:

$1 \quad \frac{1}{2} \quad \frac{1}{3} \quad \frac{1}{4} \quad \frac{1}{5} \ldots$

b Describe how you got your answer to **a**.

c Add the first seven terms of the sequence together and express your answer as a mixed number.

N5.7 Converting fractions, decimals and percentages

This spread will show you how to:

- Recognise the equivalence of fractions, decimals and percentages.

KEYWORDS

Convert	Fraction
Decimal	Numerator
Denominator	Percentage

You can convert between fractions, decimals and percentages.

Mental methods

This number line shows some common conversions:

$0.1 = \frac{1}{10}$ $0.2 = \frac{1}{5}$ $0.25 = \frac{1}{4}$ $0.\dot{3} = \frac{1}{3}$ $0.5 = \frac{1}{2}$ $0.75 = \frac{3}{4}$

0 10% 20% 25% 33.3% 50% 75% 1

If you know these by heart you can work out more difficult conversions:

$\frac{3}{5} = 3 \times \frac{1}{5} = 3 \times 0.2 = 0.6$, or 60% $\quad \frac{2}{3} = 2 \times \frac{1}{3} = 2 \times 0.3333\ldots = 0.6666\ldots = 66.7\%$

Written and calculator methods

You can convert from a decimal to a fraction by using your knowledge of place value.

example

Convert 0.345 to a fraction in its lowest terms.

Write 0.345 in a place value table:

Units	.	$\frac{1}{10}$	$\frac{1}{100}$	$\frac{1}{1000}$
0	.	3	4	5

$$\text{So } 0.345 = \frac{3}{10} + \frac{4}{100} + \frac{5}{1000}$$
$$= \frac{300}{1000} + \frac{40}{1000} + \frac{5}{1000}$$
$$= \frac{345}{1000}$$
$$0.345 = \frac{345}{1000} = \frac{69}{200}$$

- You can convert a fraction to a decimal by dividing the numerator by the denominator ...
- ... and you can convert a decimal to a percentage by multiplying by 100:

$\frac{3}{8} = 3 \div 8 = 0.375$ — You should know this conversion

$0.375 \times 100\% = 37.5\%$ — You should be able to ×100 in your head

Exercise N5.7

1 What is the decimal equivalent of each of these fractions?
a $\frac{1}{2}$ **b** $\frac{1}{4}$ **c** $\frac{3}{4}$ **d** $\frac{2}{4}$ **e** $\frac{6}{8}$

2 Which of the fractions in question 1 are equivalent?

3 Write each of the fractions in question 1 as percentages.

4 By looking at your answers to questions 2 and 3 describe an easy method for
a changing decimals into percentages
b percentages into decimals.

5 The probability of rain tomorrow is 40%. Express this as:
a a decimal **b** a fraction.

6 Convert each of these decimals to fractions in their lowest terms:
a 0.45 **b** 0.567 **c** 0.4 **d** 0.243
e 0.003 **f** 1.1 **g** 2.01 **h** 0.05

7 Repeat question 6 but convert the decimals into percentages.

8 **a** Copy and complete this fraction wall for $\frac{1}{2}$s, $\frac{1}{5}$s, $\frac{1}{9}$s, and $\frac{1}{12}$s,

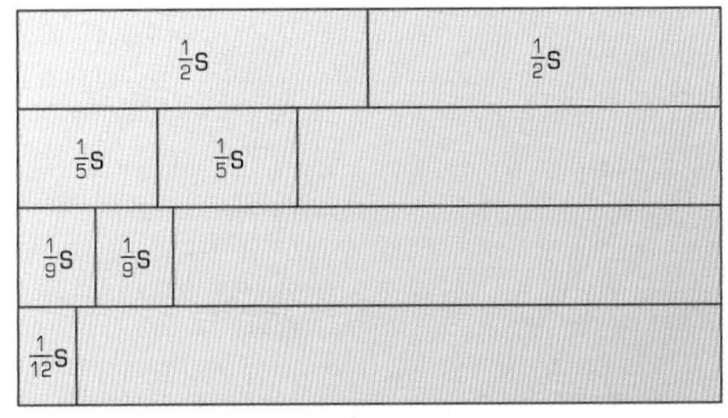

b Redraw the wall labelling decimals instead of fractions.
c Which is bigger: 0.4 or $\frac{5}{12}$?

9 Look at this sequence:
$\frac{1}{2}, \frac{1}{3}, \frac{1}{4}, \frac{1}{5}, \frac{1}{6}, \ldots$
a What are the next two fractions in this sequence?
b Write the sequence as decimals (use the division method if needed).
c What did you do with the $\frac{1}{7}$ fraction for part **b**?
d If you didn't already do it, write the fraction $\frac{1}{7}$ correct to 3 decimal places.

10 Use a calculator to investigate the following statement:
'Do fractions that have a denominator of 11 produce decimals with a predictable pattern?'

11 **a** Fred's teacher had a strange way of giving out the test marks to his class. Instead of giving them their scores as a simple number, he would tell them what fraction of the total they got. Work out the scores each person scored as a percentage correct to 1 decimal place:

Fred	$\frac{3}{8}$
Mary	$\frac{13}{60}$
Harry	$\frac{4}{9}$
Bill	$\frac{7}{11}$
Carrie	$\frac{10}{17}$
Tom	$\frac{2}{3}$
Barry	$\frac{4}{5}$
Alison	$\frac{5}{6}$

b Who scored the highest on the test?

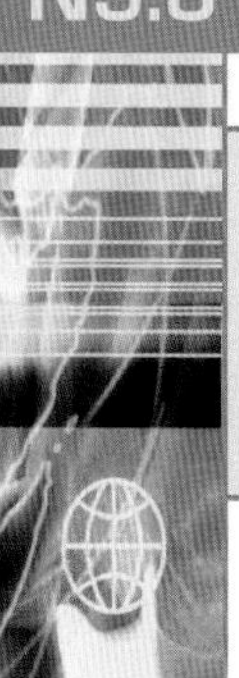

N5.8 Calculating parts of quantities

This spread will show you how to:

- Calculate fractions of numbers, quantities or measurements.
- Multiply and divide a fraction by an integer.
- Calculate percentages of numbers, quantities and measurements.

KEYWORDS

Amount	Fraction
Decimal	Increase
Decrease	Percentage

You can use different methods to find a fraction of an amount.

- You can often work it out **mentally** ...

  For example, $\frac{1}{7}$ of 63 kg

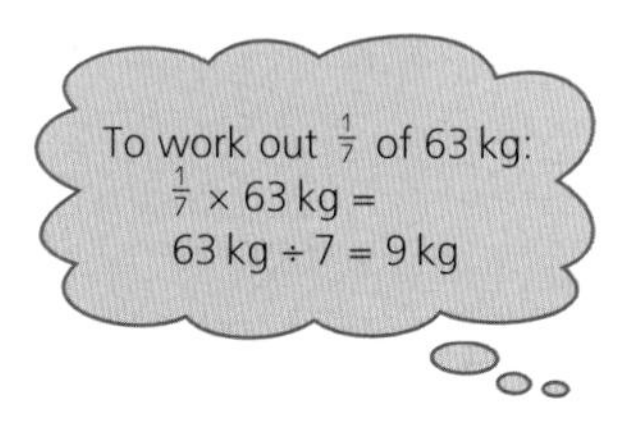

- You could use a **written** method ...

  For example, $\frac{3}{7}$ of 30 km

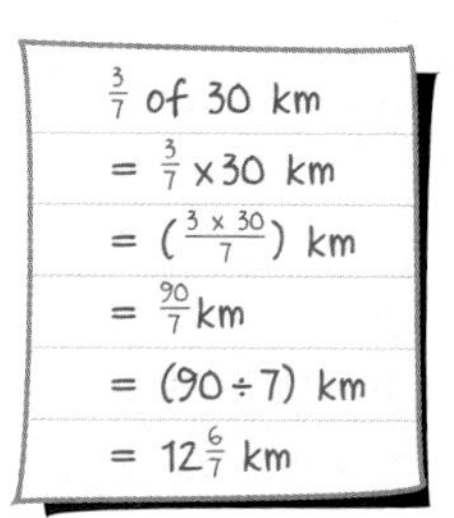

You can divide a fraction by an integer like this:

$\frac{1}{3} \div 4 =$

$1 \div 3 \div 4$

$= 1 \div 12$

$= \frac{1}{12}$

- You could use a **calculator** and convert the fraction to a decimal ...

  For example, $\frac{13}{20}$ of £716

  13 ÷ 20 × £716
  = £465.40

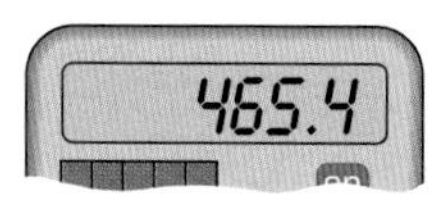

You often need to calculate a percentage increase or decrease.

example

Natalie earns £5.50 per hour. Her employer gives her an 8% pay rise. How much does Natalie earn per hour now?

First estimate: 10% of £5 is 50p

Now calculate 8% of £5.50:

$\frac{8}{100} \times £5.50$

$= 8 \times £\frac{5.50}{100}$

$= £\frac{44}{100}$

$= £0.44$ (close to the estimate of 50p)

Add the increase to the original amount: £5.50 + £0.44 = £5.94

Natalie now earns £5.94.

If the problem involved a percentage decrease, you would have to **subtract** from the original amount.

Exercise N5.8

1 A mathematics test was out of 40, but the teacher gave the pupils their marks as percentages. Work out their actual scores.

Adam 20%	Bill 45%
Carl 60%	David 35%
Eric 45%	Freda 80%
Geoff 75%	Harriett 90%

2 These fractions relate to the fraction of a chocolate bar each of three children received:
Mandy $\frac{3}{5}$
Susan $\frac{1}{4}$
Alex $\frac{3}{20}$
Each person had to share their amounts with 4 other people, so Mandy's $\frac{3}{5}$ had to be shared between 5 people altogether.
Work out what fraction each of the three children actually ended up with.

3 Multiply each of these fractions by 7 and write your answers in their simplest form:
a $\frac{1}{2}$ **b** $\frac{1}{4}$ **c** $\frac{3}{4}$
d $\frac{2}{4}$ **e** $\frac{6}{8}$

4 Calculate:
a $\frac{2}{3} \div 5$ **b** $\frac{3}{4} \div 6$ **c** $\frac{5}{7} \div 8$
d $\frac{1}{4} \div 4$ **e** $\frac{3}{8} \div 9$

5 Harry was 160 cm tall.
Eddie was $\frac{3}{5}$ as tall as Harry.
Paul was $\frac{7}{8}$ as tall as Eddie.
How tall were Eddie and Paul?

6 Find:
a $\frac{1}{60}$ of \$40
b $\frac{3}{50}$ of £4 000 000
c $\frac{11}{15}$ of 600 m
d $\frac{13}{300}$ of 900 litres

7 A car salesperson has the cars listed in the car showroom. It is estimated that each car will lose 18% of its value every year.
Range Rover £30 000
Ford Mondeo £13 000
Ford Escort £9000
Jaguar XJ6 £35 000
Rolls Royce £190 000
Vauxhall Vectra £11 500
The questions below relate to the information in the list above.
a Work out the individual value of each vehicle after 1 year and after 2 years.
b What fraction of their value do the cars lose each year?

8 This is an advert for a house:

HOUSE FOR SALE
3 Bedroom Detached Property
Nice views, good location.
£95 000

a Assuming the house increases in value by 23% in the next year, what will its new value be?
b The year after this the 'new price' (the price at the end of the first year) decreased by 12%. How much was the house worth at the end of the second year?

9 **Investigation**
- Start with 20
- Find 40% of 20
- Now find 40% of your answer above.
- Repeat this process

Do you eventually reach 0?
Repeat starting with other numbers and choosing a different percentage to work with.
Discuss your findings with a partner.

Summary

You should know how to ...

1 Calculate percentages and find the outcome of a given increase or decrease.

2 Use standard column procedures for multiplication and division of integers and decimals.

3 Understand where to position the decimal point by considering equivalent calculations.

4 Solve more complex problems by breaking them into smaller steps or tasks.

Check out

1 a Add 20% to £19.50
  b Decrease £35 by 20%
  c Find the cost of a TV at £790 + VAT when VAT is 17.5%

2 Show all your working to calculate:
  a 97×86 b 125×31
  c 8.62×3.4 d 16.7×8.36
  e $513 \div 19$ f $667 \div 23$
  g $111.8 \div 4.3$ h $36.3 \div 2.2$

3 Given that $23 \times 7 = 161$, write down:
  a 23×0.1 b 23×0.7
  c 23×0.07 d 2.3×0.07
  e $23 \div 0.1$ f $23 \div 0.01$
  g $2.3 \div 0.01$ h $0.023 \div 0.1$

4 **20% off!**
In a sale there is 20% off all prices.
  a What single number can you multiply by to find the sale price of an item?
  b After a month the shop reduces all its sale items by a further 20%.
  What is the percentage reduction from the original price? Illustrate your answer with examples.

13 Analysing statistics

This unit will show you how to:

- Decide which data to collect to answer a question.
- Plan how to collect the data, including sample size.
- Collect data using a suitable method, such as a data collection sheet.
- Construct diagrams to represent data, including pie charts, bar charts, frequency diagrams and simple line graphs for time series.
- Identify which diagrams are most useful in the context of the problem.
- Interpret tables, graphs and diagrams for both discrete and continous data.
- Relate summarised data to the questions being explored.
- Recognise when it is appropriate to use the range, mean, median, and mode and, for grouped data, the modal class.
- Calculate a mean using an assumed mean.
- Communicate the results of a statistical enquiry and the methods used.
- Justify the choice of what is presented.
- Identify the necessary information to solve a problem.
- Suggest extensions to problems, conjecture and generalise.

Market research can help to make products and services better.

Before you start

You should know how to ...

1 Sort data into discrete, continuous and categorical.

2 Choose the most appropriate graph to display data.

3 Calculate the mean, median, mode and range of a set of numbers.

Check in

1 Which parts of the following statement describe discrete, continuous and categorical data?
A parcel wrapped in brown paper weighed 432g and cost £1.41 to post.

2 Ten different parcels are posted.
What type of graph could you draw to illustrate:
a the colour of the wrapping paper; **b** their weights; **c** how much it costs to post the parcels?

3 Calculate the mean, median, mode and range of 4, 5, 5, 3, 8.

Planning the data collection

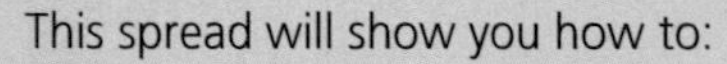

This spread will show you how to:

- Decide which data to collect to answer a question, and identify possible sources.
- Decide how to collect the data.

KEYWORDS

Data
Secondary
Primary

Richie Vanson, an entrepreneur, decides to invest in a new theme park at a seaside resort. He has to choose between Beach Town and Sandy Town.

Each town wants to convince Richie to invest in their resort.

Beach Town collect data on the cleanliness of their beaches and the water quality of the sea.

Sandy Town collect data on hotel availability and the nationality of visitors to the resort.

They both decide to include information on the typical weather throughout the year and the availability of transport to their town.

They decide to collect this data for each month:

- Rainfall (cm)
- Hours of sunshine
- Maximum and minimum temperature (°C)
- Snow (cm)
- Wind (average mph)
- Cloud cover

Month	Rainfall (cm)	Hours of sunshine per day
January	12.2	3.2
February	11.1	4.1
March	9.7	5.3
April	10.8	6.0

The data is collected using the internet and local newspaper records.

Two sets of data are collected, for 2000 and 1999, so that there is less chance of bias.

To collect the right data for your enquiry:

- Decide which data is relevant.
- Research possible sources of the data.
- Plan and design a data collection sheet.

Exercise D3.1

1 **a** A sports teacher believes that the taller you are the faster you can run.
What data would you need to collect to find out if this is true?

b Another teacher thinks this may be true for boys, but not for girls.
Design a data collection sheet for this teacher.
(You do not have to collect any data).

2 If it rains on St Swithins day (July 15th) it will rain for the next 40 days.

What information would you need to collect to test if this is true?

3 Gareth wants to have a holiday in the first week of August in Wales or Scotland.
He believes that it always rains in Wales and that it is never sunny in Scotland.

Choose a town or location in each of Wales and Scotland and design a data collection sheet to help Gareth decide where he should have a holiday.
Think carefully about how much data you need to collect and give a reason for your choice.

4

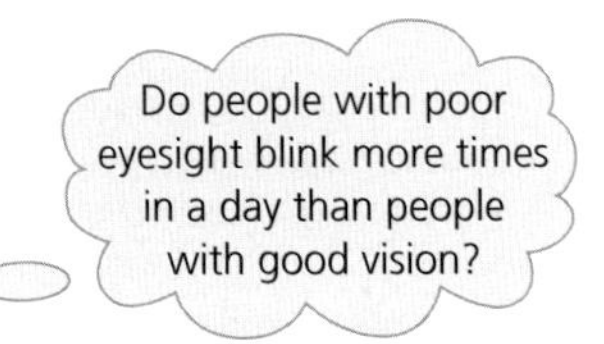

Jennifer decides to carry out a survey to see if this is true.
This is part of her data collection sheet.

Number of blinks in 10 seconds	Good eyesight	Poor eyesight

a What type of data is Jennifer collecting?

Natalie thinks that the question depends upon whether a person is short-sighted or long-sighted and whether they wear contact lenses or glasses.

b Design a data collection sheet for Natalie to carry out her survey.
(You do not have to collect any data).

D3.2 Constructing statistical diagrams

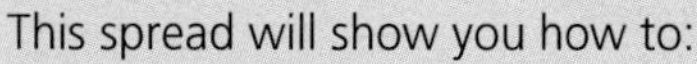

This spread will show you how to:

- Construct graphs and diagrams, including simple line graphs for time series and pie charts for categorical data, to represent data and identify key features.

KEYWORDS

Time series   Pie chart

You use a diagram to represent data as it is easier to read at a glance.

- **Always identify the key features of a statistical diagram and explain what they mean.**

Becci, at Beach Town, chose to present the data she collected on the seawater quality for the summer of 2001 in a line graph.

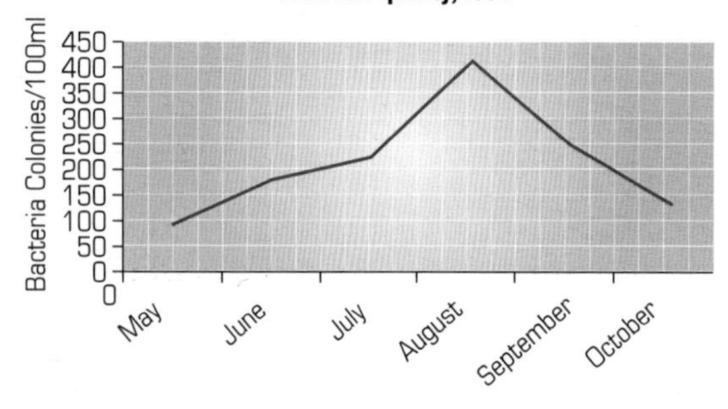

Time is always given on the horizontal axis.

Her graph shows that there were more bacteria colonies per 100 ml in August.

Sandra, at Sandy Town, chose to present the data she collected on the number and types of accommodation for visitors in a pie chart.

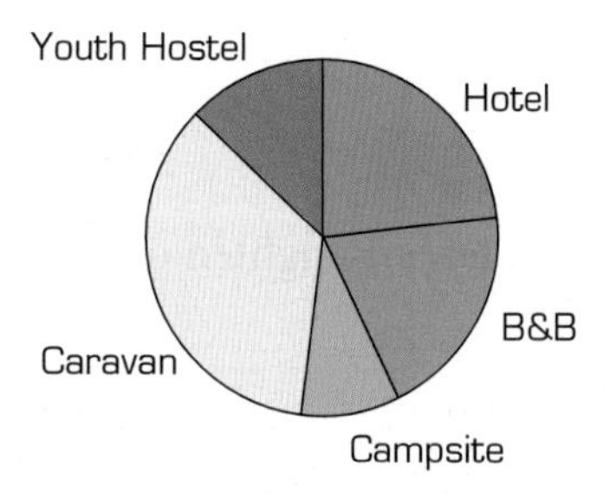

A pie chart is good for displaying categories rather than numerical data.

The sectors represent the proportion in each category.

Her pie chart shows that there is a greater proportion of caravans to stay in than other types of accomodation.

If you have too many categories, a pie chart is not useful.

Exercise D3.2

1 This table gives the world air speed record, in kilometers per hour, at five-year intervals.

Year	1910	1915	1920	1925	1930	1935	1940	1945	1950
Speed (km/h)	110	200	310	450	570	700	750	970	1070

a Draw a time series graph to represent these data (draw the x-axis up to 1960).
b Comment on the trend shown by your graph.
c What do you think the world air speed record would be in 1955 and in 1960?

The world air speed record in 1955 was 1320 km/h and in 1960 was 2440 km/h.
d Add these data to your graph and comment on your answer to part **c**.

2 **a** This table gives the world population in millions of people (to the nearest 10 million) recorded at 10-year intervals.

Year	1920	1930	1940	1950	1960	1970	1980	1990
Population (millions)	1860	2070	2300	2520	3020	3700	4450	5290

Draw a time series graph to represent these data.
Comment on the trend shown by your graph.

b This table gives the world population in millions of people (to the nearest 10 million) recorded each year from 1985 to 1992.

Year	1985	1986	1987	1988	1989	1990	1991	1992
Population (millions)	4850	4940	5020	5110	5200	5290	5390	5480

Draw a time series graph to represent these data.
Comment on the trend shown by your graph.

c Compare the comment on the graphs you have drawn in parts **a** and **b**.

3 This table describes the time allocated to Anoushka's daily activities during each weekday.

Activity	Time (hours)
School	7
Sleeping	8
Playing & hobbies	3
Watching TV	2
Eating & drinking	1
Other	3
Total	24

a There are 360° in a circular pie chart. How many degrees represent 1 hour?
b What angle corresponds to sleeping?
c Find the angles for all the other activities and draw a pie chart for Anoushka's day.

D3.3 Comparing data using diagrams

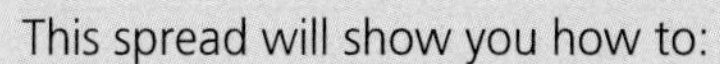

This spread will show you how to:

- Interpret tables, graphs and diagrams and draw inferences related to the problem.

KEYWORDS
Composite bar chart
Compare
Interpret
Trend

Becci and Sandra collect data for Beach Town and Sandy Town respectively.
They draw graphs to represent their findings and present them to Richie Vanson.
Richie studies their graphs for maximum and minimum temperature for 2000 and 1999.

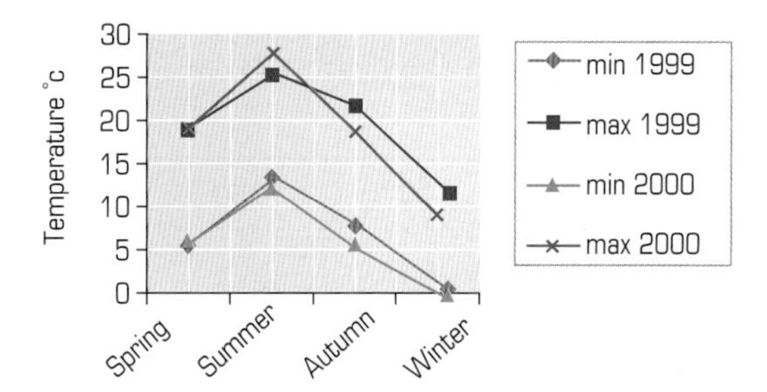

This time series graph makes it easy to generalise and see the trend for each year. It is not easy to compare the two years.

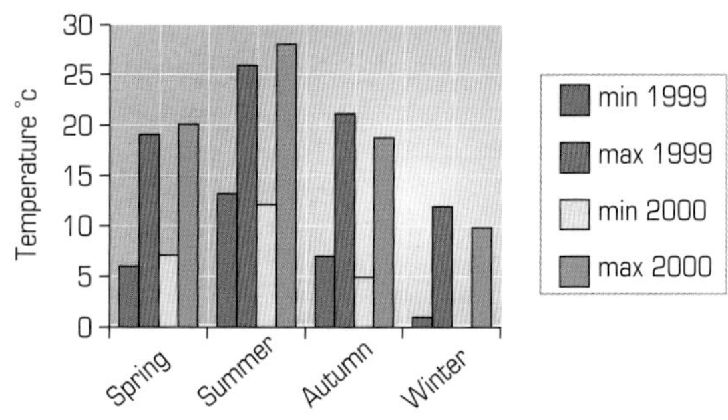

This comparative bar chart makes it easy to compare the two years. It is difficult to see the overall trend.

- Data displayed using appropriate diagrams is often easier to interpret.
- The shape of a graph shows trends in the data and helps you to draw conclusions.
- To compare graphs easily you need to use the same scale on each diagram.

Richie Vanson conducted his own survey into the number of visitors to the two towns over the past four years. He drew a composite bar chart to display the data.

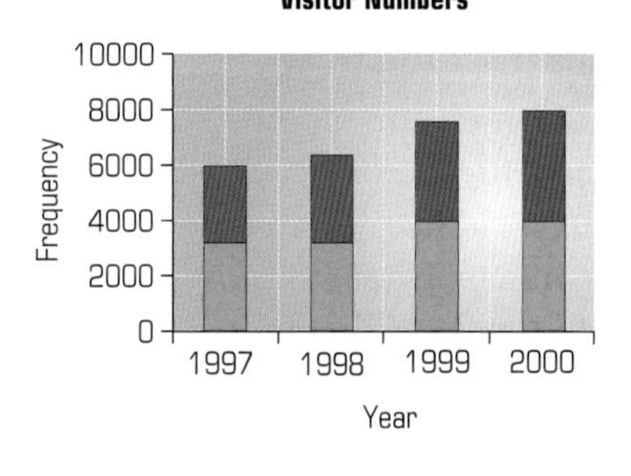

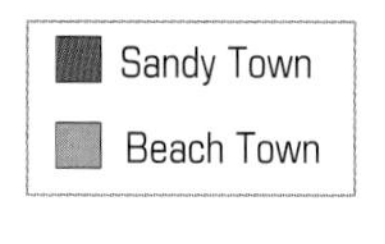

The graph shows that although Beach Town has more visitors each year, the number of visitors to Sandy Town is increasing steadily.

In a composite bar chart you can compare totals easily, but it is hard to see the separate data.

Exercise D3.3

1 Year 7 were asked to vote for a representative on the school council.
The pie chart shows how the votes were cast.
120 people voted.

a Who was voted as form captain?

b How many voted for Edwin?

c How may more votes did Dave get than Freddie?

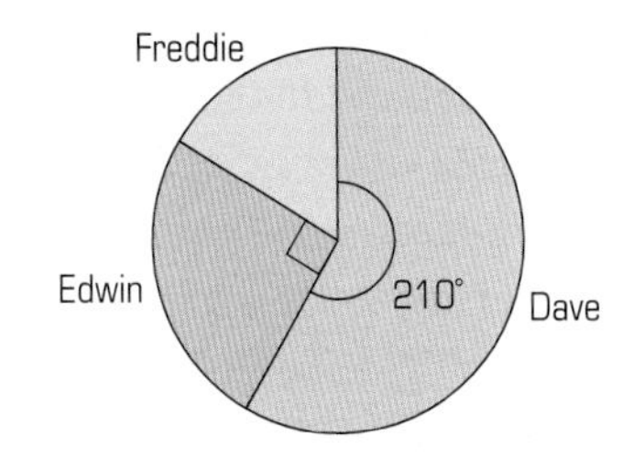

2 Sally and Kit each made four-sided coloured spinners.
They each spun their spinner 40 times.
Their results are displayed in the bar graphs.

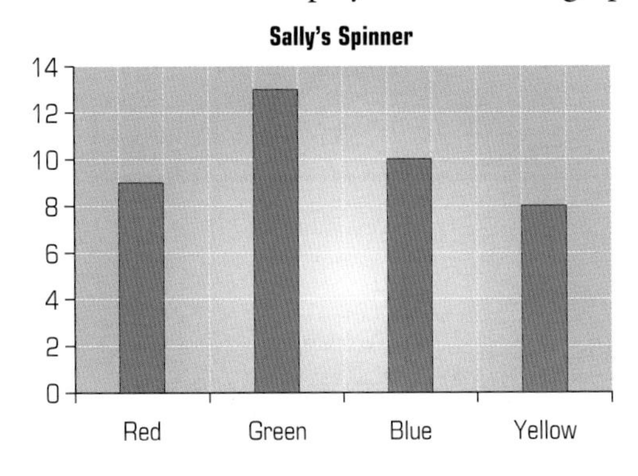

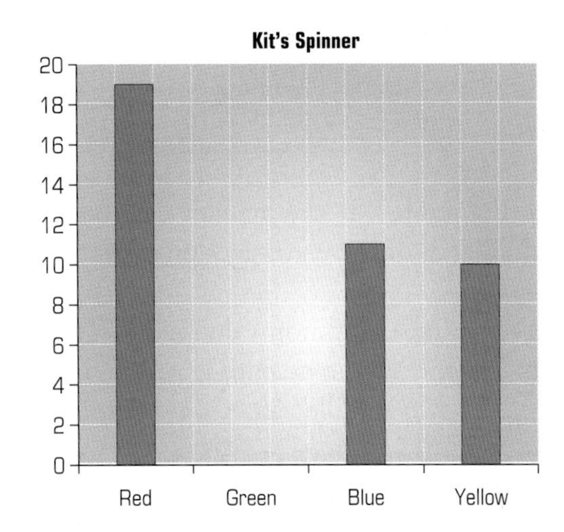

Compare the two graphs and make as many comments as you can about the similarities and differences in the two spinners.
Suggest possible reasons.

3 The noise level in two different classes, Mr G. and Ms K., was measured every ten minutes throughout a one-hour lesson.
The results are shown on the graph.

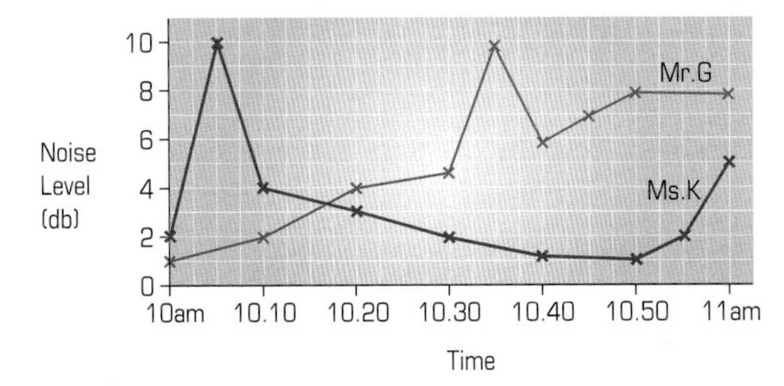

Describe what you think was happening in each of the classrooms.
Whose lesson would you like to have been in?

D3.4 Describing data using statistics

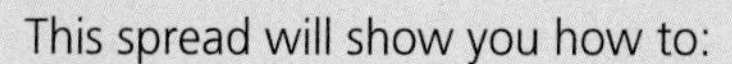

This spread will show you how to:

- Calculate the mean for a set of data.
- Find the median of a set of data.
- Calculate a mean using an assumed mean.
- Compare distributions using the range and an average.

KEYWORDS

Assumed mean, Mode, Median, Range, Mean

You can summarise data by finding an **average** and a measure of **spread**.

Each of the averages that you can choose can give a different impression.

example

The total rainfall, to the nearest cm, for each month in 2000 in Beach Town was:

5.4 3.7 3.7 3.7 4.6 4.5 5.7 5.9 4.9 5.7 6.4 4.8

Find:

a the mode **b** the median **c** the mean.

d Which of these averages gives the most typical value?

a The most common value is 3.7. The mode is 3.7 cm.

b Put the data in ascending order: 3.7, 3.7, 3.7, 4.5, 4.6, 4.8, 4.9, 5.4, 5.7, 5.7, 5.9, 6.4.
The two middle values are 4.8 and 4.9. The median is 4.85 cm.

c Add the numbers up: $5.4 + 3.7 + \cdots + 4.8 = 59$
Divide by 12: $59 \div 12 = 4.916666 \ldots$
The mean is 4.92 cm to 2 d.p.

d The mode is also the lowest value, so it is not really typical.
The mean and median both give a more typical picture.

To get a better impression of the data you can look at the spread or range of the data.

- Range = highest value − lowest value.

The range of rainfall for Beach Town is 45 cm − 2 cm = 43 cm.
This shows that there was a high variation in the amount of rainfall.

- When you summarise data you need an average to describe the data and a measure of spread to show the amount of variation.
- You can compare two data sets using the range and an average.

The mean rainfall in 2000 in Sandy Town was 14 cm. The range was 52 cm.
This tells you that on average there is less rainfall per month in Sandy Town than Beach Town, but the variation in the amount of rain is greater.

Exercise D3.4

1 Find the mean, median, mode and range of these groups of numbers.

a 6 13 4 12 10 6 8
b 15 16 17 18 19 16 14
c 2 3 2 4 2 5 2 7 3
d 29 27 20 22 27 25

2 Twelve friends, five girls and seven boys, compared how much pocket money they each received in a month.

The amounts for the girls were: £24 £20 £60 £32 £20.

a Calculate the range of the amounts.
b Calculate the mean, median and model amount of pocket money.
c Which of these measures of average is the most appropriate to use?
d Give a reason for your answer.

The amounts for the boys were: £24 £20 £30 £28 £30 £22 £18.

e Calculate the range and the median of these amounts.
f Comment on the differences and similarities between the amount of pocket money received each month by this group of boys and girls.

3 Eight chocolate bars are weighed and their weights recorded to the nearest gram:

63 61 67 68 65 65 64 66

Follow these steps to find the mean weight of the bars of chocolate and fill in the boxes.

- Choose an **assumed mean** of 60.
- Subtract 60 from each data value:
  3, 1, 7, 8, 5, 5, 4, 6
- Find the mean of these data:
  $3 + 1 + 7 + 8 + 5 + 5 + 4 + 6 = \square$
  $39 \div \square = 4.875$
- Add to the assumed mean:
  $\square + 60 = 64.875$
- The mean is 64.875g

The manufacturer of the chocolate claims that the average weight is 65 grams.
Is the manufacturer justified in making this claim?

4 Kilogram bags of flour were weighed and their weights recorded to the nearest gram.

1008 1003 998 999 1000 1002 1007 996 1003 1005.

Use an assumed mean of 1000 to calculate the mean average weight of the bags of flour.

D3.5 Communicating results

This spread will show you how to:

- Communicate methods and results.
- Write a short report of a statistical enquiry, illustrated with appropriate diagrams, graphs and charts. Justify the choice of presentation.

KEYWORDS
Appropriate
Statistical enquiry
Summary

Becci collected data on Beach Town to show why Beach Town would be a good place to invest in. She compiled graphs to represent the data and calculated statistics.
Becci produced a report of her findings of Beach Town to present to Richie Vanson.

A statistical report should:

- specify the reason for the enquiry
- describe how the data was collected
- illustrate the data with appropriate diagrams
- calculate relevant statistics to summarise the data
- interpret the findings and justify the results
- consider whether any further enquiry is needed.

Here is a section of Becci's report.

Becci describes how the data were collected and illustrated the data with an appropriate diagram.

She calculates statistics to summarise the data. She uses the median as it is lower than the mean average so that the water appears to be better quality.

Becci has interpreted her findings, but has not said which ranges of bacteria colonies/100 ml would define water as excellent, good, or poor. She has not said how many were good and how many were excellent or how close the value 1600 is to being classified as poor water quality.

Becci has collected data for 1997 which is a long time ago. You need to question why she is not using data that is more up to date.

Bathing water quality was tested every two weeks from May to September in 1997 by taking a sample of sea water and counting the number of bacteria colonies/100 ml.

Date	12/5	26/5	9/6	23/6	7/7	21/7	4/8	18/8	1/9	15/9
Bacteria 100ml	50	200	40	160	55	450	950	1600	25	260

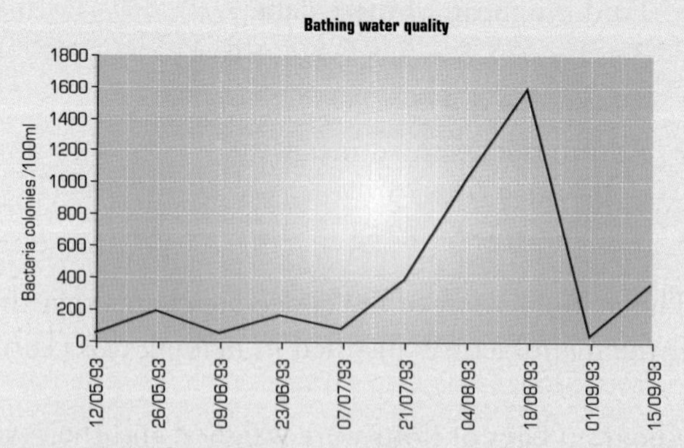

The range in bacteria/100 ml is 1575.
The median amount of bacteria/100 ml is 180.
Water quality is defined as excellent, good or poor.
The water quality in Beach Town was excellent or good throughout 1997.

Exercise D3.5

Richie Vanson decided to collect his own data about the visitors from the United Kingdom and overseas to Beach Town and Sandy Town in 2001.
The data he collected are given in these tables.

Beach Town

Season Visited	UK %	Overseas %
Jan-March	13	18
Apr-June	26	27
July-Sept	42	36
Oct-Dec	19	19

Age (Years)	UK %	Overseas %
Under 15	7	10
15–24	13	11
25–34	20	18
35–44	24	21
45–54	14	21
55–64	12	11
65+	10	6

	UK %	Overseas %
Visits (thousands)	3.7	0.52
Nights spent (thousands)	13.8	4.5
Spending (thousands)	557	198

Day Visits: 292 000, spending £14 560 000

Sandy Town

Season Visited	UK %	Overseas %
Jan-March	12	14
Apr-June	28	29
July-Sept	41	40
Oct-Dec	19	17

Age (Years)	UK %	Overseas %
Under 15	9	9
15–24	18	12
25–34	21	17
35–44	24	21
45–54	16	20
55–64	8	13
65+	4	7

	UK %	Overseas %
Visits (thousands)	4.2	0.26
Nights spent (thousands)	22.2	2.0
Spending (thousands)	866	77

Day Visits: 188 000, spending £7 340 000

Number of tourists and spending in 2001 for the whole of the United Kingdom

	UK	Overseas
Visits (millions of people)	175.4	25.2
Nights spent (millions)	576.4	203.8
Spending (£ millions)	26 132	12 672

Use the data to decide which resort, Beach Town or Sandy Town, Richie Vanson should invest in to build a theme park.
Calculate statistics and draw suitable diagrams to support your decision.
Comment also on whether sufficient information has been collected and/or whether the data are useful.

Summary

You should know how to ...

1 Compare two distributions using the range and one or more of the mean, median and mode.

2 Construct pie charts for categorical data.

3 Construct bar charts and frequency diagrams for discrete and continuous data.

4 Construct simple line graphs for time series.

5 Identify which diagrams are most useful in the context of the problem.

Check out

1 A survey on the average amount of time, in hours, spent on homework each week was carried out.
The results were:
Boys: 2, 4, 3, 5, 6, 5, 6, 7, 5, 3, 7, 6
Girls: 7, 5, 5, 5, 6, 8, 10, 6, 7, 10, 9, 6

Calculate the range and the mean and comment on the differences and similarities shown between the homework habits of the boys and the girls in this survey.

2 Construct a pie chart to display the results for the girls in question 1.

3 Construct:

a a comparative bar chart

b a composite bar chart

to display the results for both boys and girls in question 1.

4 One of the boys chose to compile a match report of the times when goals were scored in the premier league one Saturday.

Time minutes	Frequency
0–15	3
15–30	2
30–45	7
45–60	4
60–75	6
75–90	5

Draw a time series graph to represent this data.

5 For the bar charts in question 3, which is the most useful for comparing the homework habits between boys and girls?

4 Probability experiments

This unit will show you how to:

- Find and record all possible mutually exclusive outcomes for single events and two successive events in a systematic way, using diagrams and tables.
- Estimate probabilities from experimental data.
- Understand that if an experiment is repeated there may be, and usually will be, different outcomes.
- Understand that increasing the number of times an experiment is repeated generally leads to better estimates of probability.
- Compare experimental and theoretical probabilities in different contexts.
- Solve problems and investigate in the context of probability.

Outcomes like this are not impossible but they are highly improbable.

Before you start

You should know how to ...

1 Record outcomes for single event.

2 Find simple probabilities.

3 Construct a frequency table.

Check in

An eight sided spinner is spun. It has 3 orange sides (O), 1 red side (R) and 4 green sides (G). Joel spins the spinner once.

1 What colours could the spinner land on?

2 What is the probability that the spinner lands on
a orange **b** red or green?

3 The spinner is spun 12 times. It lands on:
G, G, R, O, O, G, O, G, O, R, G, G
Draw a frequency table to illustrate the colours the spinner lands on.

Theoretical probability

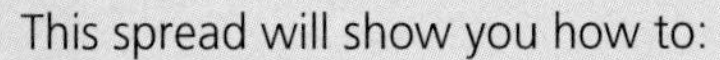

This spread will show you how to:
- Find and record outcomes for two events in a systematic way.
- Solve problems involving probability.

KEYWORDS
Outcome
Probability
Sample space diagram

Before you can calculate probabilities you need to know all the possible outcomes.

example

At a fete there are two silver money trees.
There are five envelopes left on each tree.
The amounts in each of the envelopes on each tree are: 50p, 25p, 20p, 10p and 5p.

You pay 25p each time you choose an envelope.
Reuben chooses two envelopes, one from each tree.
Will Reuben receive more money than he paid?

First list all possible outcomes.
When there are two events, one way to show the outcomes is in a **sample space diagram**.

	5p	10p	20p	25p	50p
5p	10p	15p	25p	30p	55p
10p	15p	20p	30p	35p	60p
20p	25p	30p	40p	45p	70p
25p	30p	35p	45p	50p	75p
50p	55p	60p	70p	75p	100p

The list of possible amounts that he could get, 10p, 15p, 20p, ..., is called the set of **all possible outcomes**.

Reuben paid a total of 50p and chose two envelopes, one from each of the two money trees.
There are 25 possible outcomes. Nine of these would give a total amount greater than 50p.

The probability that Reuben would receive more than he paid is $\frac{9}{25}$.
This is less than an even chance. He would not be wise to play.

To solve problems involving probability you need to:
- identify the different equally likely outcomes
- count how many of these are favourable.

Exercise D4.1

1 Dave has two packs of ordinary playing cards.
He chooses one card from each pack.
What is the probability that he has chosen

a a club and a diamond

b two red cards?

2 Joshua is playing a game with a fair six-sided dice.
If he throws an odd number or a prime number he scores a point.
What is the probability that he scores a point on the next throw?

3 A coin is flipped twice. What the probability that a tail will appear at least once?

4 Holly and Ivy are playing a game of Beetle Drive.

(To play Beetle Drive you need to throw a 6 to draw the body, 5 for the head, 4 for each antenna, 3 for each of the legs, 2 for each eye and a 1 for the mouth. You have to draw the body first. You cannot add mouth, eyes or antennae without a head. The first complete beetle wins.)

Holly's first six throws are 6, 5, 5, 3, 3, 3.

a What is the probability that she throws a three on her next throw?
Explain your answer.

Ivy throws 2, 3, 1, 4, 6, 5 on her first six throws.

b How much of her beetle has she been able to draw? Explain your answer.

c After their first six throws who is furthest ahead in the game, Holly or Ivy?

5 Simi has two dice.

- Dice A has the numbers 1, $^{-}2$, 3, $^{-}4$, 5, 6 on its faces.
- Dice B has the numbers 1, 2, $^{-}3$, 4, 5, 6 on its faces.

Simi throws both dice and finds the sum of the numbers shown.

a Draw a sample space diagram to show all the possible outcomes.

b What is the probability that the sum is zero?

c What is the probability that the sum is negative?

D4.2 Experimental probability

This spread will show you how to:

- Estimate probabilities based on experimental data.

KEYWORDS

Estimate

Experiment

Outcome

If a probability is unknown, you may be able to carry out an experiment to estimate it:

$$\text{Estimated probability of an event} = \frac{\text{number of times the event occurs}}{\text{number of trials in the experiment}}$$

A tube of sweets has different coloured lids.
The colours could be red, blue, green, orange or purple.
The manufacturers of the sweets do not produce equal amounts of each colour.

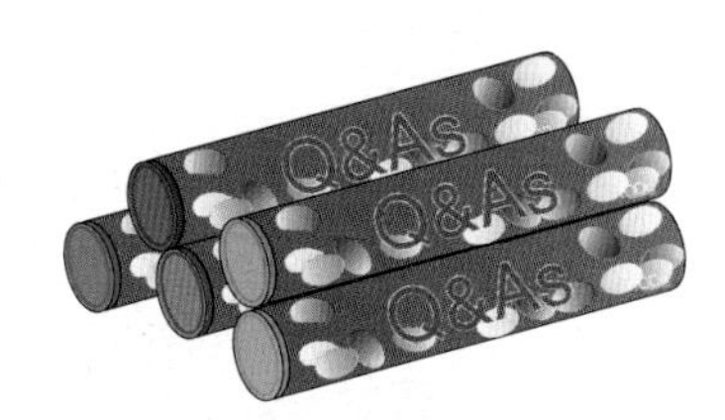

Alfie wanted to find out the probability of choosing a tube with a green lid.

He decided to carry out an experiment.

His friend Freddie suggested they buy tubes of sweets and note the colour on each lid.
Alfie bought 30 tubes of sweets and Freddie bought 50 tubes.
Their results are given in the frequency table.

Colour	Alfie	Freddie
Red	14	9
Blue	8	12
Green	1	8
Orange	5	14
Purple	2	7

Alfie said that he would be unlikely to get a green lid as it has a probability of $\frac{1}{30}$.

Freddie said that he was more likely to get a green lid as it has a probability of $\frac{8}{50}$.

Their friend Bertie combined their results and said that the probability of a green lid is $\frac{9}{80}$.

Bertie suggested that they should carry out the experiment many times and combine all the results to get the best estimate.

- **Repeating an experiment leads to better estimates of probability.**

Exercise D4.2

1 In the game 'Paper, Scissors, Stone':

- Scissors cut paper so scissors beats paper.
- Stone blunts scissors so stone beats scissors.
- Stones can be wrapped in paper so paper beats stone.
- Other combinations produce a draw.

Robert and Luke play the game 10 times. Here are their results.

Game	1	2	3	4	5	6	7	8	9	10
Robert	scissors	stone	scissors	stone	paper	paper	scissors	stone	scissors	paper
Luke	stone	scissors	paper	stone	scissors	paper	stone	scissors	paper	stone

Use their results to:

a Estimate the probability that the next game will result in a draw.
b Estimate the probability that Robert will win the next game.

2 In the first hour at a school fete 40 people play on the 'roll a penny' stall.
Six people win a prize.
It costs 10p to play and the prize is 50p.

a Estimate the probability of winning a prize.
b Does the stall make money, lose money or break even in the first hour?

3 Sarah and Alex are trying to find out if an old coin they have is biased.
Sarah flips the coin 400 times. Alex flips the coin 100 times.

	Head	Tail
Sarah	276	124
Alex	75	25

a Estimate the probability of obtaining a Tail using Sarah's results.
b Estimate the probability of obtaining a Tail using Alex's results.
c Which of your answers, **a** or **b**, is more reliable?
Give a reason for your answer.
d Without throwing the coin any more times suggest how you might find a better estimate of the probability of obtaining a tail.
Write down this probability.
e Comment on whether you think the coin is biased.

4 Rishma has a large jar of blue (B), green (G) and orange (O) marbles. She picks ten at random from the jar and then returns them to the jar.
On her first go she picks B, B, G, O, B, O, B, G, O, B

a Estimate the probability of obtaining a green marble.

On her second go she picks B, O, B, G, G, G, O, G, B, B

b Use all 20 marbles to estimate the probability of obtaining a green marble.
c If there are 1000 marbles altogether, estimate how many are green.

D4.3 Comparing experiment with theory

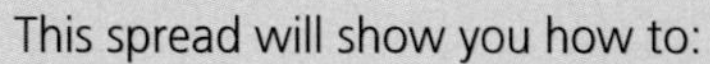

This spread will show you how to:

- Estimate probabilities based on experimental data and use relative frequency as an estimate of probability.
- Compare experimental and theoretical probabilities.

KEYWORDS

Bias
Experiment
Fair
Outcome
Theoretical probability

Edina and Patsy each make a three-sided spinner using card and a pencil.
Each sector is exactly the same shape and size.
Each spinner is equally likely to land on any one of the sectors.

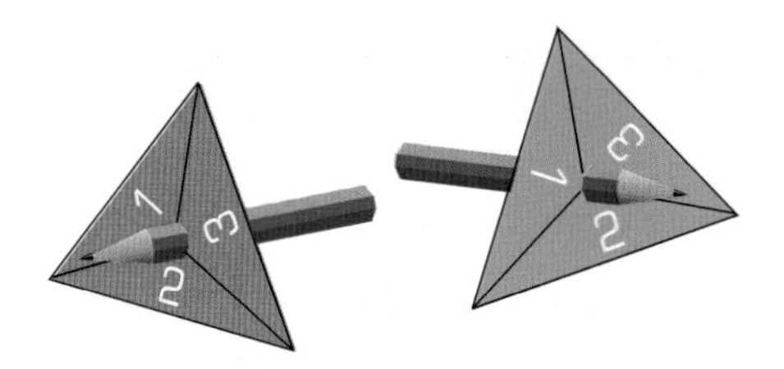

They draw a table to show the possible outcomes if they add the scores shown on both spinners.

	1	**2**	**3**
1	2	3	4
2	3	4	5
3	4	5	6

In theory, the probability of a total score of 5 is $\frac{2}{9}$ or 0.222 ...

In theory, the probability of getting a score of 2 is the same as for a score of 6.

Edina and Patsy experiment to find out if their spinners are fair.
They each spin them 100 times adding up their scores each time.
Here are their results:

Score	Frequency
2	9
3	17
4	38
5	23
6	13

Edina says the probability of getting a score of 5 is $\frac{23}{100}$ or 0.23, and this is very close to the theoretical probability 0.222 ... so the spinners must be fair.

Patsy says that the chance of getting a score of 2 is not the same as getting a score of 6, so the spinners must be **biased**.

- Experimental and theoretical probabilities are not always exactly the same, but if there is no bias the probabilities should be close.

Exercise D4.3

1 **a** Throw a dice 10 times and record the numbers that you get in a table.

Dice number	Tally	Frequency
1		
2		
3		
4		
5		
6		

Comment on whether or not you think your dice is fair.

b Throw the dice 10 more times and record the numbers. You can extend your table. Has the same set of numbers been thrown as before? Comment on your answer.

c Write how many of each number you would expect if you threw the dice 30 times.

d Throw the dice a further 10 times and record the numbers, extending the table again. Combine the results you have obtained from all 30 throws of the dice in a separate table. Compare this experimental data with the amount of each number that you expected to get. Do you think your dice is fair?
Comment on your answer.

e Compare your 30 results with 30 results from another student.
Are they the same or similar or different or very different?
Comment on your answer.

f Write down the theoretical probabilities of obtaining each number on a dice.

g Combine your 30 results with 30 results of another student.
Use these 60 results to calculate an estimate of the probability of getting each number.
You can use a table like this:

Dice number	Frequency	Estimated probability = Frequency ÷ 60
1		
2		
3		
4		
5		
6		
	Total = 60	Total =

Estimated probability is also called **relative frequency**.

Compare your experimental probabilities with the theoretical probabilities.
Comment on anything you notice.

Summary

You should know how to ...

1 Find and record all possible mutually exclusive outcomes for two successive events in a systematic way.

2 Estimate probabilities from experimental data.

Check out

1 Abigail had a jelly and ice cream party.
There were:
- strawberry
- lime or
- orange flavour jellies

There were:
- vanilla
- chocolate
- strawberry and
- toffee-fudge flavour ice creams.

Find all the possible combinations of jelly and ice cream you could have at Abigail's party.

2 A coin is thrown 10 times and the results recorded.

Throw	1	2	3	4	5
Result	H	T	H	T	T

Throw	6	7	8	9	10
Result	T	H	H	H	H

Estimate the probability of throwing a head based on

a the first five throws

b all ten throws

Which is the best estimate and why?

5 Equations and graphs

This unit will show you how to:

- Begin to distinguish the different roles played by the letter symbols in equations, formulae and functions.
- Construct and solve linear equations with integer coefficients.
- Use formulae from mathematics and other subjects.
- Derive simple formulae.
- Multiply a single term over a bracket.
- Generate terms of a linear sequence using position-to-term difinitions.
- Recognise that equations of the form $y = mx + c$ correspond to straight-line graphs.
- Construct linear functions arising from real-life problems and plot their graphs.
- Discuss and interpret graphs arising from real situations.
- Solve word problems in the context of algebra.
- Identify the necessary information to solve a problem.
- Represent problems and interpret solutions in algebraic or graphical form.

Reading a graph can be quicker and easier than making a calculation.

Before you start

You should know how to ...

1 Substitute integers into simple formulae.

2 Expand single brackets.

3 Appreciate position-to-term rules for sequences, especially of linear sequences.

Check in

1 Evaluate the following when $x = 2$:

a $3x^2$ **b** $\frac{10 - 2x}{2}$ **c** $\frac{8x}{4} + 1$

2 Expand and simplify if possible:

a $3(x + 4) + 4$

b $x(x + 3) + 10$

3 **a** Write an expression for the nth term of:

i 6, 9, 12, 15, 18, ... **ii** 4, 13, 22, 31, 40, ...

b Generate the first five terms of:

i $7n - 2$ **ii** $10 - 3n$

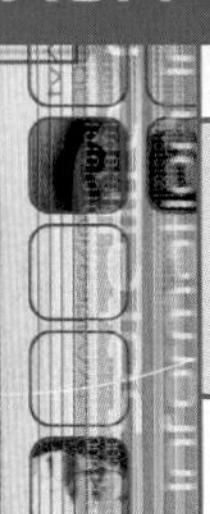

A5.1 Solving linear equations

This spread will show you how to:

- Solve linear equations by transforming both sides in the same way.

KEYWORDS
Cross-multiply
Solve
Equation
Fraction

Sometimes equations are made up of fractions.

For example: $\frac{x+3}{7} = \frac{x-5}{5}$

Fractions can be difficult to calculate with, so it is best to remove them.

You can do this by **cross-multiplying**:

$$\frac{x+3}{7} = \frac{x-5}{5}$$

$$5 \times (x+3) = 7 \times (x-5)$$

The ÷5 has 'become' a ×5   The ÷7 has 'become' a ×7

- **You can remove fractions in an equation by cross-multiplying. You multiply both sides by the denominator, or denominators.**

example

Solve the equation $\frac{x+4}{7} = \frac{x-4}{3}$ by first removing the fractions.

Cross-multiply:	$3(x+4) = 7(x-4)$
Expand the brackets:	$3x + 12 = 7x - 28$
Collect like terms:	$12 + 28 = 7x - 3x$
Simplify:	$40 = 4x$
Divide by 4:	$10 = x$

Note: you don't need to write the × sign.

The solution is $x = 10$

The same rules apply to algebraic denominators.

example

Solve the equations:

a $\frac{4}{x-2} = 8$ **b** $\frac{p+2}{p-6} = \frac{30}{6}$

a

$$4 = 8(x-2)$$
$$4 = 8x - 16$$
$$4 + 16 = 8x$$
$$20 = 8x$$
$$x = \tfrac{20}{8} = 2\tfrac{1}{2}$$

b

$$6(p+2) = 30(p-6)$$
$$6p + 12 = 30p - 180$$
$$12 + 180 = 30p - 6p$$
$$192 = 24p$$
$$p = 192 \div 24 = 8$$

Remember to use brackets.

Exercise A5.1

1 Solve these equations:

a $\dfrac{2(x-7)}{5} = 4$ **b** $\dfrac{4x-7}{9} + 2 = 11$

c $20 - \dfrac{x}{4} = 8$ **d** $\dfrac{x-23}{^-7} = 2$

2 Solve the following equations:

a $\dfrac{x+5}{4} = \dfrac{x+20}{7}$ **b** $\dfrac{8}{y+2} = \dfrac{24}{y+4}$

c $\dfrac{16}{z+1} = 2$ **d** $\dfrac{p+4}{p-8} = \dfrac{30}{6}$

3 Spot the odd one out!
Which equation has a different solution to the other two?

a $3x+11 = 6x-13$

b $20-2x = 3x-15$

c $\dfrac{x-2}{3} = \dfrac{2x+4}{10}$

4 In the flowchart, you can take either route from w and you will get the same solution.
What number do you need to begin with?

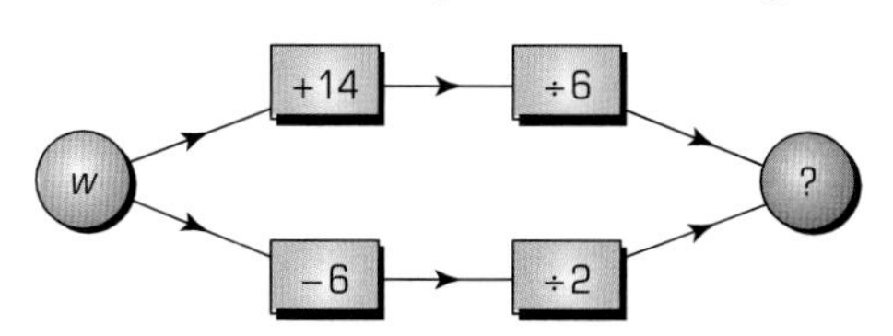

5 What value of x would make the area of the triangle the same as that of the splodge?

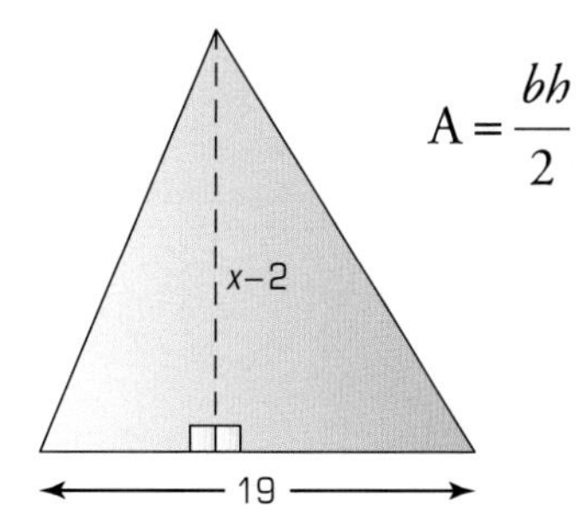

$A = \dfrac{bh}{2}$

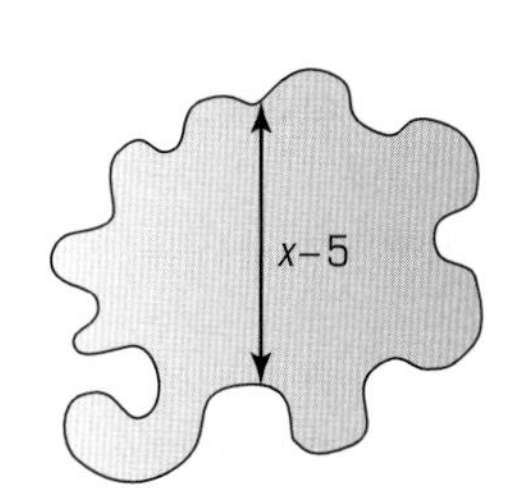

$A = \dfrac{h}{3}$

A5.2 Using formulae

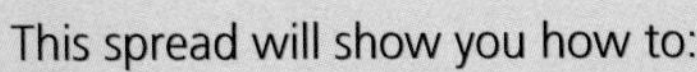

This spread will show you how to:

- Understand and use inverse operations.
- Explain the meaning of and substitute numbers into formulae.
- Derive algebraic formulae.

KEYWORDS

Formula Subject
Rearrange Substitute

A formula is a statement that links variables.

example

In the rectangle, x is the length and y is the width.

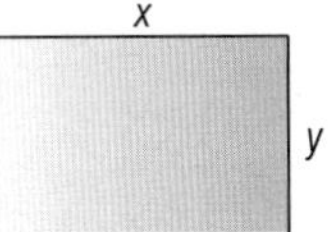

a Derive a formula for the perimeter, P.
b If $x = 5$ cm and $y = 2$ cm, find P.
c If $y = 3$ cm and $P = 16$ cm, find x.

a Perimeter $= x + y + x + y = 2x + 2y$
$P = 2x + 2y$
b $P = 2 \times 5 \text{ cm} + 2 \times 2 \text{ cm} = 14 \text{ cm}$
c $16 = 2x + 2 \times 3$
$16 = 2x + 6$ so $2x = 10$
$x = 5$ cm

In a formula, there is usually a single letter on the left-hand side.
This letter is called the **subject of the formula**.
In $P = 2x + 2y$, P is the subject. In $A = \frac{1}{2}bh$, A is the subject.

- **You can change the subject of a formula by rearranging it.**

example

This formula connects temperature in degrees Celsius with degrees Fahrenheit.

$$C = 5\left(\frac{F - 32}{9}\right)$$

a Make F the subject of the formula.
b On a warm day in Ibiza, the temperature is 30°C.
Find the temperature in Fahrenheit.

a
$$C = 5\left(\frac{F - 32}{9}\right)$$
$$9C = 5(F - 32)$$
$$\frac{9C}{5} = F - 32$$
$$\frac{9C}{5} + 32 = F$$

b $C = 30$
$F = \frac{9C}{5} + 32$
$F = 9 \times \frac{30}{5} + 32$
$= \frac{270}{5} + 32 \quad = 54 + 32$
$= 86$ The temperature is 86°F

Exercise A5.2

1 Use the given formula to find the unknown values:

a This is the formula for change (C) from a £50 note when b books, at £9 each, are bought.

$C = 50 - 9b$

Find the change when 4 books are purchased.

b This is the formula for finding the number of matches (m) in a triangular array with R rows.

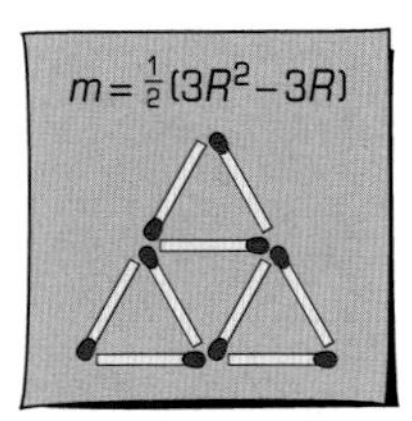

Find m when $R = 4$.

c This is Hero's formula for finding the area, A, of any triangle with sides a, b and c.

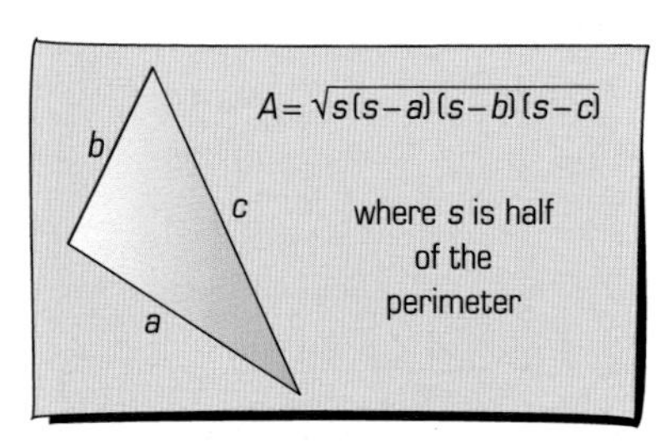

Find the area of a triangle with sides 5 cm, 10 cm and 14 cm.

2 If $p = 5$, find the value of the subject of these formulae:

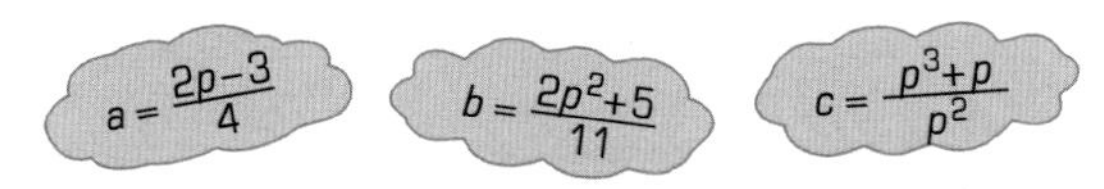

3 The formula for finding the area of a circle with radius r is given by $A = \frac{22r^2}{7}$.

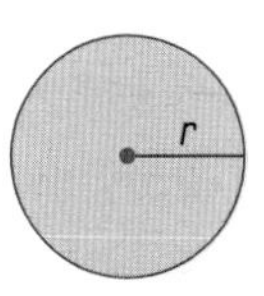

a Find the value of A when $r = 10$.

b Rearrange the formula to make r the subject.

c Find, using your rearranged formula, the value of r when $A = 500$.

4 A regular polygon has n sides.

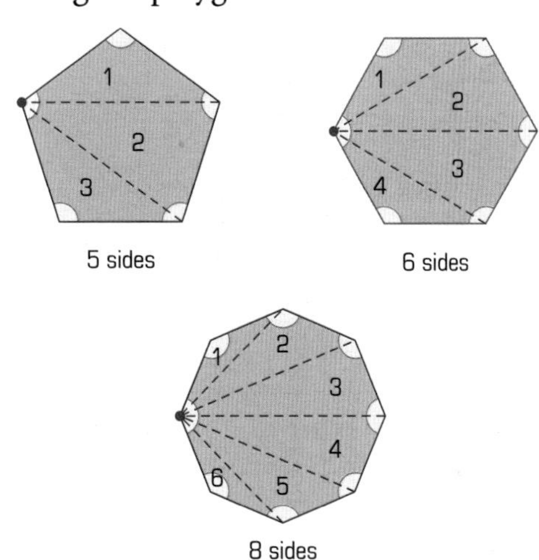

Find a formula to connect the number of sides with the:

a number of diagonals (D) from a vertex

b number of triangles (T) the polygon is split into when all the diagonals are drawn from one vertex

c total sum of the angles (A) inside the polygon.

Hint: Each triangle has 180° inside.

A5.3 Brackets and negatives

This spread will show you how to:
- Simplify expressions by collecting like terms.
- Begin to multiply a single term over a bracket.

KEYWORDS
Equation, Simplify, Expand, Solve, Expression

Brackets tell you the order of operations in an expression:
$3(x + 6)$ means you multiply everything inside the bracket by 3.

So $3(x + 6) = 3 \times x + 3 \times 6$
$= 3x + 18$

Removing a bracket is called **expanding**.

You need to be careful if you multiply a bracket by a negative number.

example

Expand the expression $^{-}3(x + 6)$.

Multiply by $^{-}3$: $^{-}3(x + 6) = {}^{-}3 \times x + {}^{-}3 \times 6$

A plus and a minus together make a minus $= {}^{-}3x + {}^{-}18$

$= {}^{-}3x - 18$

▶ **A negative outside a bracket changes the signs of the terms inside.**

You may need to expand brackets in order to solve an equation.

example

Solve the equation $8(y - 4) - 3(y - 12) = 19$ by first expanding the brackets.

First simplify the left-hand side: $8(y - 4) - 3(y - 12) = 8y - 32 - 3y + 36$

Collect like terms $= 8y - 3y - 32 + 36$

$= 5y + 4$

Write the simplified equation: $5y + 4 = 19$

$5y = 15$

$y = 3$

Sometimes you can have a minus on its own outside a bracket. This is really a $^{-}1$.

example

Find the value of the missing piece in the diagram.

$? = (5x + 3) - (2x - 6)$

$? = 5x + 3 - 2x + 6$

$? = 5x - 2x + 3 + 6$

$? = 3x + 9$

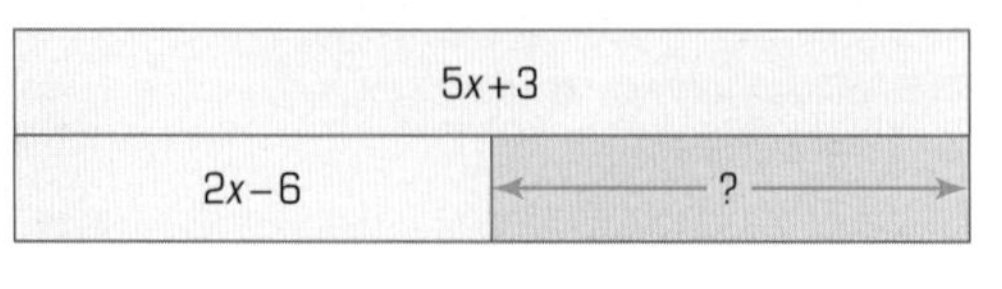

Remember to check: $2x - 6 + 3x + 9 = 5x + 3$ ✓

Exercise A5.3

1 Expand the following

a $^-5(x+2)$
b $^-3(y-2)$
c $^-8(11-z)$
d $^-4x(x-2)$

2 Show that the following expressions are all equivalent to $3p-30$:

a $6(p-4)-3(p+2)$
b $8(p+5)-5(p+14)$
c $10(p-10)-7(p-10)$

3 Write down expressions for the unknown pieces:

a

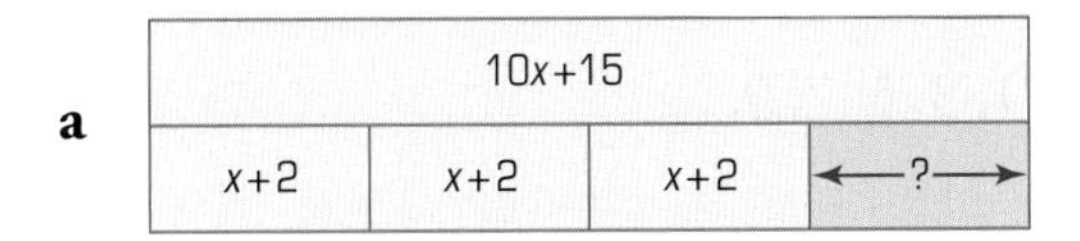

b

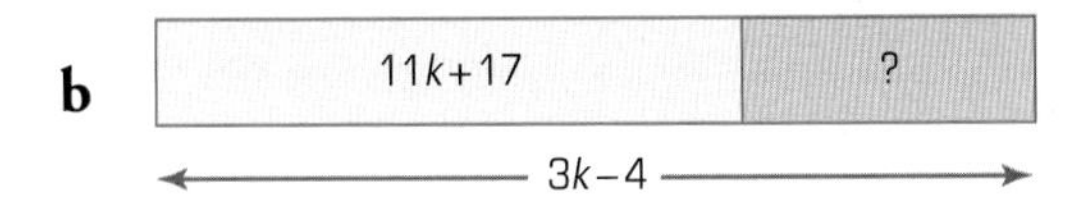

4 a

A

B

Write an expression for the nth term of each of the sequences above.

b Find an expression for the difference between the nth term of sequence *A* and the nth term of sequence *B*.

5 Solve the following equations, putting your solutions in ascending order:

a $5(x-4)-3(x+2)=15$
b $8(y+5)-2(y-3)=9$
c $9z(z+3)-3(3z^2-2)=11$

6 a Prove that the difference between a multiple of 4 and a multiple of 6 is always an even number.

b Prove that the difference between a number, plus 2, doubled and a number, plus 1 and multiplied by 4 is always an even number.

7 A father and son have their heights measured in metres.

Find the difference in their heights.

8 Show that the differences between the two sets of diagonal cells are equal.

$3(p-2)$	$p-30$
$6(p+4)$	$8(p+6)$

9 Copy and complete:
$6(p-4)-?(?-?p)=9p-30$

10 The expressions on each stepping stone shows the distance from shore A. If each jump is 6 metres, what is the value of x on each step?

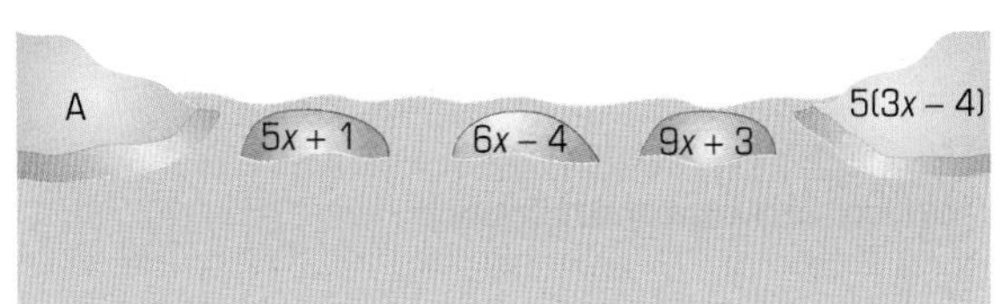

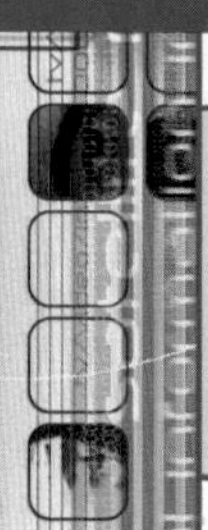

A5.4 Generating sequences

This spread will show you how to:

- Generate a sequence given a rule for finding each term from its position in the sequence.
- Begin to generate a quadratic sequence.

KEYWORDS

Difference, Sequence, Linear, Term, Quadratic

A sequence is a set of numbers that follow a rule.

Remember that a sequence is linear if each term differs from the previous one by a constant amount.

These hut numbers form a linear sequence

example

a Find a position-to-term rule for the sequence 1, 3, 5, 7, 9, ...

b What is the 85th term of this sequence?

a The sequence 1, 3, 5, 7, 9, ... is linear.
It goes up in 2s, so look at the 2 times table.

Position, n	1	2	3	4	5
Term $T(n)$	1	3	5	7	9
2 × table	2	4	6	8	10

Each term is 1 less than the corresponding multiple of 2.

The rule is: 'multiply n by 2 then subtract 1'.
Using algebra, you can write this as: $T(n) = 2n - 1$

b The 85th term of the sequence is given by $n = 85$.

$T(85) = 2 \times 85 - 1$
$= 169$

Another sequence you know is the square numbers, $T(n) = n^2$:

Position, n	1	2	3	4	5	6
Term, $T(n)$	1	4	9	16	25	36
You can look at the differences:	3	5	7	9	11	
Now look at the **second** difference:	2	2	2	2		

For the square numbers, the second difference is constant.
The square numbers form a **quadratic sequence**.

▶ A sequence in which the second difference is constant is called a quadratic sequence. It includes an n^2 term.

Exercise A5.4

1 Find a position-to-term rule for these linear sequences:

a 13, 15, 17, 19, 21, ...

b 16, 29, 42, 55, 68, ...

c $^-21$, $^-18$, $^-15$, $^-12$, $^-9$, ...

d 19.5, 17, 14.5, 12, 9.5, ...

2 Generate the quadratic sequences described by these function boxes:

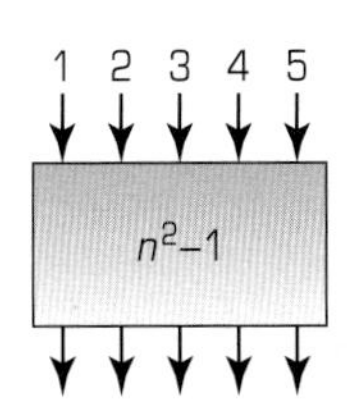

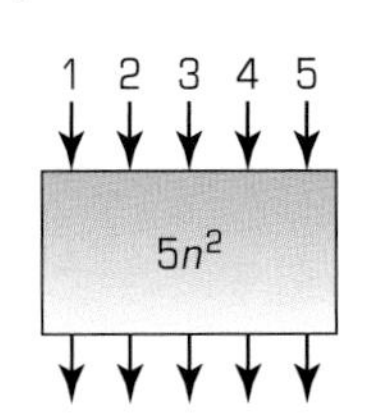

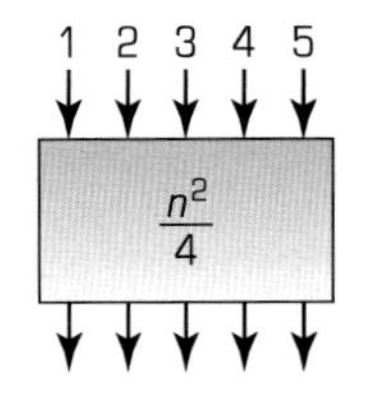

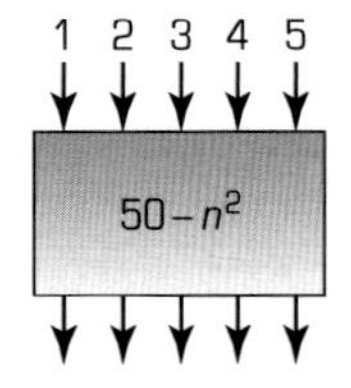

3 Write down the first five terms of these sequences:

$T(n) = n^2 + 5$

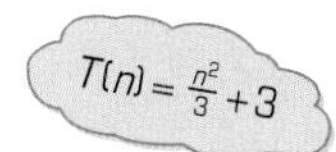

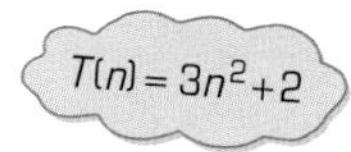

4 Find the missing function from these boxes:

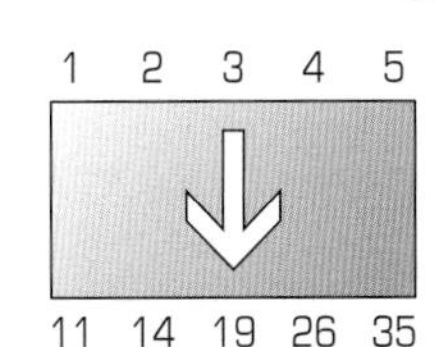

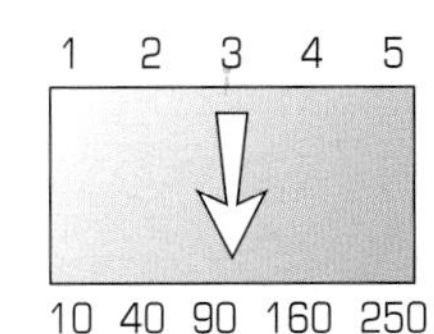

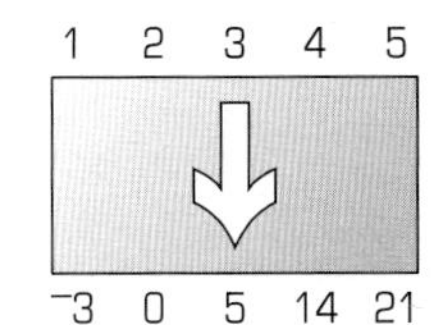

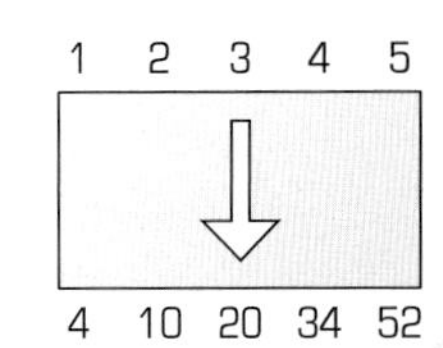

5 Find a formula for $T(n)$, the general term of each sequence, in terms of n, the term number.

a 4, 7, 12, 19, 28, ...

b 4, 16, 36, 64, 100, ...

c $^-2$, 1, 6, 13, 22, ...

d 2, 6, 12, 20, 30, ...

6 **a** Copy and complete the table below:

Sequence	Value of second difference	Formula
2, 8, 18, 32, 50, ...		
3, 12, 27, 48, 75, ...		
4, 16, 36, 64, 100, ...		
5, 20, 45, 80, 150, ...		

b What do you notice?

c Use your observation to find a formula for 9, 36, 81, 144, 225, ...

A5.5 Spot the sequence

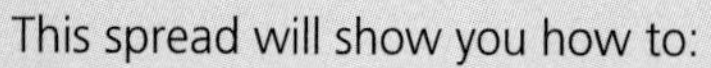

This spread will show you how to:

- Generate sequences from practical contexts.
- Describe the general term, and justify the generalisation by referring to the context.

KEYWORDS

Cube
Even
Integer
Odd
Sequence
Square
Triangular

There are different ways to describe a sequence such as 3, 5, 7, 9, ...
You can use:

- Words: 'double the position number and add 1'
- Algebra: $T(n) = 2n + 1$
- Pictures:

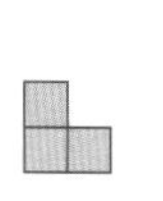
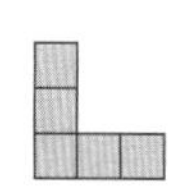
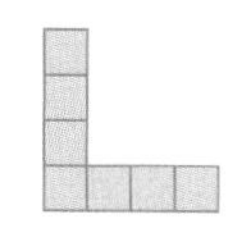
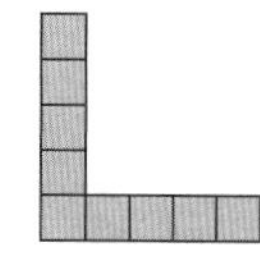

There are certain sequences that you should be familiar with.

Square numbers

1, 4, 9, 16, 25, ... ➡ $T(n) = n^2$

A number multiplied by itself, ($n \times n$) or

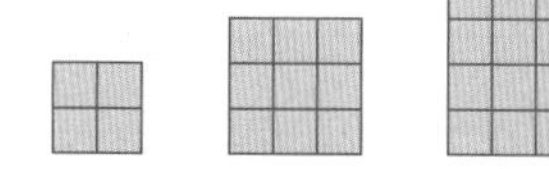

the sum of n odd numbers

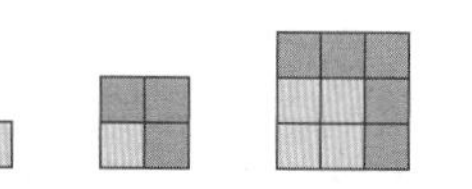

Cube numbers

1, 8, 27, 64, 125, ... ➡ $T(n) = n^3$

a number multiplied by itself three times ($n \times n \times n$)

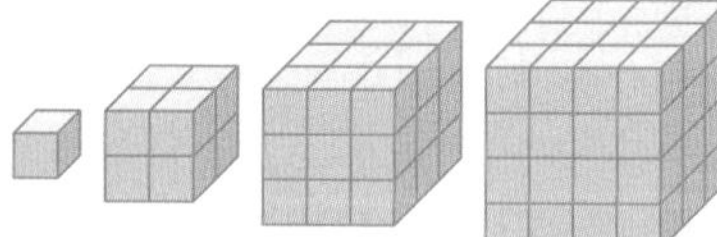
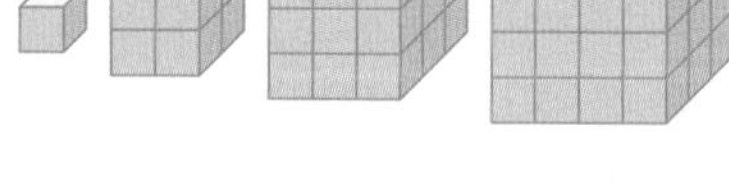

Even numbers

2, 4, 6, 8, 10, ... ➡ $T(n) = 2n$

numbers that consist of pairs

Odd numbers

1, 3, 5, 7, 9, ... ➡ $T(n) = 2n - 1$

numbers that, when paired, one is left out

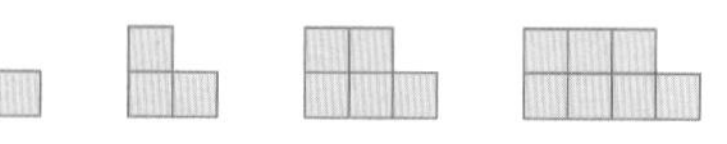

Triangular numbers

1, 3, 6, 10, 15 ... ➡ $T(n) = \frac{n(n+1)}{2}$

the sum of n positive integers

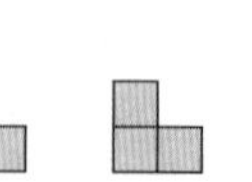
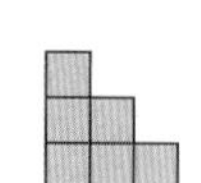
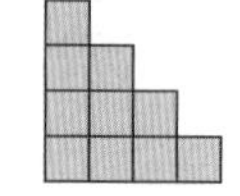

Exercise A5.5

1 Explain why the formulae fit these diagrams:

a 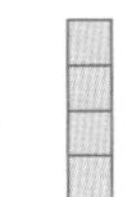 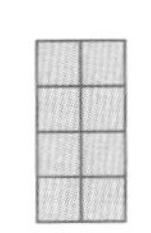 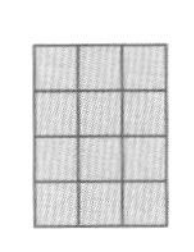 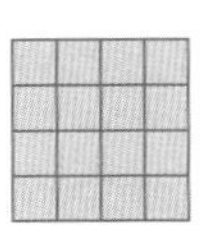 $T(n) = 4n$

b 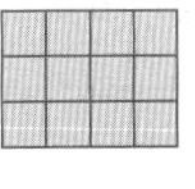 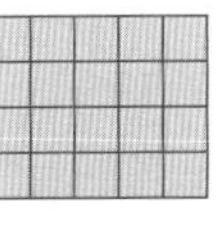 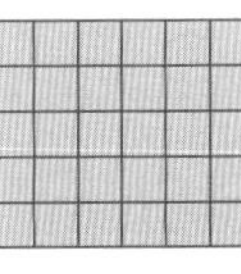 $T(n) = (n+1)(n+2)$

c 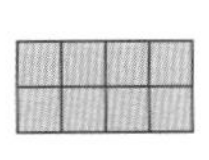 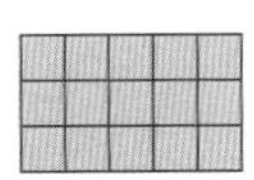 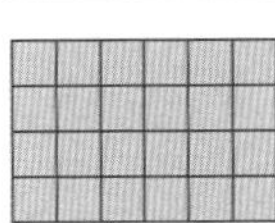 $T(n) = n^2 + 2n$

2 There are two formulae that can be used to describe this pattern:
$T(n) = n^2 + 3n$ and $T(n) = n(n+3)$.

 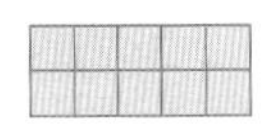 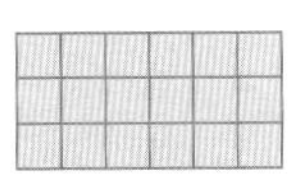 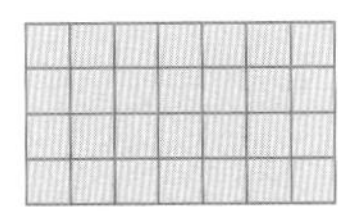

a Explain why each formula works.
b Show that each formula is exactly the same.

3 Draw diagrams to represent the following:
a Multiples of 6
b $T(n) = n^2 + 1$
c $T(n) = (n+1)^2$
d $T(n) = (n+1)^3$

4 Staircases come in 'up' varieties and 'up and down' varieties.

Up and down staircases:

 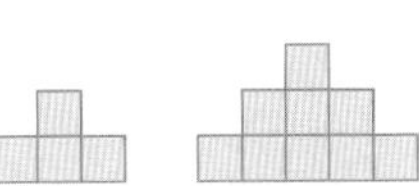 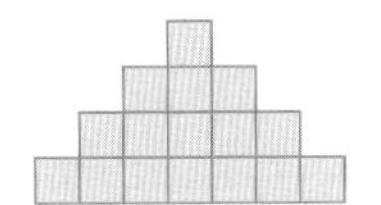

Up staircases:

 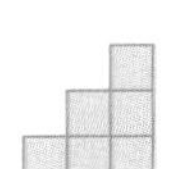 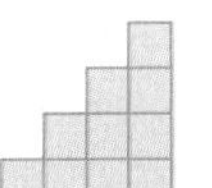

How many blocks are needed to make a staircase 100 steps high of:
a the up and down type
b the up type?

5 Here is a sequence of tile patterns:

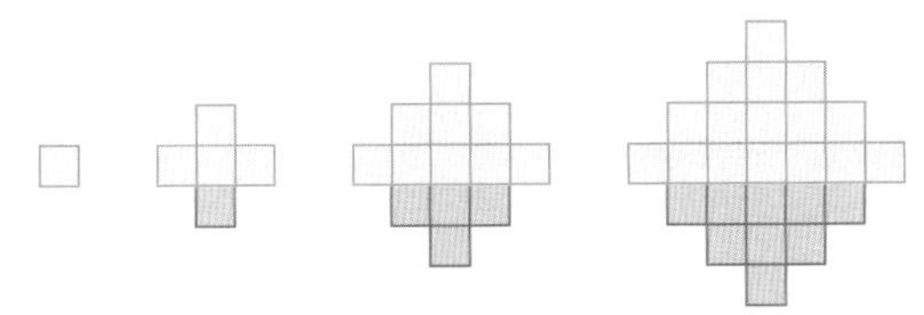

a How many yellow tiles are in each pattern?
b How many blue tiles are in each pattern?
c Write a formula for the total number of tiles in the nth pattern.

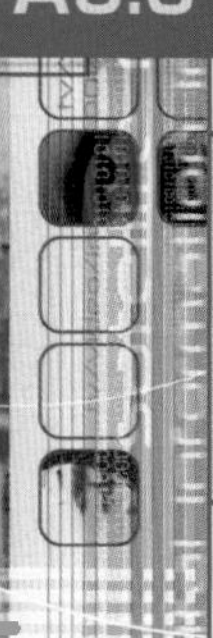

A5.6 Solving problems

This spread will show you how to:
- Solve problems using algebra.
- Suggest extensions to problems, conjecture and generalise.
- Identify exceptional cases or counter-examples.

KEYWORDS
Formula
Generalise
Sequence

Algebra allows you to solve problems by **generalising**.

Callum is trying to solve a problem.
He wants to find a formula for the total of any 2×2 square on a calendar page.

For example,

Total = 36

5	6
12	13

Total = 96

20	21
27	28

JULY

M	T	W	Th	F	Sa	Su
					1	2
3	4	5	6	7	8	9
10	11	12	13	14	15	16
17	18	19	20	21	22	23
24	25	26	27	28	29	30
31						

Callum generalises by choosing a square that starts with x:

x	$x+1$
$x+7$	$x+8$

$$\begin{aligned} \text{Total, } T &= x + (x+1) + (x+7) + (x+8) \\ &= x + x + x + x + 1 + 7 + 8 \\ &= 4x + 16 \end{aligned}$$

So for any number x, the total $T = 4x + 16$

Check:

5	6
12	13

$$\begin{aligned} T &= 4 \times 5 + 16 \\ &= 36 \end{aligned}$$

example

Here is a sequence of T-shapes, where the area is measured in squares.

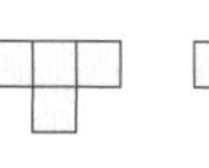

a Find a formula for the area of a general shape in this sequence.
b Referring to the pattern, describe why the formula works.

a

Pattern number, n	1	2	3	4
Area, A	4	7	10	13
3 × table	3	6	9	12

Area = 3 × pattern number + 1
$A = 3n + 1$.

b Consider pattern 4:

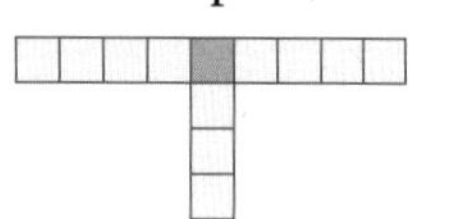

The formula $3n + 1$ works because:

There are 3 arms with 4 squares each

Plus 1 in the middle

Exercise A5.6

Consider the following problems using an algebraic approach:

1 What numbers did we put into these function machines?

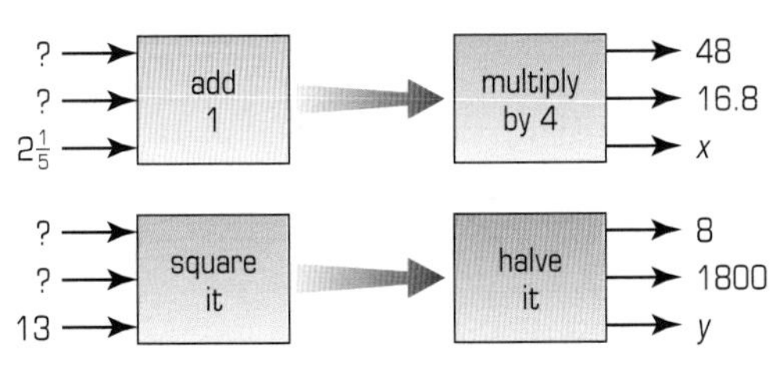

2

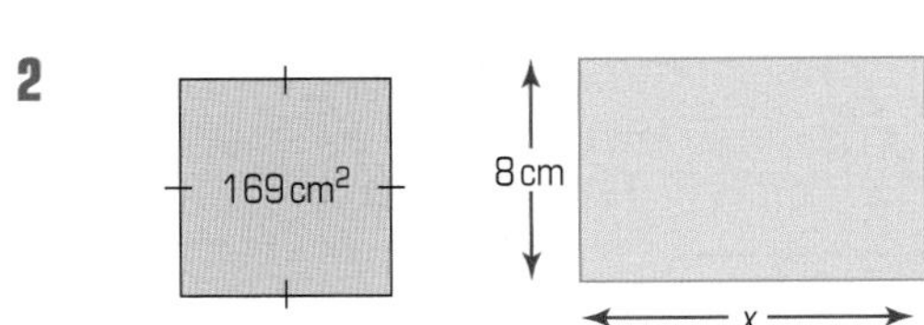

The perimeter of the square and rectangle are equivalent.
What is the value of x?

3 In each sequence, the same amount is added each time (they are linear sequences). Find the missing numbers:

a 3, __, __, __, 35

b 10, __, __, __, __, __, 46

4 If postcards cost 13p and I buy a bundle, is it possible for me to pay with a £20 note and receive £11.41 change?

5 Symmetrical T-shapes are made with square counters.

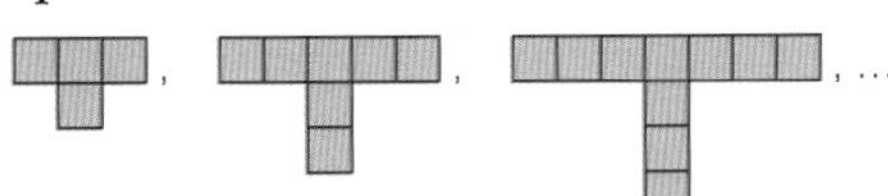

a Find a formula connecting the area of the T-shape with its perimeter. (**Hint**: use a table of values and differences).

b If a T-shape has area of 90, what is its perimeter?

c If a T-shape has perimeter 362, what is its area?

d Why does this formula work?

6 Take a multiplication grid.

2	4	6	8	10
3	6	9	12	15
4	8	12	16	20
5	10	15	20	25
6	12	18	24	
7	14	21	28	

From the grid, pick a 2 by 2 square of your choice. Find the product of opposite numbers, e.g.

4	6
6	9

$4 \times 9 = 36$
$6 \times 6 = 36$

Are these products always equal?
Try some shapes of your own.

7 "In 10 years time you will be twice as old as you were 5 years ago".

Is this possible at any point in your life?

8 **Investigation**
Take a 10 by 10 grid and pick an L-shape in this. Total the numbers.

1	2	3	4	5	
11	12	13	14	15	16
21	22	23	24	25	26
31	32	33	34	35	36
41	42	43	44	45	

How is the total connected to the corner number?
Repeat for grids of different sizes.

A5.7 Graphs of equations

This spread will show you how to:

- Plot the graphs of linear functions in the form $y = mx + c$.
- Recognise that equations of the form $y = mx + c$ correspond to straight-line graphs.

KEYWORDS
Equation
Graph
Intercept
Parallel
Gradient

A straight-line graph is made up of an infinite number of points.

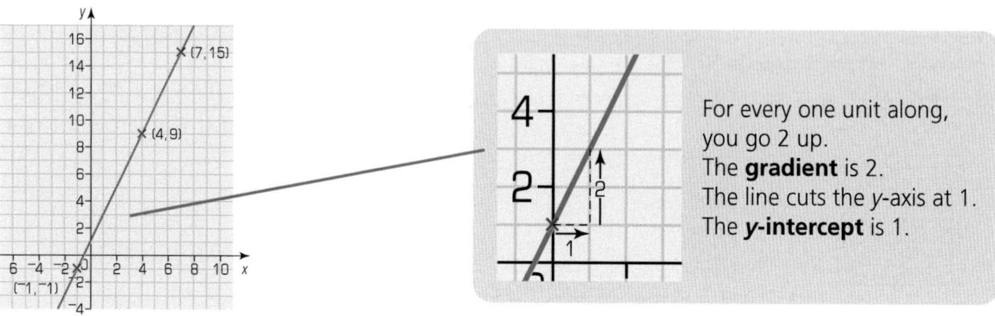

The y-coordinate is always 1 more than double the x-coordinate. The equation of the line is $y = 2x + 1$

Here are some other equations of straight-line graphs:
$y = 3x - 2$, $y = 4x + 1$, $y = 2x + 4$

▶ **The equation of a straight line is usually written as $y = mx + c$, where m and c are numbers. m is the gradient and c is the y-intercept.**

If you know the equation, you can draw the graph.

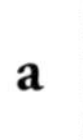

For each of these equations, complete the table and draw a graph.

a $y = 3x + 1$ **b** $y = 3x + 2$

x	−2	0	2
y			

Use the same axes for each graph.
Comment on the features of each graph, particularly where each one crosses the y-axis.

a

x	−2	0	2
y	−5	1	7

b

x	−2	0	2
y	−4	2	8

The graphs are parallel straight lines.
$y = 3x + 1$ crosses the y-axis at $y = 1$.
$y = 3x + 2$ crosses the y-axis at $y = 2$.

Exercise A5.7

1 Match the equation cards with the graph cards, explaining how you made your choice. (If unsure, use your grid and counters to plot the graphs).

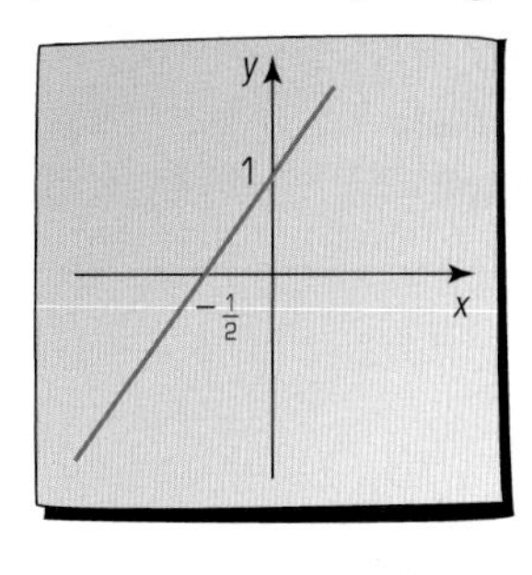

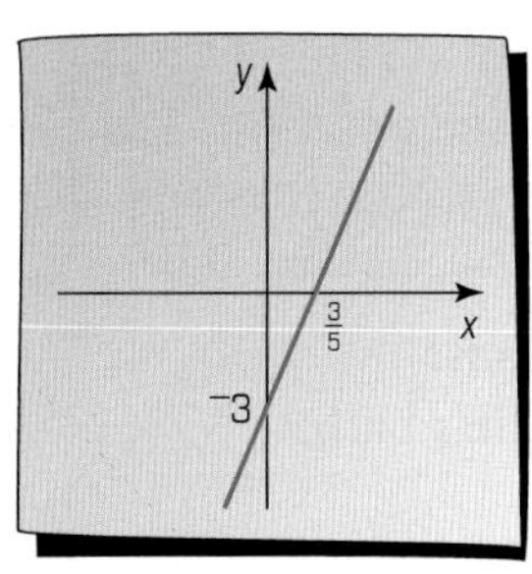

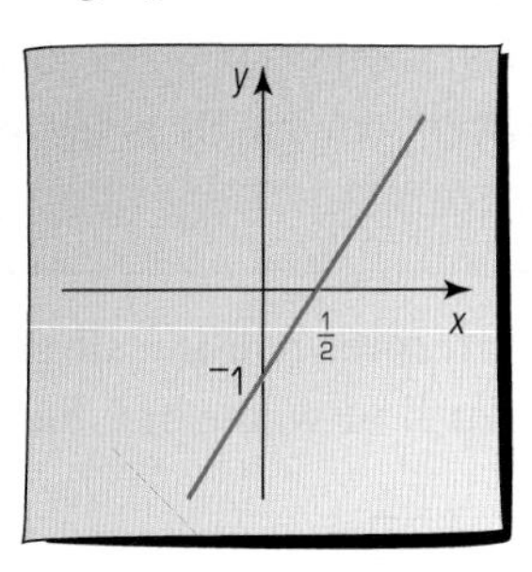

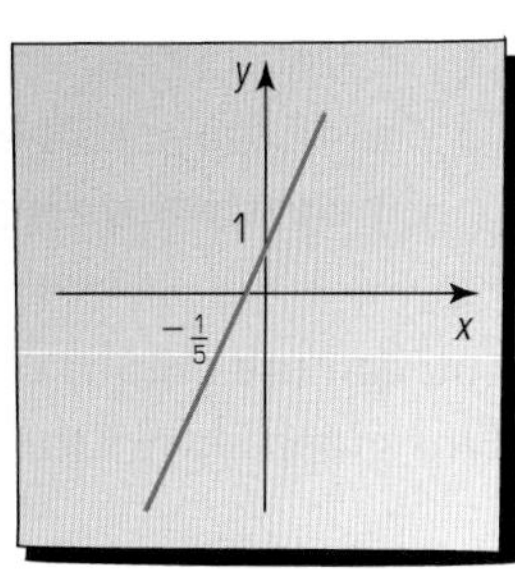

$y = 2x - 1$ | $y = 5x + 1$ | $y = 5x - 3$ | $y = 2x + 1$

2 Answer 'true' or 'false' to the following:

a $y = 7x$ is steeper than $y = 3x$
b $y = 7x$ is steeper than $y = x + 7$
c $y = 3x$ is parallel to $y = 3x + 2$
d $y = 3x$ is parallel to $y = x + 3$.

3 Write the equations of the following straight-line graphs:

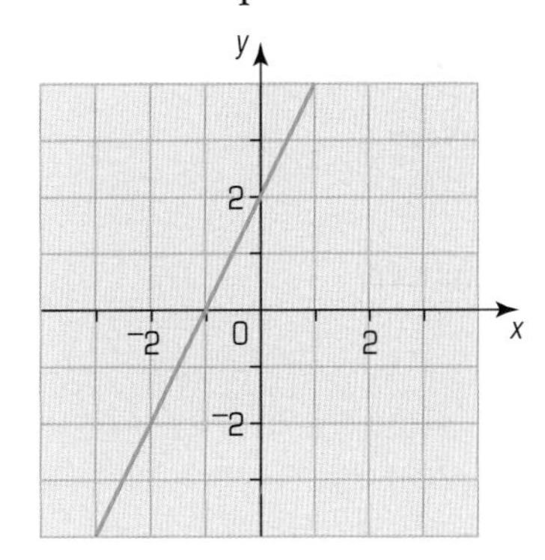

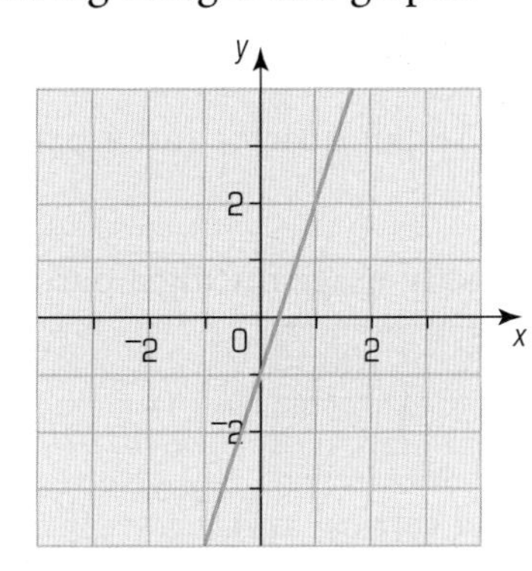

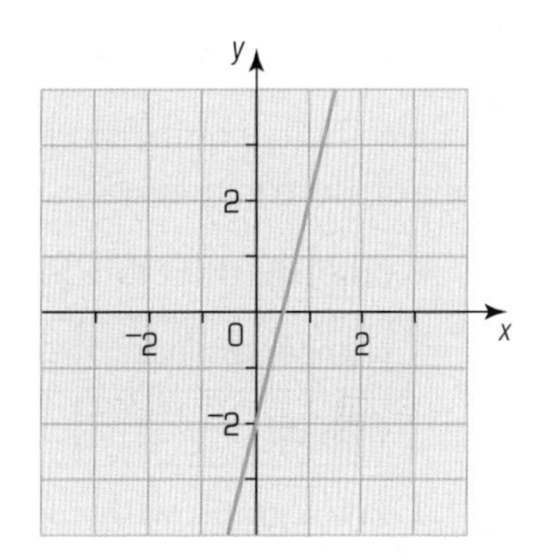

4 Write the equations of the following straight-line graphs:

a Gradient is 6 and cuts the y-axis at (0, 2)
b Gradient is $\frac{1}{2}$ and cuts the y-axis at (0, – 1)
c Parallel to $y = 3x + 5$ and cuts the y-axis at (0, 8)
d Twice as steep as $y = 2x$ and cuts the y-axis at the same place as $y = x + 6$
e Cuts the y-axis at (0, 1) and then goes through the point (1, 10)
f Goes through the points (1, 5) and (2, 9).

5 Select the equations that produce diagonal graphs and then make a sketch of them, indicating where they cut the y-axis.

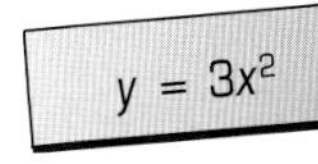

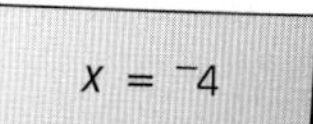

$y = 5x - 4$

$y = x - 7$

$y = 3x + 1$

$y = 2$

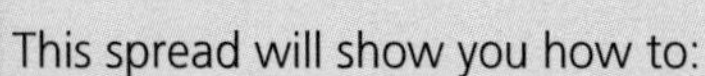

A5.8 Real-life graphs

This spread will show you how to:

- Discuss and interpret graphs from a range of sources.
- Construct functions arising from real-life problems and plot their corresponding graphs.

KEYWORDS
Axis
Formula
Graph

You can use graphs to interpret real-life situations.
This graph shows the depth of water in Chanelle's bath:

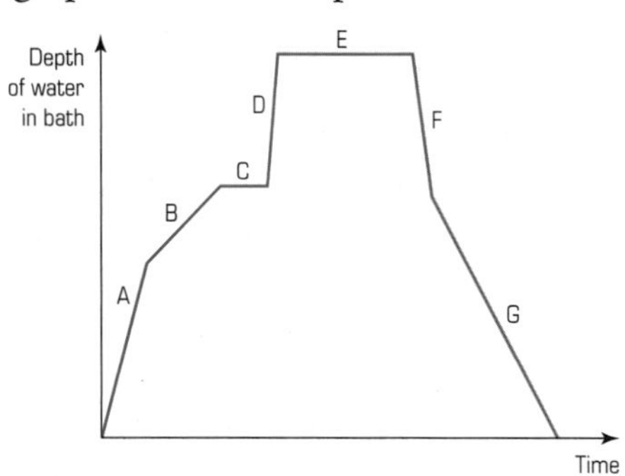

During this sequence of events, Chanelle ...
A: turns on the hot and cold taps
B: just keeps on the cold tap
C: undresses
D: gets into the bath
E: relaxes in the bath
F: gets out of the bath
G: pulls the plug

Graphs can also be a useful tool for reading information.

example

An electricity company charges £18 per quarter and 4 pence for every unit of electricity used.

a Complete this table for electricity usage during a particular quarter.

Number of units	10	50	100
Cost (£)			

b Draw a graph with cost on the vertical axis.
c Use the graph to estimate the cost of using 80 units of electricity.

a

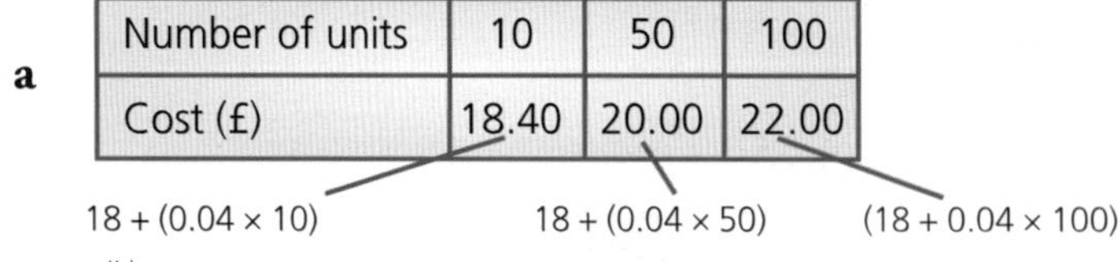

Number of units	10	50	100
Cost (£)	18.40	20.00	22.00

18 + (0.04 × 10) 18 + (0.04 × 50) (18 + 0.04 × 100)

b

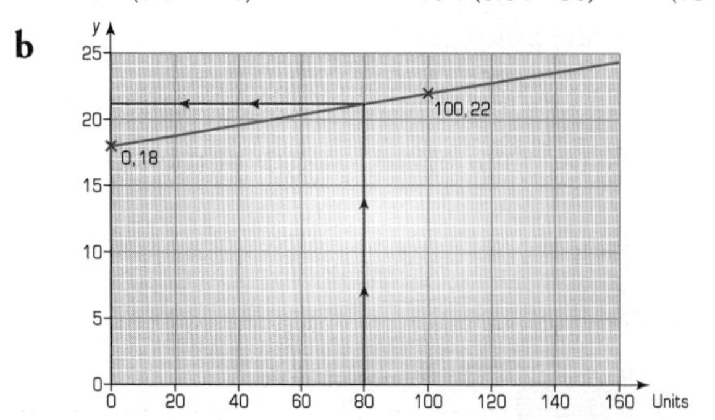

Note:
You could write a formula: $C = 18 + 0.04x$, where x is the number of units and C is the cost in £. Then if $x = 80$, $C = 18 + (0.04 \times 80) = £21.20$.

c From the graph, the cost is approximately £21.

Exercise A5.8

1 A coach uses 1 gallon of petrol for every 5 miles it travels.

a Copy and complete this table:

Miles	5	10		40
Gallons			6	

b Plot a graph to show the coach's petrol consumption. You must make the graph large enough to show a journey of 300 miles.

c Write an equation to represent your graph, stating clearly the meaning of any letters you use.

2 A campsite charges £10 per night per tent, plus an extra £2 per person who stays in the tent.

a Construct a table of values and, hence, a graph to show the cost of the campsite depending upon how many people stay in the tent. (The biggest tent available is one that sleeps 12 people.)

b Use your graph to calculate how many people stayed in the tent if the total cost was £28.

c Suggest an equation for your graph, stating clearly the meaning of any letters you use.

d Use your equation to work out the cost of pitching a supertent sleeping 23 people.

3 **a** Construct a conversion graph to convert miles to kilometres and vice versa. (Go up to 50 miles).

5 miles = 8 kilometres

b Use your graph to convert

i 22 miles to kilometres

ii 45 kilometres to miles

4 Three brothers have a pocket money system organised with their parents.

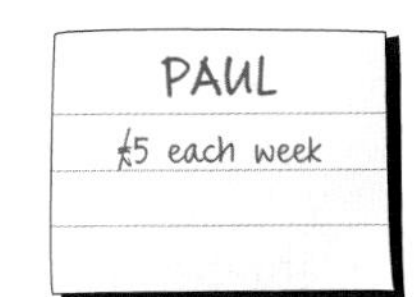

DEAN
£2 each week + 90 pence per house chore carried out

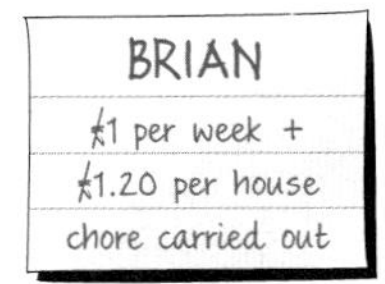

a Using one set of axes only, construct a graph for each brother. (Your axes should be number of household chores carried out versus pocket money).

b Suggest an equation for each brother's pocket money.

c Write a report about who you feel gets the best deal, using your graph as evidence.

5 This container is to be filled with water from a tap dripping at a constant rate.

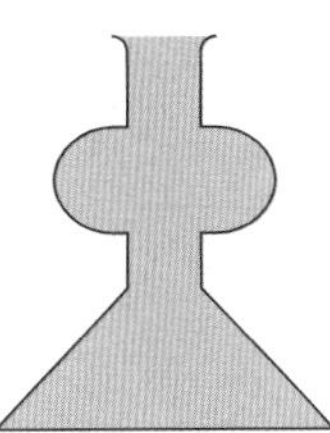

Copy and complete these sketch graphs for the container:

a

Depth

Time

b

Volume

Time

A5 Summary

You should know how to ...

1 Construct and solve linear equations with integer coefficients.

2 Substitute integers into simple formulae.

3 Simplify or transform linear equations by collecting like terms; multiply a single term over a bracket.

4 Plot the graph of linear functions.

5 Recognise that graphs of the form $y = mx + c$ correspond to straight-line graphs.

6 Discuss and interpret graphs arising from real-life situations.

Check out

1 Solve the following:

a $\frac{x+3}{9} = \frac{x-4}{7}$ b $\frac{y-4}{y+5} = \frac{2}{3}$

2 The approximate surface area of a sphere is given by the formula:

$A = 12r^2$

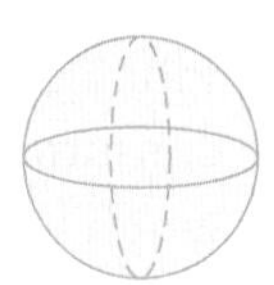

a Find the surface area of a sphere with radius 15 cm.

b Find the diameter of a sphere whose surface area is 200 cm^2.

3 a Expand, simplifying your answer if possible:

i $3(x + 4) - 2(x - 4)$

ii $8(2x - 6) - 4(3x + 8)$

b Find an expression for the difference in weights of these two elephants:

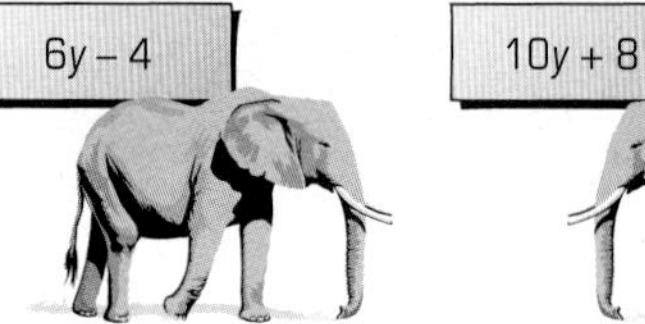

4 Plot the graphs $y = 2x - 1$ and $y = 3$.
At what point do they intersect?

5 a Which is a straight-line graph:
$y = 3x + 4$ or $y = 3x^2 + 4$?

b Which is steeper:
$y = 7x + 1$ or $y = 6x - 2$?

c Find the equation of a line with gradient 5 which cuts the y-axis at (0, 2).

6 Explain what this graph shows.

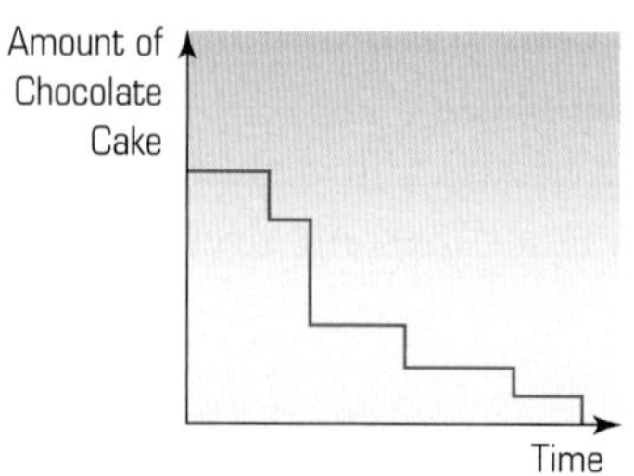

5 Polygons

This unit will show you how to:

- Use straight edge and compasses to construct:
  - the midpoint and perpendicular bisector of a line segment
  - the bisector of an angle
  - the perpendicular from a point to a line
  - the perpendicular from a point on a line
  - a triangle, given three sides (SSS)
- Identify all the symmetries of 2-D shapes.
- Solve geometrical problems using sides and angle properties of special polygons.
- Solve problems and investigate in the context of shape and space.
- Suggest extensions to problems, conjecture and generalise.
- Identify exceptional cases or counter-examples.

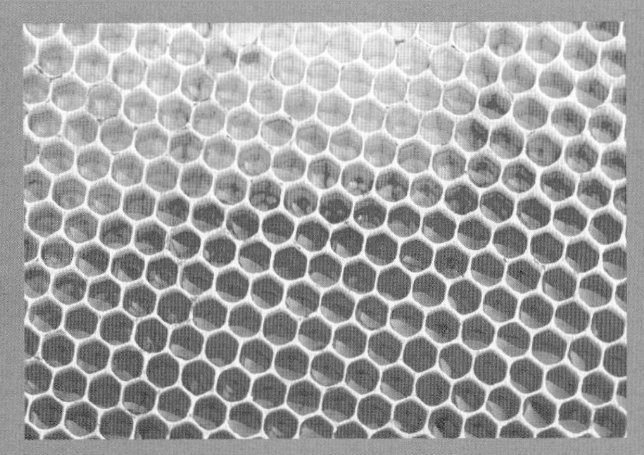

You can find a tessellation of regular hexagons in a honeycomb.

Before you start

You should know how to ...

1 Recognise and name polygons.

2 Recognise symmetry in polygons.

Check in

1 Name these shapes:

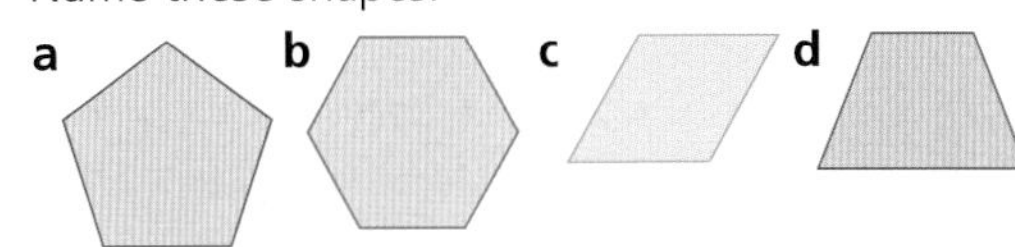

2 Describe the symmetry properties of:

a a kite **b** a parallelogram

S5.1 Constructing triangles

This spread will show you how to:

- Construct a triangle, given three sides.

KEYWORDS

Arc, Compasses, Construct, Intersection, Triangle

To construct a triangle you need to know three facts. Either:

Two angles and a side

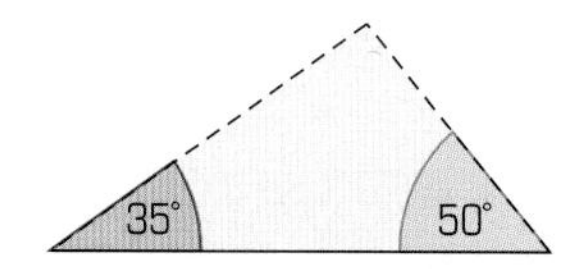

Two sides and the included angle

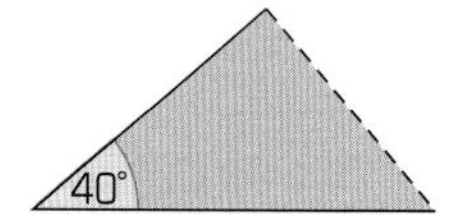

All three sides

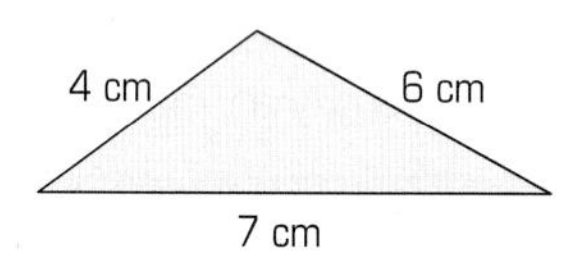

► **You can construct a triangle given all three sides just using a ruler, pencil and compasses.**

example

Construct triangle PQR where PQ = 8 cm, PR = 7 cm and QR = 6 cm.

1 Draw the longest line PQ 8 cm long

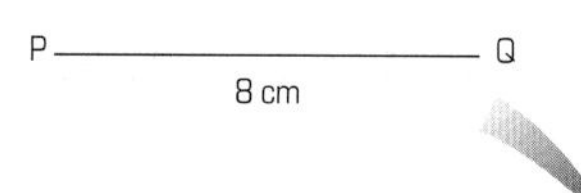

Leave enough room above the line to complete the shape.

2 Opening your compasses to 7 cm, place the point on P and draw an arc above the line.

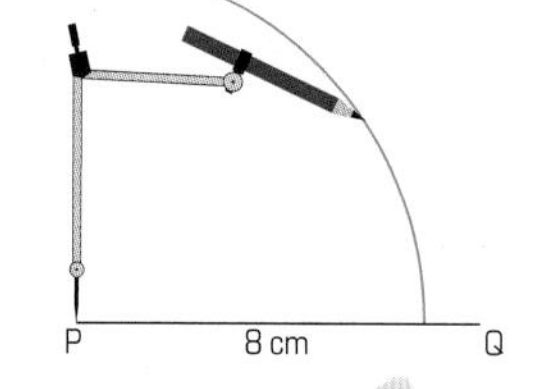

3 Opening your compasses to 6 cm, place the point on Q and draw an arc that cuts the first arc.

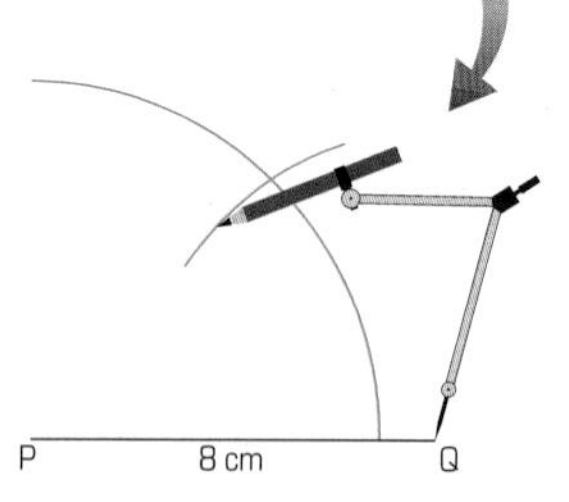

4 Label the point of intersection R and join up the points.

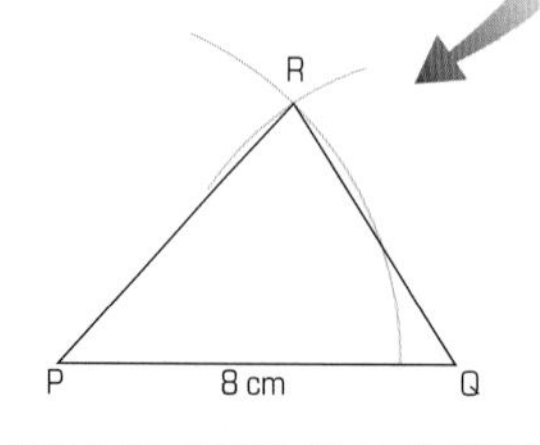

Do not rub out your construction lines. They show your method.

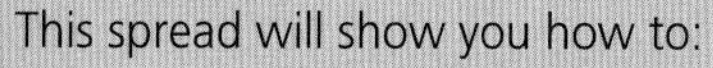

Exercise S5.1

1 Construct a triangle ABC where AB = 4 cm, AC = 5 cm and CB = 6 cm.
Measure the angles of the triangle.

2 **a** Construct a triangle LMN where LM = 3 cm, MN = 5 cm and LN = 5 cm.
Measure the angles of the triangle.
b What type of triangle is it?

3 Construct a triangle PQR where PQ = 3.5 cm, QR = 5.5 cm and RP = 6 cm.
Measure the angles of the triangle.

4 **a** Construct a triangle EFG where EF = 8 cm, FG = 8 cm and FE = 8 cm.
Measure the angles of the triangle.
b What type of triangle is it?

5 **a** Construct a triangle XYZ where XY = 6 cm, XZ = 8 cm and YZ = 10 cm.
Measure the angles of the triangle.
b What type of triangle is it?

6 **a** Construct a triangle PQR where PQ = 10cm, PR = 4 cm, QR = 5 cm.
b What goes wrong? Explain why it doesn't work.

7 **a** Can you construct a triangle with sides 10 cm, 5 cm, 6 cm?
b Give a reason for your answer.

8 **a** Can you construct a triangle with sides 10 cm, 5 cm, 5 cm?
b Give a reason for your answer.

9 **a** Can you construct a triangle with sides 12 cm, 10 cm, 20 cm?
b What type of triangle is it?

10 **a** Construct a triangle with sides 3 cm, 4 cm and 5 cm.
b What type of triangle is this?

11 A flagpole is 10 metres high.
It casts a shadow of 6 metres on the ground.

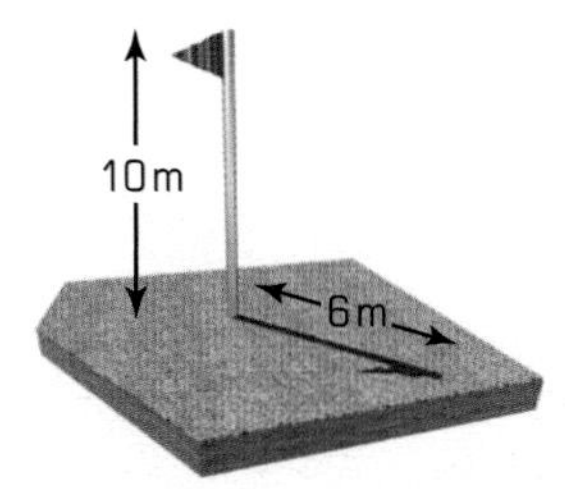

a Construct a triangle to represent this.
Use a scale of 1 cm to represent 2 m.
b Measure the angle that the light from the sun makes with the ground.

S5.2 Constructing quadrilaterals

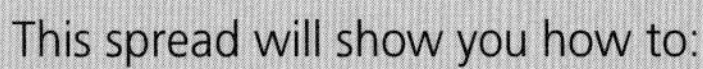

This spread will show you how to:

- Construct a quadrilateral.

KEYWORDS
Compasses, Protractor, Construct, Quadrilateral

You can construct a quadrilateral using a pair of compasses and constructing two triangles.

example

Construct a quadrilateral ABCD where AB = 5 cm, BC = 4.7 cm, CA = 4.5 cm, AD = 3 cm and DC = 3.5 cm.

1 Sketch the shape that you are trying to draw.

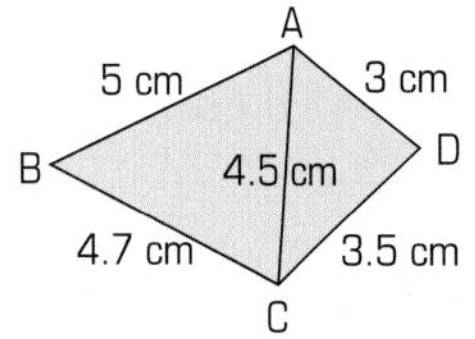

2 Draw the common line AC 4.5 cm long.

A
4.5 cm
C

The letters go around the shape in order.

3 Now use your compasses to construct triangle ACD.

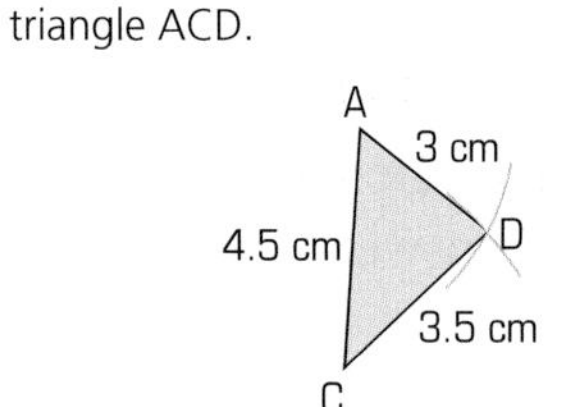

4 Use your compasses to construct triangle ABC, as described in S5.1.
This is the completed quadrilateral ABCD.

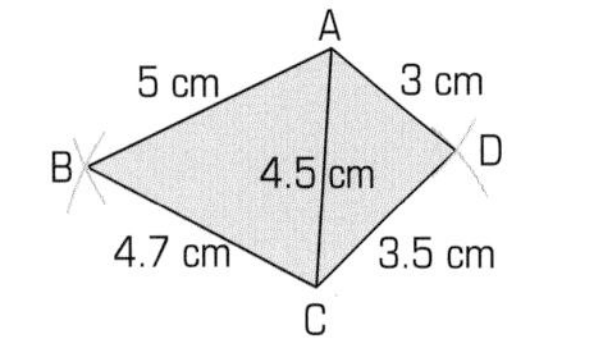

You should not rub out your construction lines.

You can also construct a quadrilateral if you are given information about angles.

example

Construct PQRS where PQ = 8 cm, PS = 6 cm, QR = 5 cm, ∠P = 100° and ∠Q = 95°.

1 Draw the line PQ.

P —— Q
8 cm

2 Draw angles P and Q using a protractor.

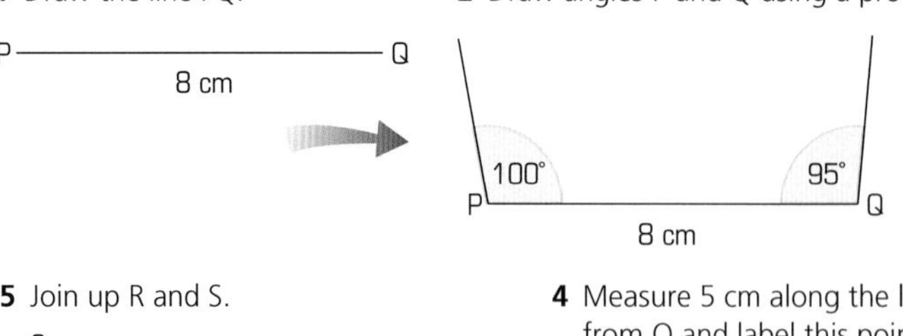

3 Measure 6 cm along the line from P and label this point S.

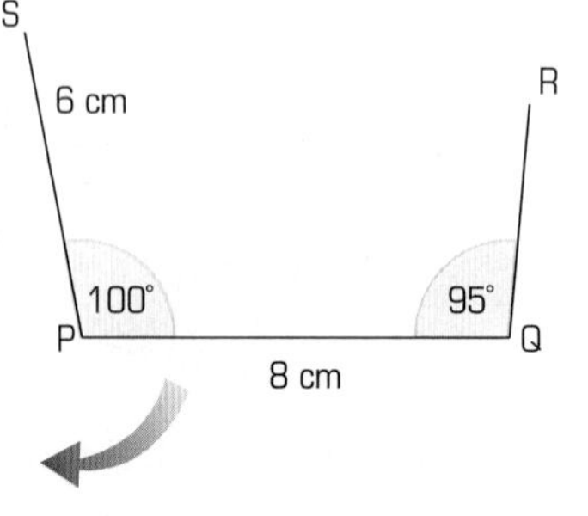

4 Measure 5 cm along the line from Q and label this point R.

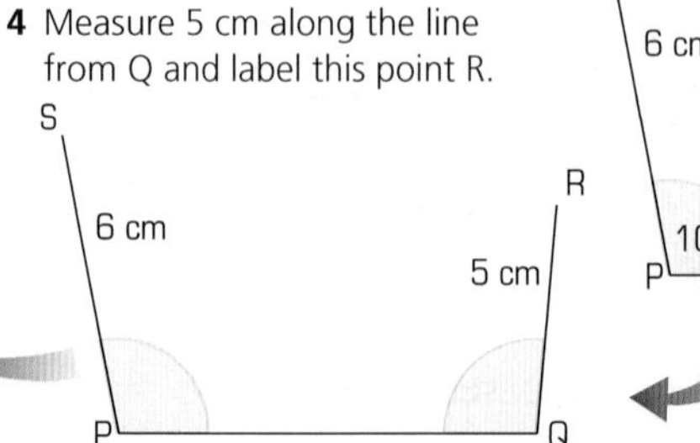

5 Join up R and S.

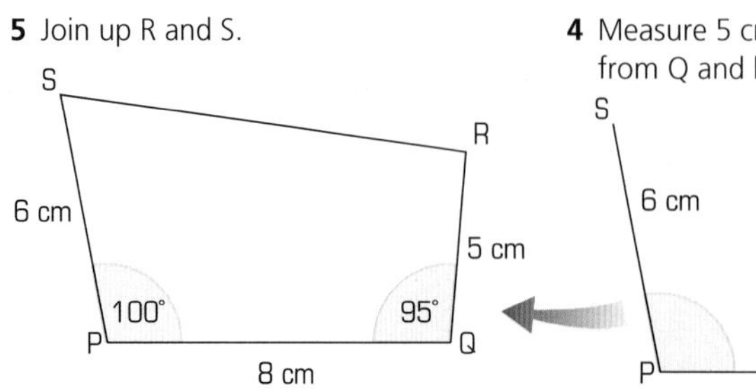

Exercise S5.2

1 Construct this quadrilateral:

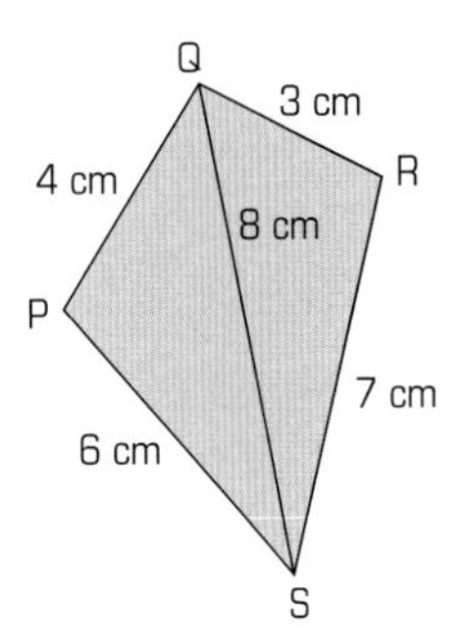

2 **a** Construct this quadrilateral:
b Measure AB.

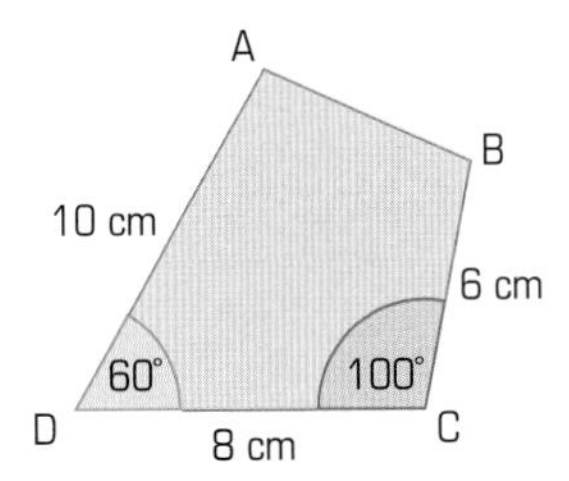

3 Construct a quadrilateral ABCD where AB = 6.3 cm, BC = 7.7 cm, CA = 5.5 cm, AD = 4.5 cm and DC = 7.5 cm.

4 **a** Construct the quadrilateral PQRS where PQ = 7.5 cm, PS = 6.5 cm, QR = 5.5 cm, angle P = 105° and angle Q = 85°.
b Measure RS.

5 Here is a field ABCD.

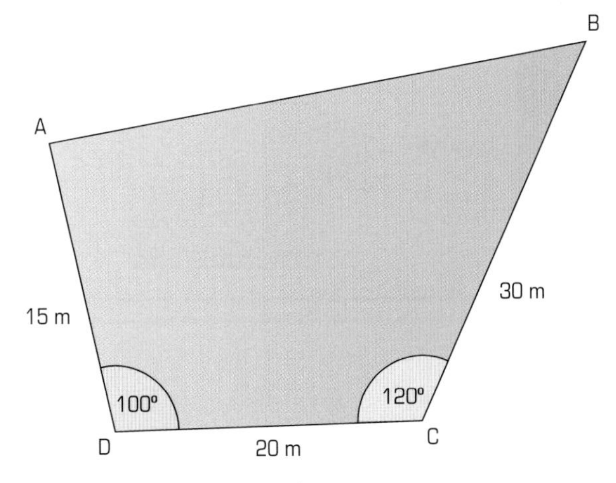

a Draw a plan of the field using the scale 1 cm to 2 m (so 15 m would be drawn as $7\frac{1}{2}$ cm).
b Measure AC.
How far in metres would it be to walk across the field from A to C?

S5.3 Constructing 3-D shapes

This spread will show you how to:

- Identify solid shapes.
- Use ruler and compasses to construct nets of solid shapes.

KEYWORDS
Cross-section Prism
Net Tetrahedron
Pentagonal

You can make a solid shape from its net.

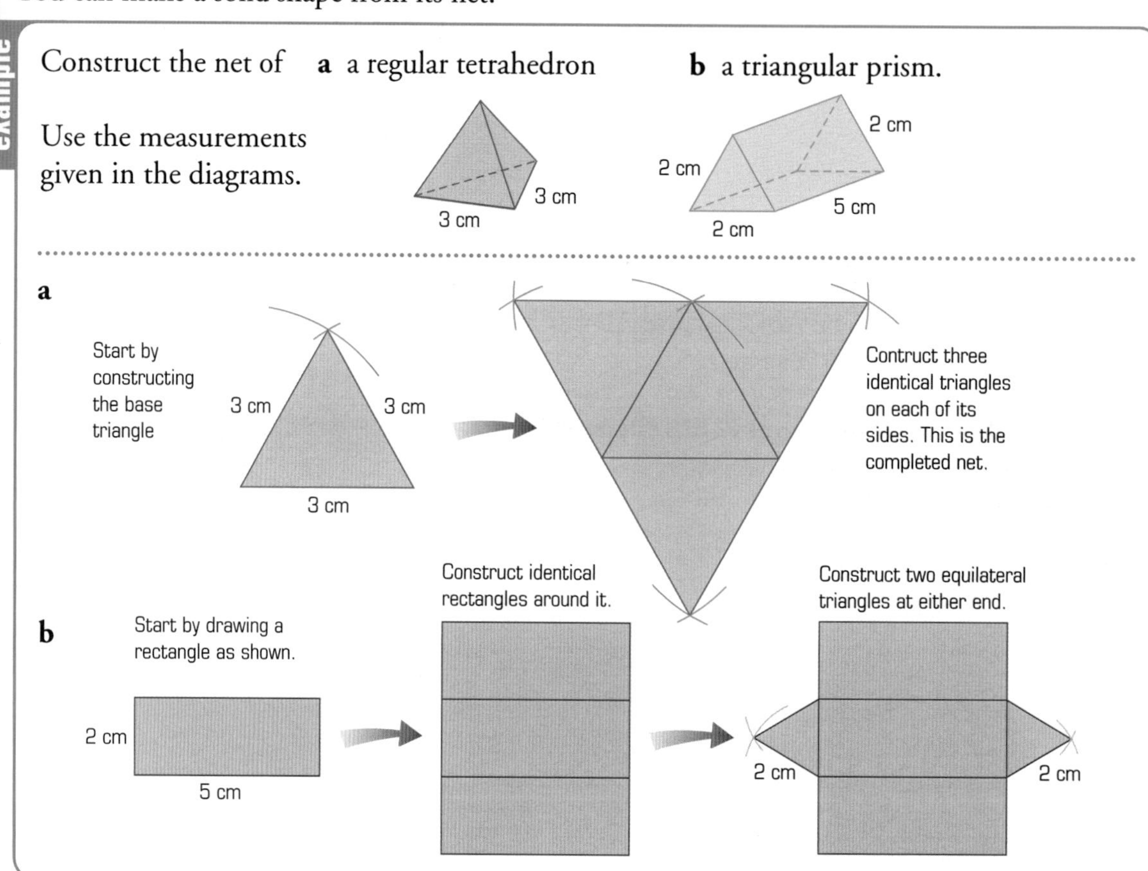

A triangular prism is an example of a **prism**.

► **A prism is a solid with a uniform cross-section.**

Wherever you slice a prism it is exactly the same shape and size.

A pentagonal prism has a pentagonal cross-section.

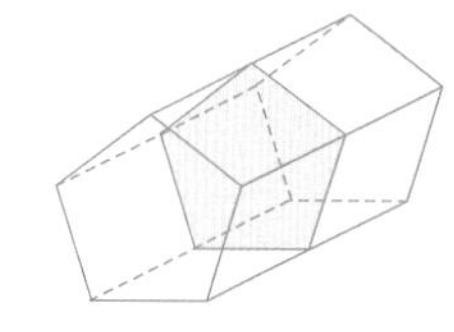

A cuboid is a prism with a rectangular cross-section.

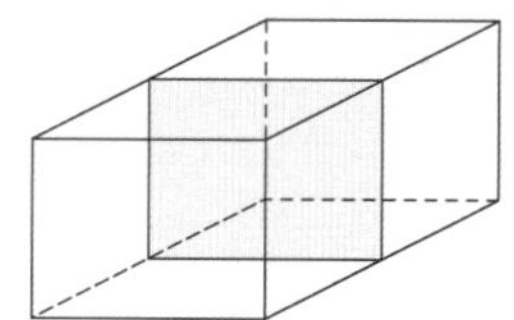

Exercise S5.3

1 Construct the nets of these solids.
Which of the solids are prisms?

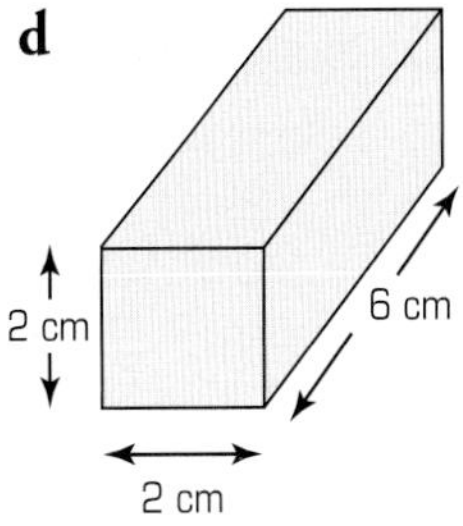

2 **a** Construct the nets for two identical square-based pyramids.

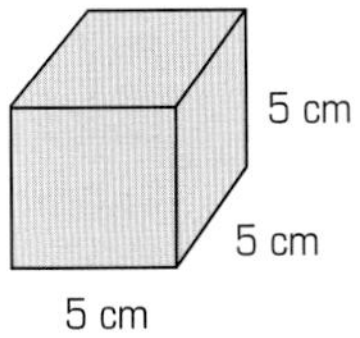

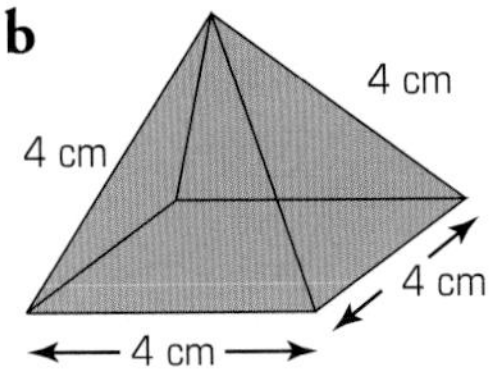

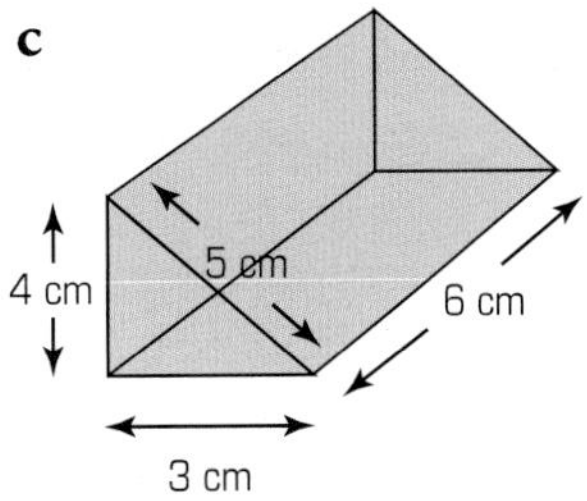

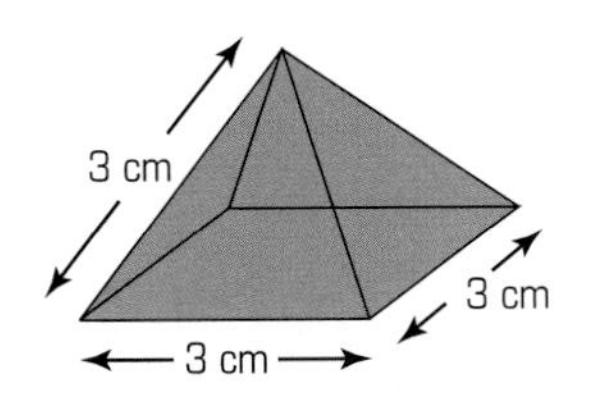

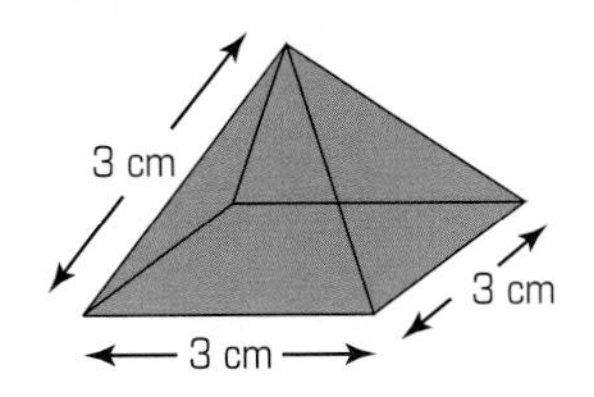

b You can join the two solids together to make an octahedron: (eight triangular faces).
Construct the net of an octahedron.

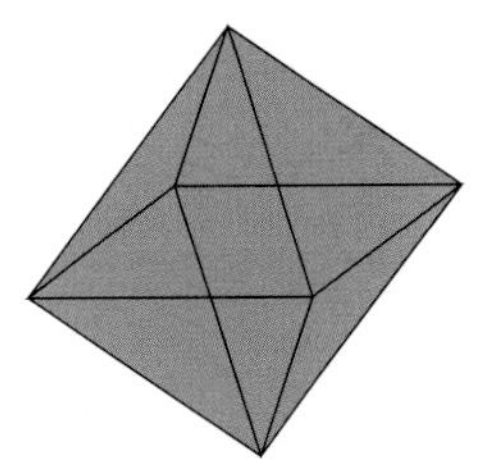

3 Here is the net of a cuboid.

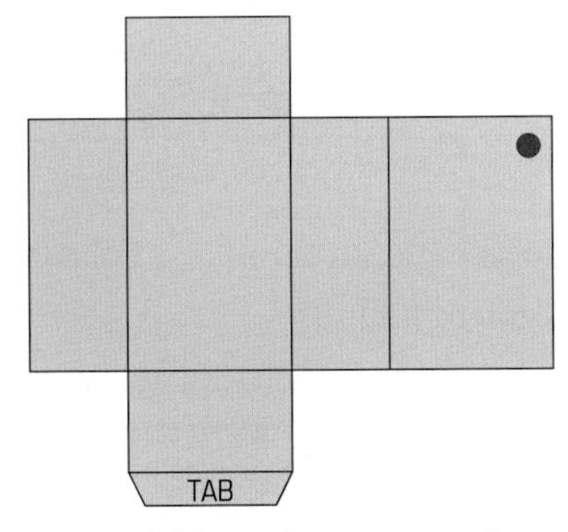

Imagine folding the net to make a cuboid.

a Copy the net and write the letter T on the edge that the tab will stick to.

b On your sketch, put a ● on the two corners that will meet the corner with the ● on.

S5.4 Constructing perpendiculars

This spread will show you how to:

- Construct the perpendicular from a point to a line.
- Construct the perpendicular from a point on a line.

KEYWORDS

Angle	Perpendicular
Arc	Protractor
Bisector	Quadrilateral
Compasses	Rhombus

Here are two useful constructions that you should know:

example

Construct the perpendicular from point P to line segment AB.

P

A ——————— B

1. Open out your compasses and draw two arcs on AB as shown.

P

A ——————— B

2. Putting the compass point at A and keeping the setting the same, draw an arc on the other side of AB

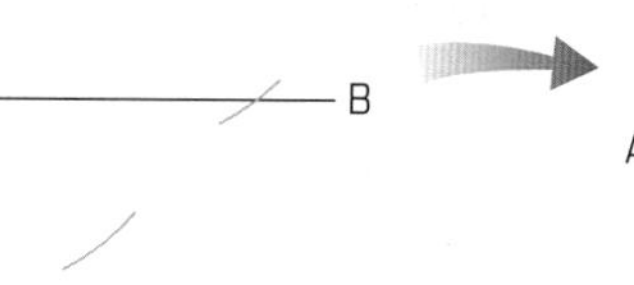

3. Put the point at B and do the same. Call the intersection Q. Join PQ. This is the perpendicular from P to AB.

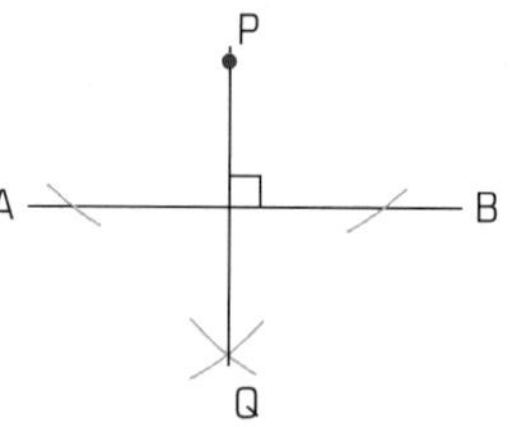

example

Construct the perpendicular from the point Q on line segment CD.

1. Open out your compasses and draw two arcs with Q in the centre.

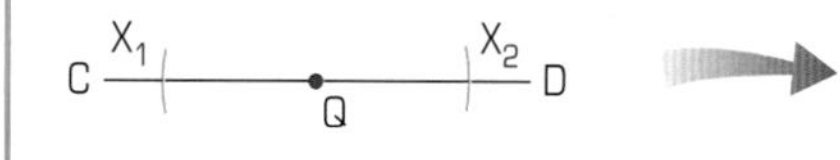

2. Putting the point at X_1, draw arcs above and below CD.

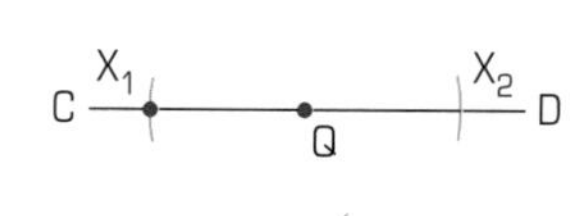

3. Put your compass point at X_2 and do the same. Join the two intersections. This is the perpendicular from Q on the line CD.

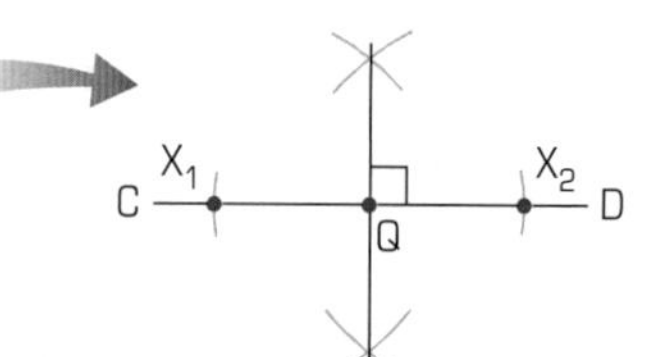

Exercise S5.4

1 Accurately construct this diagram.

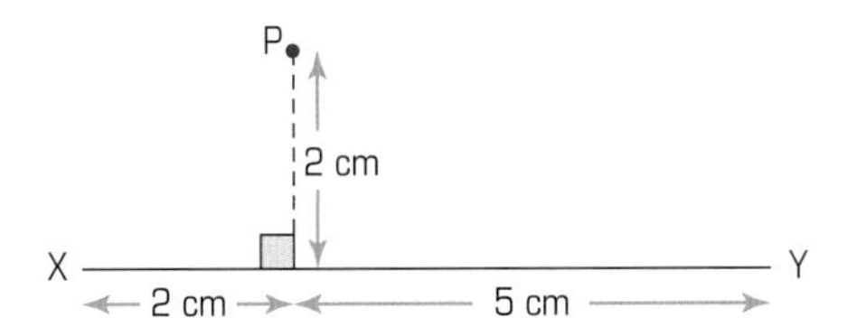

2 **a** Accurately construct this diagram.

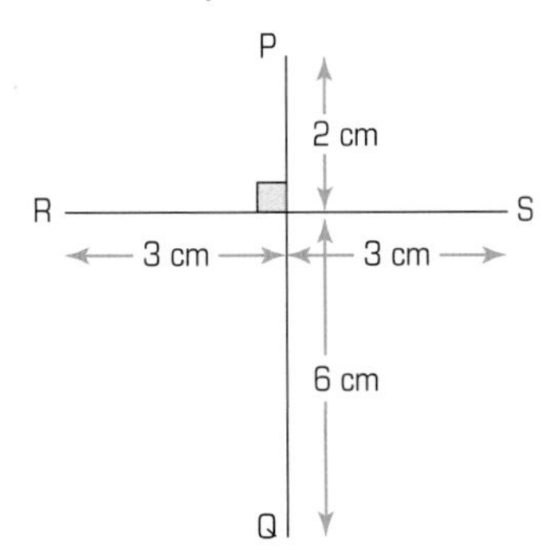

b Join together the vertices PSQR with straight lines.
What shape do they form?

c Find the area of the shape PSQR.

3 Town B is 6 km north of Town A and there is a straight road between them. Leila travels $\frac{2}{3}$ of the distance from A to B, then takes a road due East for 3 km to Town C.

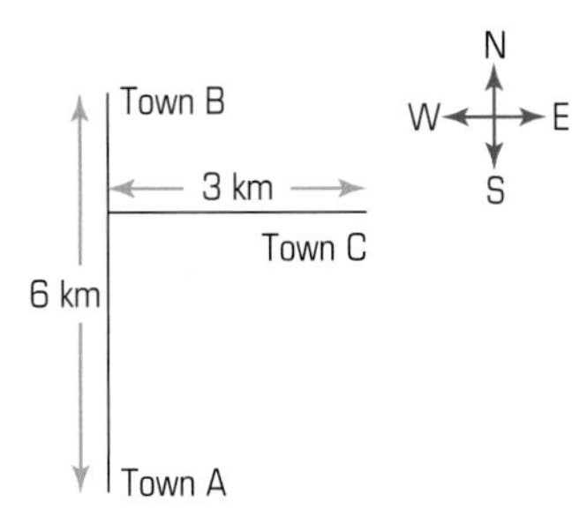

a Construct a map showing the three towns on a scale of 1 cm = 1 km.

b Measure the straight-line distance between Town A and Town C.
What is this distance in kilometres?

4 **a** Draw a straight line AC = 5 cm.

b Construct the rhombus ABCD, with AC as one diagonal, and BD = 4 cm as the other diagonal.

c Measure the angle BAD.

d Measure the angle BAC, and compare with your answer to **c**.
What does this tell you about the diagonals of a rhombus?

4 Draw axes on squared paper from 0 to 10.

a Plot points A(2, 5) and B(6, 5) and join the line AB.

b Mark on your diagram the midpoint of AB and label it C.
What are the coordinates of C?

c Mark on your diagram the point D(6, 1) and locate the midpoint of the line AD by measuring. Label the midpoint E.
What are the coordinates of E?

d Calculate the mean average coordinates of A and D like this:

$$\left(\frac{2+6}{2}, \frac{5+1}{2}\right)$$

What do you notice?

e **Investigation**
Choose pairs of coordinates and find their midpoint. Try to find a general rule.

f Copy and complete this statement:

> The midpoint of the line segment joining (x_1, y_1) to (x_2, y_2) is given by:
> $$\left(\frac{x_1 + x_2}{2}, \frac{y_1 + \square}{\square}\right)$$

S5.5 Properties of polygons

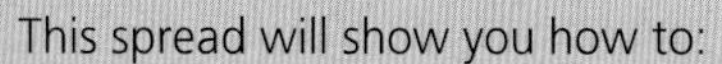

This spread will show you how to:

- Recognise and visualise symmetries of 2-D shapes.

KEYWORDS

Diagonal, Equilateral, Hexagon, Reflective, Regular, Rotational, Parallel, Symmetry

A **polygon** is a shape with straight sides.
A **regular polygon** has all sides and all angles equal.
A **regular hexagon** has six sides equal and six angles equal.

You can think of a regular hexagon as being made up of six regular, or equilateral, triangles.

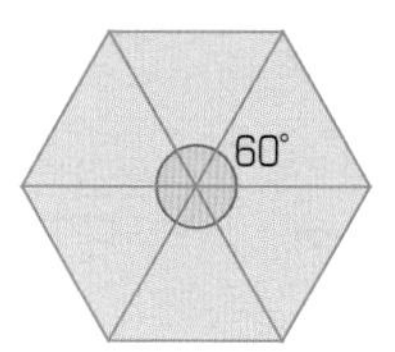

A regular hexagon has six lines of reflective symmetry.

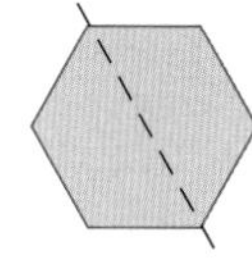

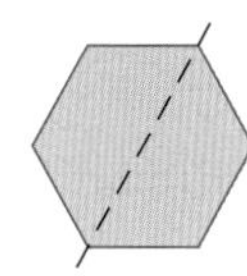

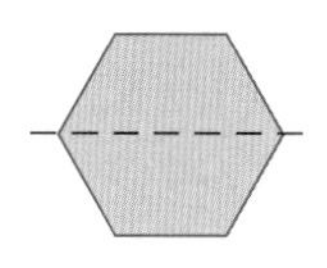

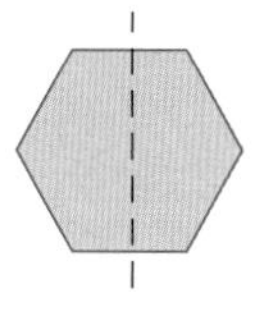

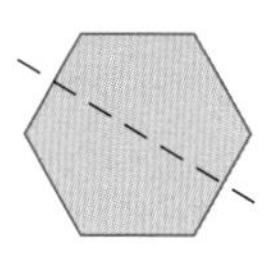

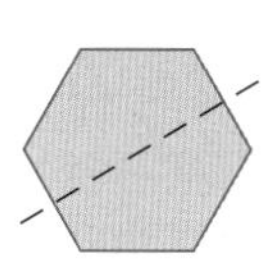

There are six lines that cut the shape into two identical halves.

The order of rotational symmetry of a regular hexagon is 6.

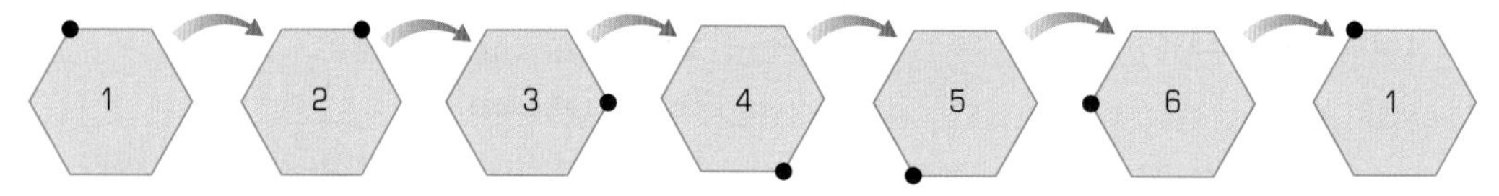

It has six identical positions when rotated through 360°.

- **A regular *n*-sided polygon has *n* lines of symmetry and order of rotational symmetry *n*.**

There are other properties of a regular hexagon that you can easily see from its shape.

- opposite sides are parallel to each other
- all diagonals to an opposing vertex are the same length
- interior angles are equal
- exterior angles are equal

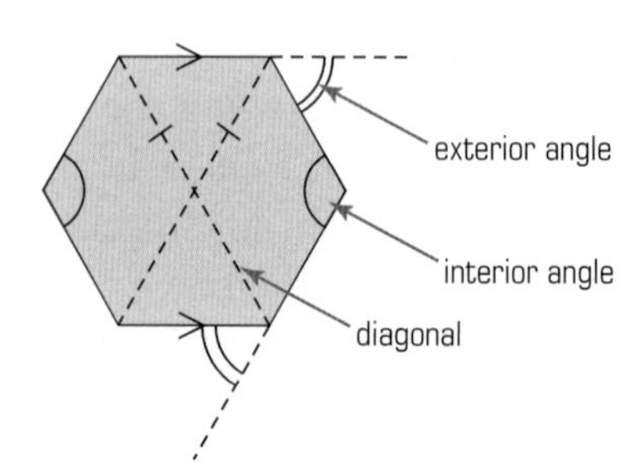

Exercise S5.5

Complete the table of properties for these polygons.

Name of Polygon	Number of Sides	Equal Angles?	Equal Sides?	Equal Diagonals?	Line of Symmetry	Order of Rotation	Parallel Sides?	Perpendicular Sides?
a								
b								
c								
d								
e								
f								
g								
h								
i								
j								
k								

Tessellating polygons

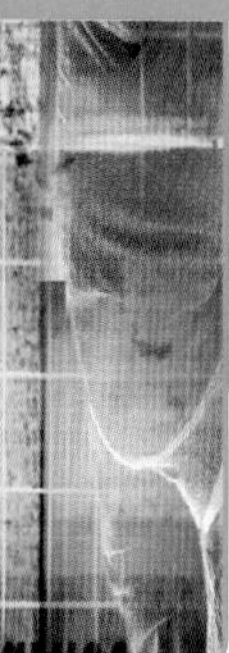

This spread will show you how to:
- Recognise transformations and symmetries of 2-D shapes.
- Identify the geometric properties of polygons to solve problems.

KEYWORDS

Equilateral	Polygon
Hexagon	Tessellation
Pentagon	Regular

Some shapes fit together ...

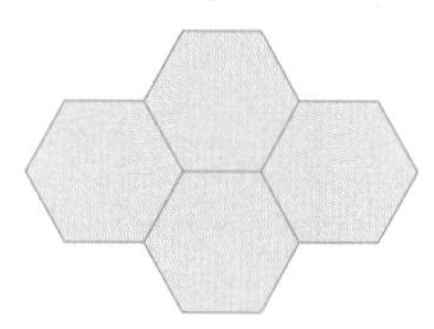

You can make a tiling pattern of regular hexagons.

... and other shapes do not

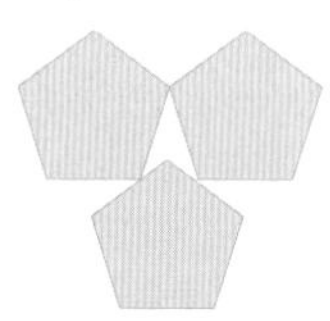

A tiling pattern of regular pentagons will leave gaps.

▶ A polygon **tessellates** if a repeated pattern can fit together without gaps or overlaps.

Here is a tessellation of equilateral triangles:

- There are six interior angles around point A.
- Each interior angle is 60°.
- The interior angles around A add up to 360°.
- Every other point in the pattern is like A.

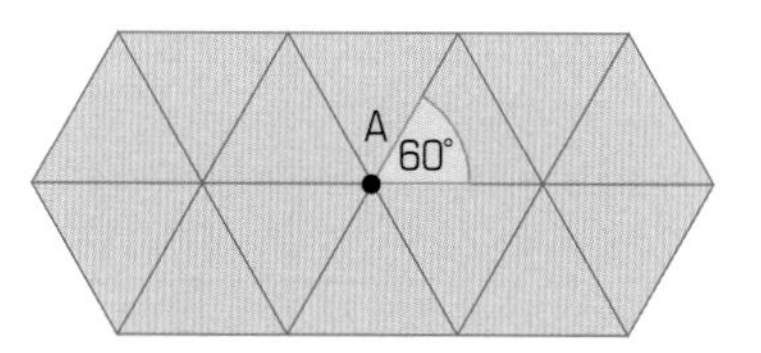

▶ A regular polygon will tessellate if its interior angle divides exactly into 360°.

You can make a tessellation out of irregular polygons as well.

example

Make a tessellation of this arrowhead.
Explain why it tessellates.

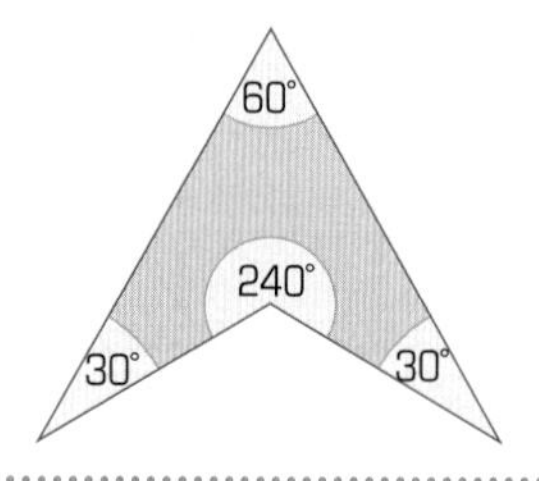

Add up the angles around point A.
30°+60°+30°+240°= 360°.
Point B is the same as A, just upside down.
The interior angles around every point add up to 360° and so the arrowhead tessellates.

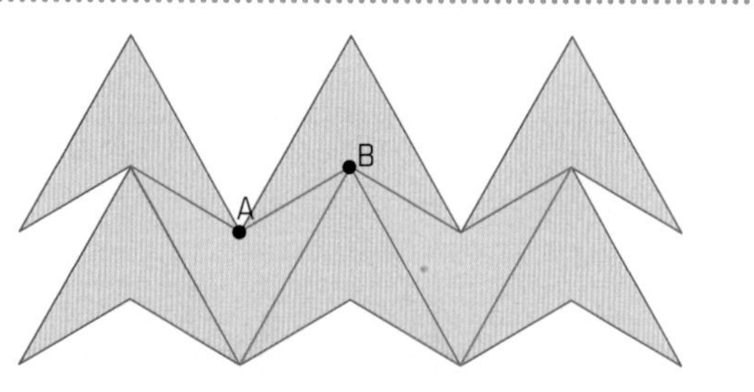

▶ A polygon will tessellate if the interior angles at each point in the pattern add up to 360°.

Exercise S5.6

Find all the shapes that will tessellate.
You may need to make accurate templates on card for some of them.
For any shapes that will not tessellate, try to explain why.

a

rectangle

b

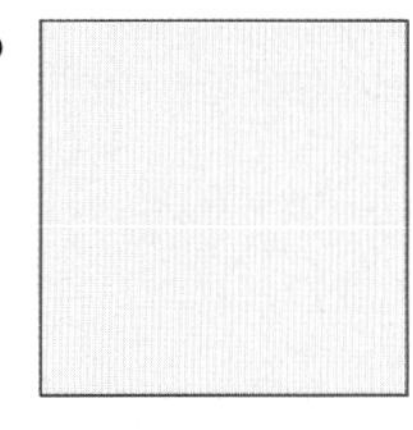

square

c

parallelogram

d

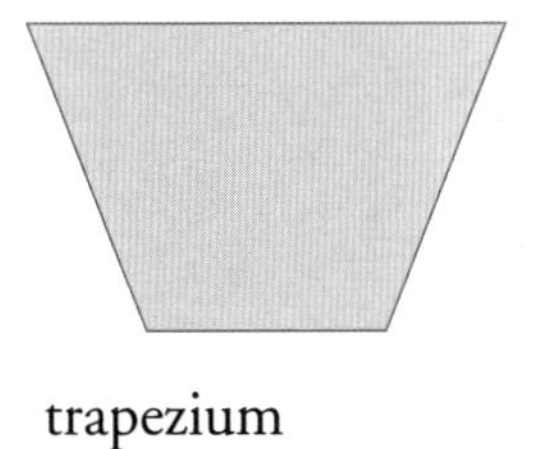

trapezium

e

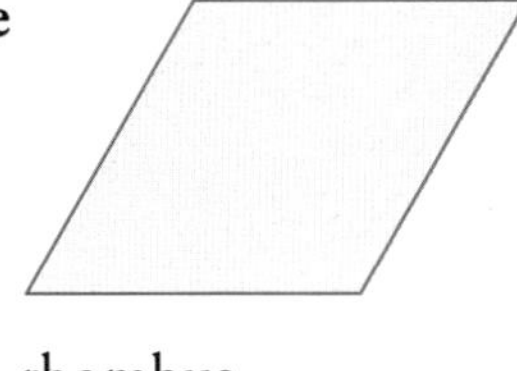

rhombus

f

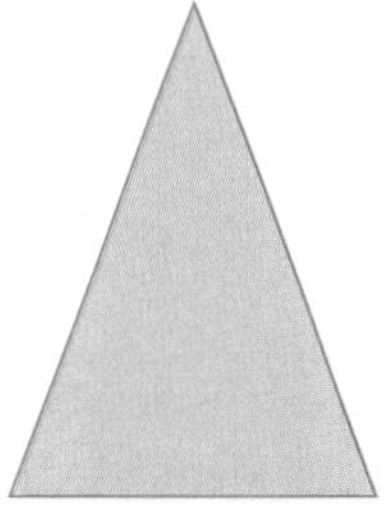

isosceles triangle

g

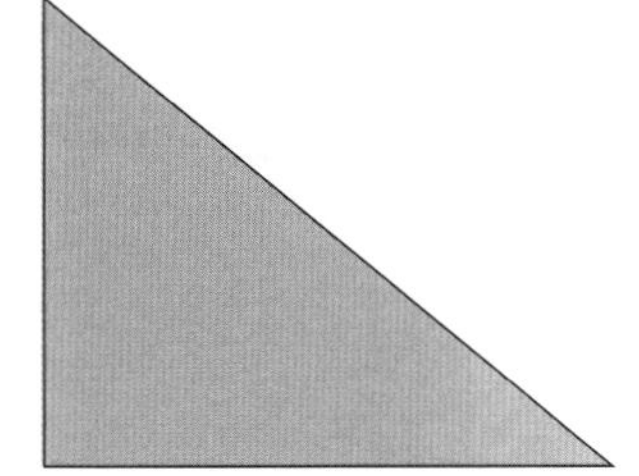

right-angled triangle

h

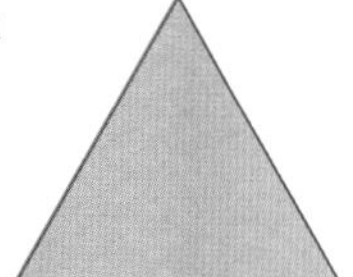

equilateral triangle

i

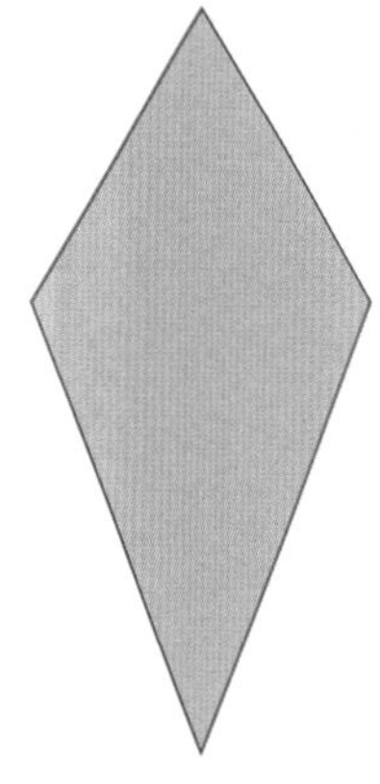

kite

j

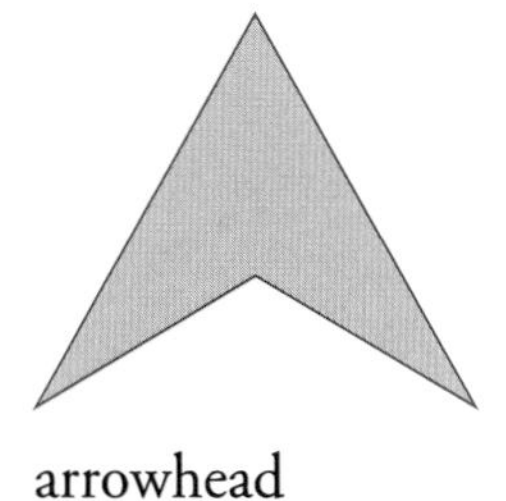

arrowhead

k

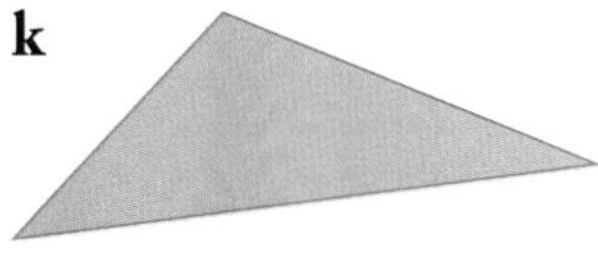

scalene triangle

l

irregular quadrilateral

Summary

You should know how to ...

1 Use a straight edge and compasses to construct:

- the midpoint and perpendicular bisector of a line segment
- the bisector of an angle
- the perpendicular from a point to a line
- the perpendicular from a point on a line

2 Solve problems in the context of shape and space.

3 Suggest extensions to problems, conjecture and generalise. Identify exceptional cases or counter-arguments.

Check out

1 Draw a line XY that is 8 cm long.
Construct the perpendicular from a point P on XY, where P lies 3 cm from X.

2 Find the missing angles:

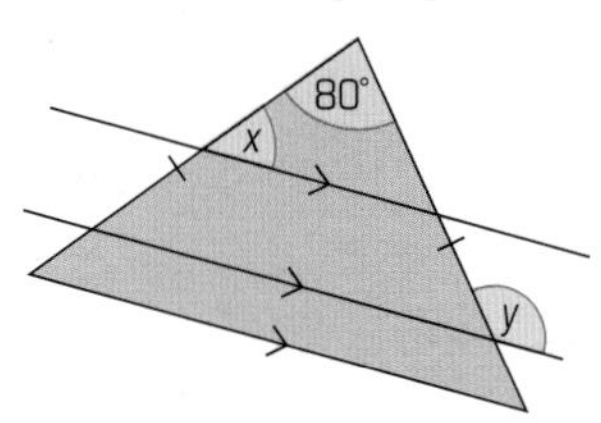

3 Use logical argument to establish whether these statements are true or false:

a A square is a rhombus but a rhombus is not a square.

b The diagonals of a rhombus bisect its angles.

c A rhombus is made up of two identical isosceles triangles.

accuracy
D2.2, N3.8

Accuracy is the degree of rounding of a value.

add, addition
N1.2, N1.5

Addition is the sum of two numbers or quantities.

adjacent (side)

Adjacent sides are next to each other and are joined by a common vertex.

algebra
A1.6, A2.1, A2.2

Algebra is the branch of mathematics where symbols or letters are used to represent numbers.

alternate
S2.2

A pair of alternate angles are formed when a line crosses a pair of parallel lines. Alternate angles are equal.

amount
N4.2, N5.8

Amount means total.

angle: acute, obtuse, right, reflex
S2.1

An angle is formed when two straight lines cross or meet each other at a point. The size of an angle is measured by the amount one line has been turned in relation to the other.

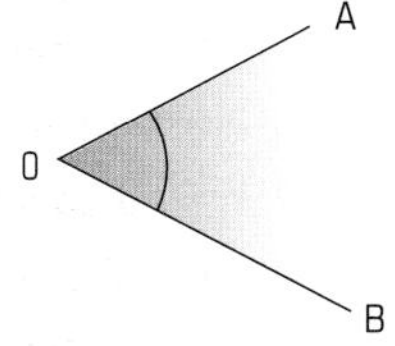

An acute angle is less than 90°.

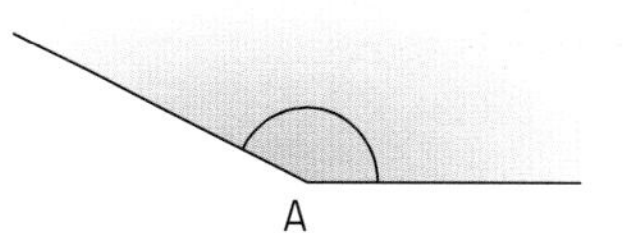

An obtuse angle is more than 90° but less than 180°.

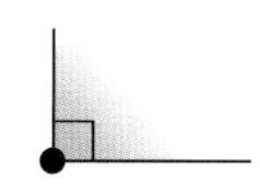

A right angle is a quarter of a turn, or 90°.

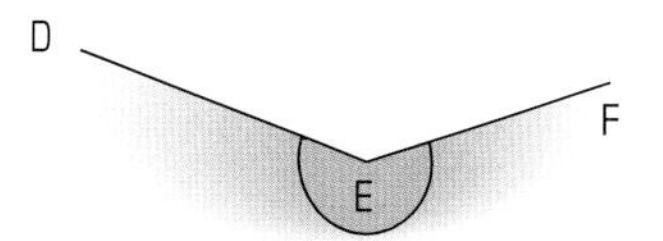

A reflex angle is more than 180° but less than 360°.

angles at a point
S2.2, S3.1

Angles at a point add up to 360°.

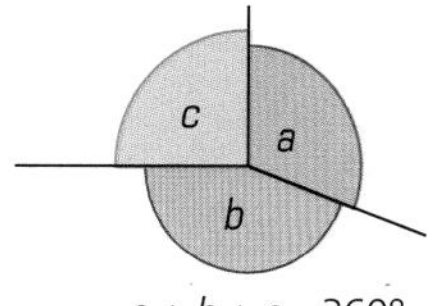

$a + b + c = 360°$

angles on a straight line
S2.2, S3.1

Angles on a straight line add up to 180°.

b a

$a + b = 180°$

anti-clockwise
S4.4

Anti-clockwise angles are positive.

approximate, approximately
N1.5, N1.6, N3.4, N3.5, N3.6, N3.7, N5.1

An approximate value is a value that is close to the actual value of a number.

approximately equal to (≈)
N3.4, N3.5, N5.4

Approximately equal to means almost the same size.

arc
S3.3, S5.1, S5.4

An arc is part of a curve.

area: square millimetre, square centimetre, square metre, square kilometre
S1.1, N2.5, S1.2, S1.3

The area of a surface is a measure of its size.

assumed mean
D3.4

An assumed mean is used to simplify the arithmetic when calculating the mean. The assumed mean is subtracted from all the data and then added on after the mean of the smaller numbers has been calculated.

average
D1.1, D1.2, D3.4, D3.5

An average is a representative value of a set of data.

axis of symmetry
S3.1, S4.5

An axis of symmetry of a shape is a line about which the shape can be folded so that one half fits exactly on top of the other half.

axis, axes
S2.3, A3.5, A5.7

An axis is one of the lines used to locate a point in a coordinate system.

bar chart
D2.4, D2.5, D3.3

A bar chart is a diagram that uses rectangles of equal width to display data. The frequency is given by the height of the rectangle.

base (number)
N3.2

The base is the number which is raised to a power, for example in 2^3, 2 is the base.

base (of plane shape or solid)
S1.2, S3.5, S5.3

The lower horizontal edge of a plane shape is usually called the base. Similarly, the base of a solid is its lower face.

between
N5.1

Between means in the space bounded by two limits.

bias
D4.3

An experiment or selection is biased if not all outcomes are equally likely.

bisector
S3.4, S5.4

A bisector is a line that divides an angle or a line in half.

brackets
N1.6, N3.2, N3.8, A4.3

Operations within brackets should be carried out first. In algebra the term outside the bracket multiplies every term inside the bracket.

calculate, calculation
N1.6

Calculate means work out using a mathematical procedure.

calculator: clear, display, enter, key, memory
N1.6, N3.8

You can use a calculator to perform calculations.

cancel, cancellation
N2.2, N4.2, N4.5

A fraction is cancelled down by dividing the numerator and denominator by a common factor.

For example, 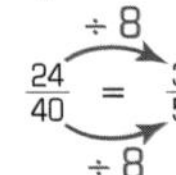

capacity: millilitre, centilitre, litre; pint, gallon
S1.3

Capacity is a measure of the amount of liquid a 3-D shape will hold.

centre of rotation
S4.4

The centre of rotation is the fixed point about which a rotation takes place.

certain
D1.4

An event that is certain will definitely happen.

chance
D1.4

Chance is the probability of something happening.

class interval
D2.3

A class interval is a group that you put data into to make it easier to handle.

closed question
D2.2

Closed questions have a limited number of particular answers.

common factor
N4.5

A common factor is a factor of two or more numbers.
For example, 2 is a common factor of 4 and 10.

commutative

An operation is commutative if the order of combining two terms does not matter. For example, addition is commutative as $4 + 3 = 3 + 4$, but subtraction is not commutative because $4 - 3 \neq 3 - 4$.

compare
N1.1, N4.4, N4.5, N5.6

Compare means to assess the similarity of.

compensation
N3.3, N5.3

The method of compensation makes calculations easier, for example some multiplications are easier if you double one of the numbers and then compensate by halving the answer.

concave
S3.2, S5.5

A concave shape bends inwards. A concave polygon has at least one interior angle greater than 180°.

This is a concave quadrilateral.

congruent
S4.6

Congruent shapes are exactly the same shape and size.

consecutive
A1.1

Consecutive means following on in order.
For example 2, 3 and 4 are consecutive integers.

construct
S3.3, S3.4, S5.1, S5.2

To construct means to draw a line, angle or shape accurately.

continuous
D2.3

Continuous data can take any values between given limits, for example less than 1 m.

convert
N2.3, N2.4, N2.6, N3.1, N4.1, N5.7

Convert means to change.

convex
S3.2, S5.5

Convex means bending outwards. A convex polygon has no interior angles greater than 180°.

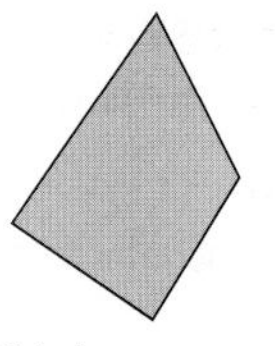

This is a convex quadrilateral.

coordinate pair
A3.5, A5.7

A coordinate pair is a pair of numbers that give the position of a point on a coordinate grid.
For example, (3, 2) means 3 units across and 2 units up.

coordinate point
A5.7

A coordinate point is the point described by a coordinate pair.

coordinates
A3.5, A3.6, S4.3

Coordinates are the numbers that make up a coordinate pair.

corresponding
S2.2

A pair of corresponding angles are formed when a straight line crosses a pair of parallel lines. Corresponding angles are equal.

cross-multiply
A5.1

Cross-multiplying is a method for removing fractions from equations.

cross-section
S1.4, S5.3

The cross-section of a solid is the plane shape you get when you slice it.

data
D1.1, D2.1, D3.1

Data are pieces of information.

database

A database is a means of storing sets of data.

data collection sheet
D2.2, D3.1

A data collection sheet is a sheet used to collect data. It is sometimes a list of questions with tick boxes for collecting answers.

decimal fraction
N1.5, N2.4

A decimal fraction shows part of a whole represented as tenths, hundredths, thousandths and so on.
For example, 0.65 and 0.3 are decimal fractions.

decimal number
N4.1, N5.7, N5.8

A decimal number is a number written using base 10 notation.

decimal place (d.p.)
N1.1, N3.5, N3.7, N5.1

Each column after the decimal point is called a decimal place.
For example, 0.65 has two decimal places (2 d.p.)

degree (°)
S2.1

Angles are measured in degrees. There are 360° in a full turn.

denominator
N2.1, N2.2, N2.3, N4.1
N5.6, N5.7

The denominator is the bottom number in a fraction. It shows how many parts there are in the whole.

diagonal
S5.5

A diagonal of a polygon is a line joining any two non-adjacent vertices.

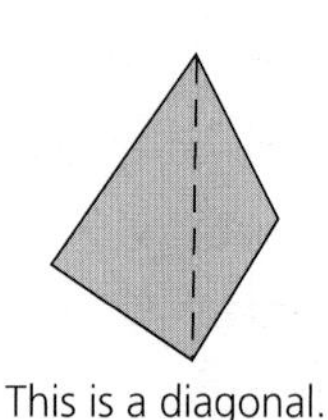

This is a diagonal.

difference
N1.2
You find the difference between two amounts by subtracting one from the other.

digit
N1.1, N5.1, N5.3
A digit is any of the numbers 0, 1, 2, 3, 4, 5, 6, 7, 8, 9.

dimension
S1.1, S1.4
The length, width or height of a shape or solid.

direction
S2.3
The direction is the orientation of a line in space.

directed number
N1.2
A number with a positive or negative sign.

discrete
D2.3
Discrete data can only take certain definite values, for example integers between 10 and 20.

distance
S1.1, S1.2
The distance between two points is the length of the line that joins them.

divide, division
N3.6, N3.7, N4.5
Divide (÷) means share equally.

divisible, divisibility
A3.1, N5.3
A whole number is divisible by another if there is no remainder after division.

divisor
N3.7, N5.5
The divisor is the number that does the dividing.
For example, in $14 \div 2 = 7$ the divisor is 2.

draw
S2.1, S5.1
Draw means create a picture or diagram

edge (of solid)
S1.4, S3.5
An edge is a line along which two faces of a solid meet.

edge

elevation
S3.5
An elevation is an accurate drawing of the side or front of a solid.

enlargement
S4.2, S4.6
An enlargement is a transformation that multiplies all the sides of a shape by the same scale factor.

equally likely
D1.5, D4.1, D4.3
Events are equally likely if they have the same probability.

equals (=)
A2.5, A3.4, A4.1
Equals means having exactly the same value or size.

equation
A3.4, A3.5, A3.6, A4.1, A4.2, A4.4, A5.1, A5.3
An equation is a statement showing that two expressions have the same value.

equation (of a graph)
A3.5, A3.6, A5.7
An equation is a statement showing the relationship between the variables on the axes.

equidistant
S2.2, S3.4
Equidistant means the same distance apart.

equivalent, equivalence
N2.2, N2.3, N2.4, N4.1, N4.2, N5.5, N5.6

Two quantities, such as fractions which are equal, but are expressed differently, are equivalent.

estimate
N1.5, D1.6, N3.4, N3.6, S3.2, D4.2, N5.8

An estimate is an approximate answer.

evaluate
A2.1, A2.4

Evaluate means find the value of an expression.

event
D1.5

An event is an activity or the result of an activity.

exact, exactly

Exact means completely accurate.
For example, three divides into six exactly.

expand
A4.3, A5.3

'Expand the expression' means remove all the brackets.

experiment
D1.6, D2.1, D4.2, D4.3

An experiment is a test or investigation to gather evidence for or against a theory.

expression
A1.6, A2.1, A2.2, A2.3, A2.4, A4.1, A4.3, A5.3

An expression is a collection of numbers and symbols linked by operations but with no equals sign.

exterior
S3.2

An exterior angle is made by extending one side of a shape.

face
S1.4, S3.5

A face is a flat surface of a solid.

face

factor
A2.4, A3.1, N5.2, N5.3, N5.6

A factor is a number that divides exactly into another number.
For example, 3 and 7 are factors of 21.

fair
D4.3

In a fair experiment there is no bias towards any particular outcome.

finite
A1.1

A finite sequence has a definite beginning and end.

formula
A2.5, A3.2, A5.2, A5.6, A5.8

A formula is a statement that links variables.

fraction
N2.1, N2.4, N2.5, N2.6, N3.7, N4.1, N4.2, N5.1, N5.6, N5.7, N5.8, A5.1

A fraction is the result of dividing an integer by another (non-zero) integer. A fraction is a part of a whole. For example $\frac{2}{5}$ of the shape shown is red.

frequency
D1.1, D2.3

Frequency is the number of times something occurs.

frequency diagram
D2.4, D2.5

A frequency diagram uses bars to display grouped data. The height of each bar gives the frequency of the group, and there is no space between the bars.

function
A1.4, A1.5, A1.6, A3.4,

A function is a rule.
For example, + 2, − 3, × 4 and ÷ 5 are all functions.

function machine A1.4, A1.5, A3.4	A function machine links an input value to an output value by performing a function.
generalise A5.6	Generalise means formulate a general statement or rule.
generate A1.2, A3.2, A3.3, A3.5	Generate means produce.
gradient A5.7	Gradient is a measure of the steepness of a line.
graph A3.5, A3.6, A5.7, A5.8	A graph is a diagram that shows a relationship between variables.
greater than (>) N1.1, N5.6	Greater than means more than. For example $4 > 3$.
grid S2.3	A grid is a repeated geometrical pattern used as a background to plot coordinate points. It is usually squared.
grouped data D2.3	Grouped data is data put into groups to make it easier to handle.
height, high S1.2	Height is the vertical distance from the base to the top of a shape.
highest common factor (HCF) N5.2	The highest common factor is the largest factor that is common to two or more numbers. For example, the HCF of 12 and 8 is 4.
horizontal A3.6, S4.3	Horizontal means flat and level with the ground.
hundredth N1.1	A hundredth is 1 out of 100. For example 0.05 has 5 hundredths.
impossible D1.4	An event is impossible if it definitely cannot happen.
improper fraction N2.3	An improper fraction is a fraction where the numerator is greater than the denominator. For example, $\frac{8}{5}$ is an improper fraction.
index, indices A2.4, N3.2	The index of a number tells you how many of the number must be multiplied together, for example 2^3 means $2 \times 2 \times 2$.
infer D2.5	Infer means to conclude from evidence.
infinite A1.1	An infinite sequence has no definite end. An infinite quantity cannot be counted or measured.
input, output A1.4, A1.5, A3.4,	Input is data fed into a machine or process. Output is the data produced by a machine or process.
intercept A5.7	The *y*-intercept is the point at which a graph crosses the *y*-axis.
integer N1.2, N2.5, N5.1, A5.5	An integer is a positive or negative whole number (including zero). The integers are: ..., $^-3$, $^-2$, $^-1$, 0, 1, 2, 3, ...
interior S3.2	An interior angle is inside a shape, between two adjacent sides.

interpret
D1.3, D2.5, N3.8, D3.3, D3.5
You interpret data whenever you make sense of it.

intersect, intersection
S2.2, S2.3
Two lines intersect at the point where they cross.

intersection

interval
D2.3
An interval is the size of a class or group in a frequency table.

inverse
N1.3, S4.5
The inverse is the opposite, for example, multiplication is the inverse of division.

isometric
S3.5
Isometric paper has a triangular grid for drawing 2-D diagrams of 3-D objects.

label
D2.4
A label is a description of a diagram or object.

length: millimetre, centimetre, metre, kilometre; mile, foot, inch
S1.1, S1.3
Length is a measure of distance.

less than (<)
N1.1
Less than means smaller than.
For example, 3 is less than 4, or $3 < 4$.

likelihood
D4.1
Likelihood is the probability of an event happening.

line graph
D1.3
Points representing frequencies are joined by straight lines on a graph.

line of symmetry
S4.5
A line of symmetry is a line about which a 2-D shape can be folded so that one half of the shape fits exactly on the other half.

line
S3.4
A line joins two points and has zero thickness. A line segment is part of a line.

linear
A3.2, A3.5, A3.6, A5.4
The terms of a linear sequence increase by the same amount each time. The graph is a straight line.

locus, loci
S3.4
A locus is the position of a set of points (usually a line) that satisfies some given conditions.

lowest common multiple (LCM)
N5.2, N5.6
The lowest common multiple is the smallest multiple that is common to two or more numbers, for example the LCM of 4 and 6 is 12.

lowest terms
N2.2
A fraction is in its lowest terms when the numerator and denominator have no common factors.

mapping
A1.5, A3.4
A mapping is a rule that can be applied to a set of numbers to give another set of numbers.

mean
D1.2, D3.4
The mean is an average value found by adding all the data values and dividing by the number of pieces of data.

measure
N3.1, S2.1, S3.4
When you measure something you find the size of it.

median
D1.1, D3.4

The median is an average which is the middle value when the data is arranged in size order.

metric
N3.1

The metric system is a decimal system of weights and measures.

midpoint
S3.4

The midpoint is halfway between two points.

mirror line
S4.1, S4.3

A mirror line is a line or axis of symmetry.

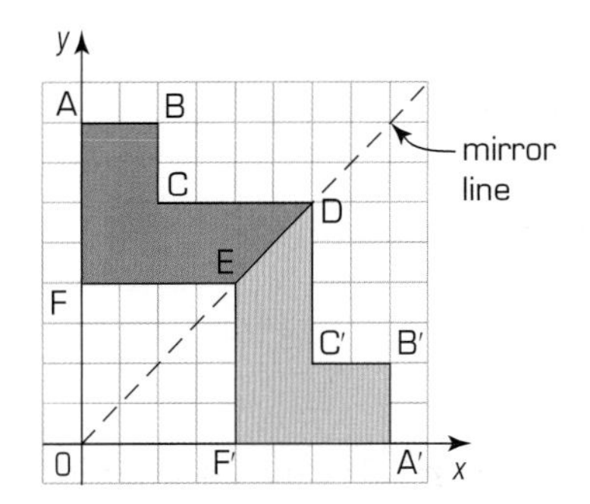

mixed number
N2.3

A mixed number has a whole number part and a fraction part, for example, $3\frac{1}{2}$ is a mixed number.

modal class
D1.1, D3.4

The modal class is the most commonly occurring class when the data is grouped. It is the class with the highest frequency.

mode
D1.1, D3.4

The mode is an average which is the data value that occurs most often.

multiple
N5.2, N5.6, A3.1

A multiple of an integer is the product of that integer and any other. For example, these are multiples of 6: $6 \times 4 = 24$ and $6 \times 12 = 72$.

multiply, multiplication
N3.1, N3.7, N5.3, N5.4

Multiplication is repeated addition, for example
$7 \times 4 = 7 + 7 + 7 + 7 = 28$

nearest
N5.1

Nearest means the closest value.

negative
N1.2, N1.3, N1.6, A4.2

A negative number is a number less than zero.

net
S1.4, S5.3

A net is a 2-D arrangement that can be folded to form a solid shape.

***n*th term**
A3.3, A3.4, A5.4

The *n*th term is the general term of a sequence.

numerator
N2.1, N2.2, N2.5, N4.1, N5.6, N5.7

The numerator is the top number in a fraction. It shows how many parts of the whole you have.

object, image
S4.1, S4.2, S4.3, S4.4

The object is the original shape before a transformation. An image is the shape after a transformation.

open question
D2.2

Open questions can have any answers.

operation
A2.4, N3.2, N3.8

An operation is a rule for processing numbers. The basic operations are addition, subtraction, multiplication and division.

opposite (sides, angles)
S2.2

When two straight lines cross, four angles are made; any pair that touch only at the crossing point are called opposite angles (also known as vertically opposite angles). In a triangle, the opposite side of an angle is the side that is not an arm of the angle. In a quadrilateral, opposite sides do not meet at any vertex.

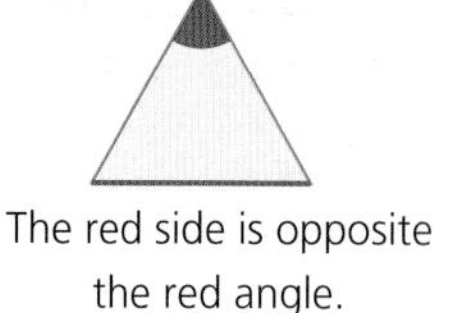
The red side is opposite the red angle.

order
N1.1

To order means to arrange according to size or importance.

order of operations
A2.4, N3.2

The conventional order of operations is:
brackets first, then indices,
then division and multiplication,
then addition and subtraction.

order of rotational symmetry
S4.5

The order of rotation symmetry is the number of times that a shape will fit on to itself during a full turn.

origin
S2.3

The origin is the point where the *x*- and *y*-axes cross, that is (0, 0).

outcome
D1.5, D1.6, D4.1, D4.2

An outcome is the result of a trial or experiment.

parallel
S2.2, S3.3, S5.6

Two lines that always stay the same distance apart are parallel. Parallel lines never cross or meet.

partition
N1.4, N3.3, N3.4, N5.3

To partition means to split a number into smaller parts.
For example, 57 could be split into 50 + 7, or 40 + 17.

percentage (%)
N2.6, N4.1, N4.2, N4.3, N5.7, N5.8

A percentage is a fraction expressed as the number of parts per hundred.

perimeter
S1.1, S1.2

The perimeter of a shape is the distance around it. It is the total length of the edges.

perpendicular
S1.2, S2.2, S3.4, S3.5, S5.4

A line or plane is perpendicular to another line or plane if they meet at a right angle.

pie chart
D1.3, D2.4, D2.5, D3.2

A pie chart uses a circle to display data. The angle at the centre of a sector is proportional to the frequency.

place value
N1.1, N3.1, N3.3

The place value is the value of a digit in a decimal number.
For example, in 3.65 the digit 6 has a value of 6 tenths.

plan
S3.5

The plan is the view of a solid from above.

plane

A plane is a flat surface.

point

A point has no size.

polygon: pentagon, hexagon, octagon
S3.2, S4.5, S5.3, S5.5, S5.6

A polygon is a closed shape with three or more straight edges.

A pentagon has five sides.

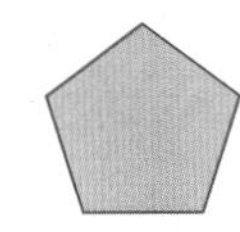

A hexagon has six sides.

An octagon has eight sides.

position
A1.2, A1.3, A1.4

The position of a term in a sequence is an integer, starting with 1, which describes its place in the sequence.

positive
N1.2

A positive number is greater than zero.

power
A2.4, N3.2, N5.1

The power of a number tells you how many of the number must be multiplied together, for example 10 to the power 4 is 10 000.

predict
A2.2, A3.3

Predict means forecast.

primary
D2.1, D3.1

Data you collect yourself is primary data.

prime
A3.1, N5.2

A prime number is a number that has exactly two different factors.

prime factor
A3.1, N5.6

A prime factor is a factor that is a prime number.

probability
D1.4, D1.5, D4.1, D4.2, D4.3

Probability is a measure of how likely an event is.

probability scale
D1.4

A probability scale is a line numbered 0 to 1 or 0% to 100% on which you place an event based on its probability.

product
N3.3, A3.1, N5.2

The product is the result of a multiplication.

proper fraction
N2.3

A proper fraction is a fraction where the numerator is smaller than the denominator.

proportion
N4.4, N4.5

Proportion compares the size of a part to the size of a whole. You can express a proportion as a fraction, decimal or percentage.

protractor (angle measurer)
S2.1, S3.2, S5.2

A protractor is an instrument for measuring angles in degrees.

quadratic
A5.4

A quadratic expression contains a square term.

quadrant
S2.3

A quadrant is one of the four sections the plane is divided into by the *x*- and *y*-axes.

quadrilateral: arrowhead (delta), kite, parallelogram, rectangle, rhombus, square, trapezium
S1.2, S2.3, S3.2, S3.3, S5.2, S5.4

A quadrilateral is a polygon with four sides.

rectangle

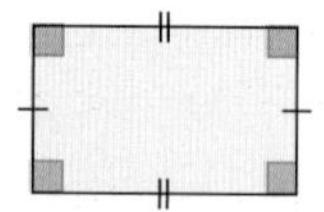

All angles are right angles. Opposite sides equal.

parallelogram

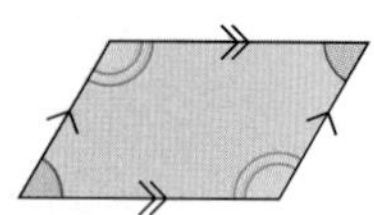

Two pairs of parallel sides.

kite

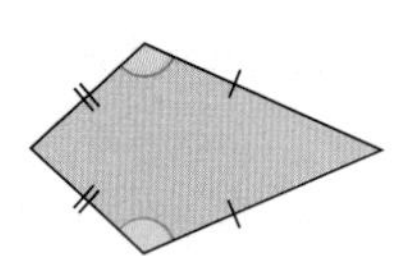

Two pairs of adjacent sides equal. No interior angle greater than 180°.

rhombus

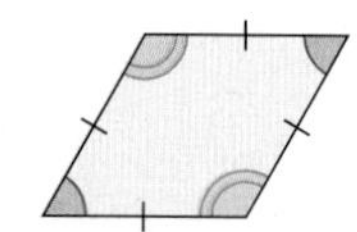

All sides the same length. Opposite angles equal.

square

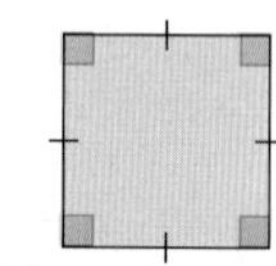

All sides and angles equal.

trapezium

One pair of parallel sides.

An arrowhead has two pairs of equal adjacent sides and one angle greater than 180°.

questionnaire
D2.2

A questionnaire is a list of questions used to gather information in a survey.

quotient
N5.5

A quotient is the result of a division, for example, the quotient of 12 ÷ 5 is $2\frac{2}{5}$, or 2.4.

random
D1.5, D4.1

A selection is random if each object or number is equally likely to be chosen.

range
D1.1, D3.4

The range is the difference between the largest and smallest values in a set of data.

ratio
N4.5

Ratio compares the size of one part with the size of another part.

recurring
N2.4, N5.1

A recurring decimal has an unlimited number of digits after the decimal points.

reflect, reflection
S4.1, S4.5, S4.6, S5.5

A reflection is a transformation in which corresponding points in the object and the image are the same distance from the mirror line.

reflection symmetry
S4.5, S5.5

A shape has reflection symmetry if it has a line of symmetry.

regular
S5.5, S5.6

A regular polygon has equal sides and equal angles.

relationship
A3.5, A5.7

A relationship is a link between objects or numbers.

remainder
N5.5

A remainder is the amount left over when one quantity is divided by another. For example, 9 ÷ 4 = 2 remainder 1.

represent
D3.2, D3.3, D3.4, D3.5

You represent data whenever you display it in the form of a diagram.

response
D2.2
A response is an answer to a question in a survey.

rotate, rotation
S4.4, S4.5, S4.6, S5.5
A rotation is a transformation in which every point in the object turns through the same angle relative to a fixed point.

rotation symmetry
S4.5, S5.5
A shape has rotation symmetry if when turned it fits onto itself more than once during a full turn.

round
N1.6, N3.4, N3.5, N5.1, N5.5
You round a number by expressing it to a given degree of accuracy. For example, 639 is 600 to the nearest 100 and 640 to the nearest 10.
To round to one decimal place means to round to the nearest tenth. For example 12.47 is 12.5 to 1 d.p.

row, column
A row is horizontal; a column is vertical.

rule
A1.1, A1.2, A1.3, A1.4, A1.5, A1.6, A3.2, A3.3, A5.4
A rule describes the link between numbers or variables.
For example, the rule linking 2 and 6 may be +4 or ×3.

sample space
D4.1
A sample space is a diagram that records the outcomes of two events.

scale
S1.3
A scale gives the ratio between the size of an object and its diagram.

scale factor
S4.2
A scale factor is the multiplier in an enlaragement.

secondary
D2.1, D3.1
Data already collected is secondary data.

sequence
A1.1, A1.2, A1.3, A3.2, A3.3, A3.4, A5.4, A5.5
A sequence is a set of numbers or objects that follow a rule.

shape
S3.2
A shape is made by a line or lines drawn on a surface, or by putting surfaces together.

side (of 2-D shape)
S3.2, S3.3
A side is a line segment joining vertices.

sign
N3.2
A sign is a symbol used to denote an operation, for example +, −, ×, ÷.

significant
N1.1
The first non-zero digit in a number is the most significant figure. For example, the most significant figure in 207 is the 2, which represents 200.

simplify, simplest form
N2.1, N2.2, N4.5
A fraction (or ratio) is in its simplest form when the numerator and denominator (or parts of the ratio) have no common factors. For example, $\frac{3}{5}$ is expressed in its simplest form.

simplify
A2.3, A2.4, A4.3, A5.3
To simplify an algebraci expression you gather all like terms together into a single term.

sketch
S3.4, S5.1, S5.2
A sketch shows the general shape of a graph or diagram.

solid (3-D) shape: cube, cuboid, cylinder,hemisphere, prism, pyramid, square-based pyramid, sphere, tetrahedron
S1.4, S3.5, S5.3

A solid is a shape formed in three-dimensional space.

cube

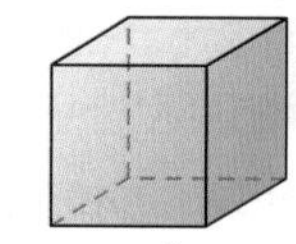

six square faces

cuboid

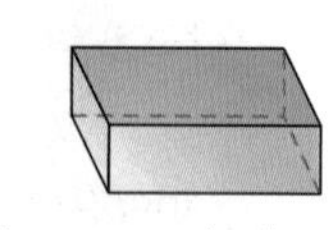

six rectangular faces

prism

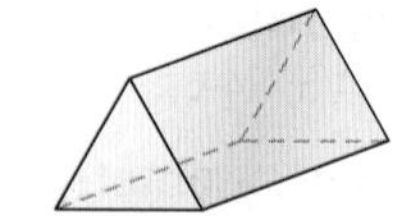

the end faces are constant

pyramid

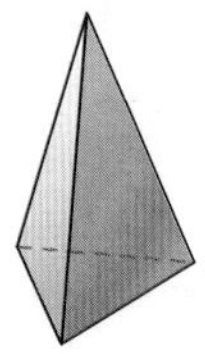

the faces meet at a common vertex

tetrahedron

all the faces are equilateral triangles

square-based pyramid

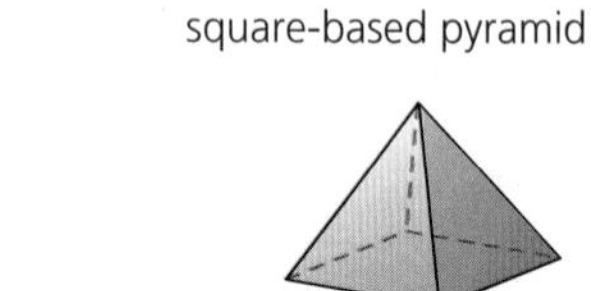

the base is a square

solution (of an equation)
A5.1

The solution of an equation is the value of the variable that makes the equation true.

solve (an equation)
A4.1, A4.2, A4.4, A5.1, A5.3

To solve an equation you need to find the value of the variable that will make the equation true.

spin, spinner
D1.5, D4.3

A spinner is an instrument for creating random outcomes, usually in probability experiments.

square number, squared
A3.4, A5.4, A5.5

If you multiply a number by itself the result is a square number, for example 25 is a square number because $5^2 = 5 \times 5 = 25$.

square root
A3.1

A square root is a number that when multiplied by itself is equal to a given number. For example $\sqrt{25} = 5$, because $5 \times 5 = 25$.

statistic, statistics
D1.3, D3.4, D3.5

Statistics is the collection, display and analysis of information.

straight-line graph
A3.5, A3.6, A5.7

A straight-line graph is the graph of a linear equation.

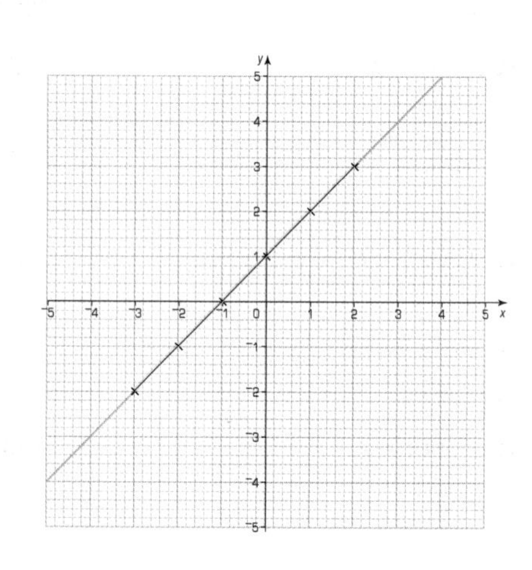

subject
A5.2
The subject of a formula is the single letter that is usually on the left-hand side of the formula.

substitute
A2.1, A2.4, A5.2
When you substitute you replace part of an expression with a value.

subtract, subtraction
N1.2, N1.3, N1.5, N3.3
Subtraction is the operation that finds the difference in size between two numbers.

sum
D1.2
The sum is the total and is the result of an addition.

summary
D3.5
A summary is a brief account of the key points of a statistical enquiry.

surface, surface area
S1.4
The surface area of a solid is the total area of its faces.

survey
D2.1, D2.2
A survey is an investigation to find information.

symbol
A2.2
A symbol is a letter, number or other mark that represents a number or an operation.

symmetry, symmetrical
S3.1, S4.5, S5.5
A shape is symmetrical if it is unchanged after a rotation or reflection.

tally
D2.3
You use a tally mark to represent an object when you collect data. Tally marks are usually made in groups of five to make it easier to count them.

temperature: degrees Celsius, degrees Fahrenheit
Temperature is a measure of how hot something is.

tenth
N1.1
A tenth is 1 out of 10 or $\frac{1}{10}$.
For example 0.5 has 5 tenths.

term
A1.1, A1.2, A1.3, A1.4, A1.6, A2.3, A2.4, A4.2
A term is a number or object in a sequence. It is also part of an expression.

terminating
N2.4
A terminating decimal has a limited number of digits after the decimal point.

tessellation
S5.6
A tessellation is a tiling pattern with no gaps.

theoretical probability
D4.3
A theoretical probability is worked out without an experiment.

thousandth
N1.1
A thousandth is 1 out of 1000 or $\frac{1}{1000}$,
for example, 0.002 has 2 thousandths.

three-dimensional (3-D)
S3.5, S5.3
Any solid shape is three-dimensional.

time series
D2.4, D3.2
A time series shows how something varies over time.

to one decimal place (to 1 d.p.)
N3.5, N5,1

To round to 1 d.p. means to round to the nearest tenth, for example 12.47 is 12.5 to 1 d.p.

transformation
S4.6

A transformation moves a shape from one place to another.

translate, translation
S4.3, S4.6

A translation is a transformation in which every point in an object moves the same distance and direction. It is a sliding movement.

triangle: equilateral, isosceles, scalene, right-angled
S1.2, S2.3, S3.1, S3.2, S3.3, S5.1, S5.6

A triangle is a polygon with three sides.

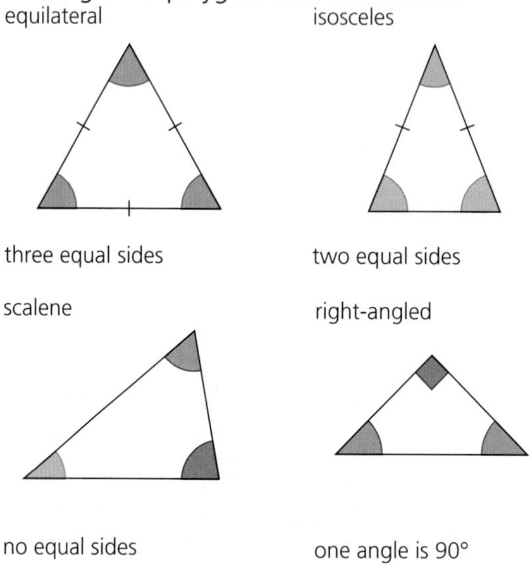

equilateral

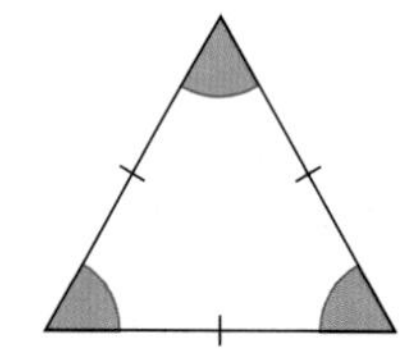

three equal sides

isosceles

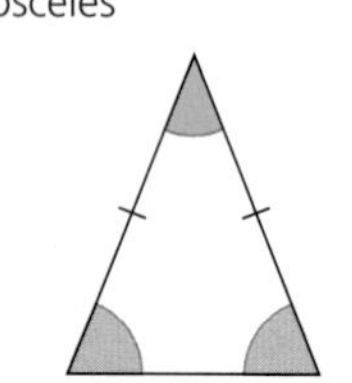

two equal sides

scalene

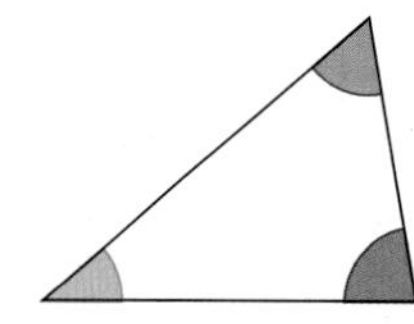

no equal sides

right-angled

one angle is 90°

triangular number
A5.5

A triangular number is the number of dots in a triangular pattern:
The numbers form the sequence
1, 3, 6, 10, 15, 21, 28 ...

two-dimensional (2-D)
S3.5

A flat shape has two dimensions, length and width or base and height.

unitary
N4.4, N4.5

In the unitary method you first work out the size of a single unit and then scale it up or down.

unknown
A4.1

An unknown is a variable. You can often find its value by solving an equation.

variable
A1.6, A2.5, A5.2

A variable is a symbol that can take a range of values.

vector
S4.3

A vector describes a translation by giving the *x*- and *y*-components of the translation.

vertex, vertices
A2.2, S1.4, S3.5

A vertex of a shape is a point at which two or more edges meet.

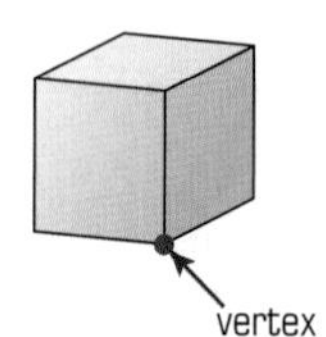

vertical
A3.6, S4.3

Vertical means straight up and down.

vertically opposite angles
S2.2, S3.1

When two straight lines cross they form two pairs of equal angles called vertically opposite angles.

$a = c \quad b = d$

whole
N2.1

The whole is the full amount.

width
S1.1, S1.3

Width is a dimension of an object describing how wide it is.

x-axis, y-axis
S2.3

On a coordinate grid, the *x*-axis is the horizontal axis and the *y*-axis is the vertical axis.

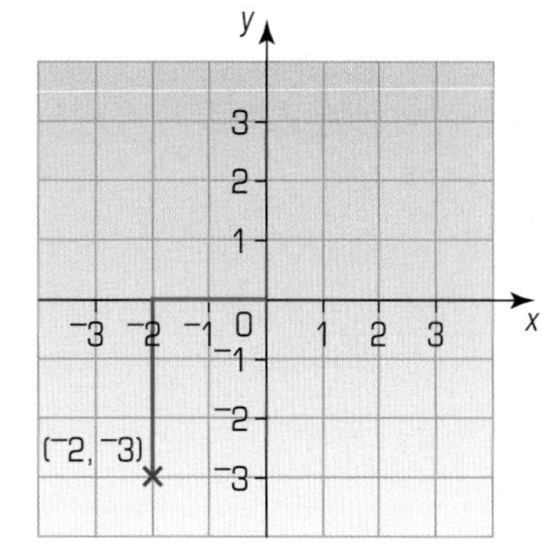

x-coordinate, y-coordinate
A5.7, S2.3

The *x*-coordinate is the distance along the *x*-axis.
The *y*-coordinate is the distance along the *y*-axis.
For example, ($^{-}2$, $^{-}3$) is $^{-}2$ along the *x*-axis and $^{-}3$ along the *y*-axis.

zero
N3.1

Zero is nought or nothing.
A zero place holder is used to show the place value of other digits in a number.
For example, in 1056 the 0 allows the 1 to stand for 1 thousand. If it wasn't there the number would be 156 and the 1 would stand for 1 hundred.

A1 Check in

1 a 5, 12, 19, 26, 33, 40, 47
  b 19, 15, 11, 7, 3, $^{-}1$, $^{-}5$
  c 12, 11.7, 11.4, 11.1, 10.8, 10.5, 10.2
2 a 54 b 84 c 11
  d 7 e 9 f 4
3 a 4, 8, 16, 32, 64
  b 10, 20, 40, 80, 160
  c 2.5, 5, 10, 20, 40
  d 6.1, 12.2, 24.4, 48.8, 97.6

A1 Check out

1 a 36, 47, 58
  b 48, 96, 192
  c 9, 16, 25
  d 5.5, 6.2, 6.9
2 a 5, 18, 31, 44, 57
  b 7, 3, $^{-}1$, $^{-}5$, $^{-}9$
  c 2, 6, 10, 14, 18
  d 1, 8, 27, 64, 125
3 a 100 triangles, each with 3 matches = 300 matches
  b Each pentagon needs 4 matches and 1 to close if off, so 100 pentagons need $4 \times 100 + 1 = 401$ matches
4 a $n \longrightarrow \frac{n}{10}$
  b $x \longrightarrow 5x - 2$

N1 Check in

1 a Two thousand and four point six
  b 80 002
2 a 0.1 b 0.01
3 a 23 kg, 230 kg, 2300 kg (or 2.3 tonnes)
  b £5.20, £0.52, 5.2p
4 a 5 b 4.6

N1 Check out

1 a 0.04 km, 200 cm, 0.5 m
  b 3 mm, 0.03 m, 30 cm
  c $^{-}2.3$, $^{-}2.06$, $^{-}2.005$
2 a $^{-}7$ b $^{-}1$ c 10 d $^{-}4$
  e 2 f $^{-}18$ g 8 h 1
3 a 6.656 b 3.686
4 £0.14. This is 14p
5 a 76, 77, 78
  b 149, 151, 153
  c 80, 85, 90, 95

S1 Check in

1 **a** 3 cm
b $a = 5$ cm, $b = 3$ cm
2 **a** = rectangle
b = parallelogram
c = equilateral triangle
d = cuboid
e = tetrahedron
3 **a** 7.68 mm^2 **b** 19.22 m

S1 Check out

1 **a** 10 cm^2 **b** 20 cm^2
c 12 cm^2 **d** 5 cm^2
3 10.51 cm^2
3 **a** 12 m^3 **b** 64 cm^3
4 volume = 72 m^3
surface area = 108 m^2

N2 Check in

1 **a** $\frac{24}{100} = \frac{6}{25}$, 24%
b $\frac{3}{10}$, 30%
c $\frac{2}{100} = \frac{1}{50}$, 2%
2 **a** $\frac{1}{7}$ **b** $\frac{1}{4}$ **c** $\frac{1}{15}$
3 0.4285714 ...

N2 Check out

1 **a** $\frac{5}{3} = 1\frac{2}{3}$ **b** $1\frac{7}{24}$
c $\frac{2}{10} = \frac{1}{5}$ **d** $1\frac{11}{21}$
2 **a** £18.60 **b** 48 g
c 25 minutes **d** 600°
3 **a** £1.90 **b** £76.50
b 30.6 kg **d** £105.78
e £16.83 **f** 721 g
4 **a** £34.50 **b** 528 m
c £66, £72.60. No, 20% of £60 = £120 so the increase would be £72.
5 55%

D1 Check in

1 a 12, 19, 24, 34, 52
  b 2.82, 2.9, 5.6, 5.65, 8.3
2 a 42 b 84 c 126
3 a depends on current weather conditions
  b $\frac{1}{2}$

D1 Check out

1 a Mean = 6
  Median = 6
  Mode = 9
  Range = 7
  b Mean = 30.4
  Median = 21
  Mode = 20
  Range = 49
2 November – it is very cloudy and there are 30 days in the month
3 impossible — A — B — C — certain
4 0.9865
5

	O	S	C	M	P
O	✓	✓	✓	✓	✓
S		✓	✓	✓	✓
C			✓	✓	✓
M				✓	✓
P					✓

A2 Check in

1 a $4x + 8$ b $\frac{y-3}{9}$ c z^2
2 a 2 b 3 c 15 d $^-3$
  e 1 f $^-42$ g $^-25$ h $^-7$
3 a 16 b 144 c 225 d $\frac{1}{4}$
  e 0.09 f 25

A2 Check out

2 a i $\frac{2x+5}{4}$ ii $7(x^2 - 2)$ iii $4x^3$
  b xy and yx
  $x + y$ and $y + x$
  $x \times x$ and x^2
  $y \div x$ and $\frac{y}{x}$
2 a i $p + 10q$ ii $2x^2 + 15x$
  iii $15ab + 12ac$ iv $11x + 9$
  b i $12xy$ ii $3b$ iii $132a^2$
  iv $32p^3$ v $2w$ vi $24x^3$
3 a i 49 ii 10 iii $^-10$
  iv 9 v 13
  b i 33 cm² ii 27 cm
4 a $x + (x + 1) + (x + 2) = 3x + 3 = 3(x + 1)$
  b $(2n + 1) + (2n + 3) = 4n + 4 = 4(n + 1)$

S2 Check in

1 **a** 39° **b** 100° **c** 285°

2 **a** (3, 3) **b** (9, 3)
c (4, 2) **d** (8, 1)

S2 Check out

1 $a = 62°$ $b = 118°$

2 **a** Rhombus
b Isosceles triangles
c 180°
d 360°

3 115°, 65° and 115°

4 **a** (2, $^{-}1$) **b** (1, 0) **c** (2, 0)

D2 Check in

1 **a** **i** 7 > 5 **ii** 3.7 > 3.64
b yes

2 **a** **i** IIII **ii** 𝍸 II **iii** 𝍸 𝍸 II
b **i** 3 **ii** 9 **iii** 16

D2 Check out

1 **a** Are you male or female?
How much time do you spend on homework each week?
b Are you: Male □ Female □
Time spent on homework each week:
<30 mins □
30–60 mins □
60–90 mins □
90–120 mins □
>120 mins □

2 **a** **i** 2 **ii** 10
b The time most students spend on homework, the mode, is the same for boys and girls.

N3 Check in

1 a 23 b 3
c 4 d 2

2 a 360 b 323
c 630 d 450

N3 Check out

1 a 3500 cm b 0.035 km
c 350 000 cm^2 d 0.000 035 km^2

2 a 79.36 b 451.03
c 1.6325 d 1.984
e 28 f 23.1
g 17.5 h 123

3 a 0.057 b 0.171
c 0.0057 d 57
e 19 f 190

4 The numbers with exact square roots are:
1, 4, 9, 16, 25
They are the square numbers.

A3 Check in

1 They form a vertical line because the *x* coordinates are all 3.

2 a 7 b 32 c 36

3 a i 27 ii 169 iii 32
iv 16 v 1000
b i $3^4 \times 4^3$ ii x^5
iii 5^3y^2 or $125y^2$

A3 Check out

1 a 12; $\frac{2}{5}$
b False. LCM of 2 and 8 is 8
c $2^2 \times 3^3 \times 5$

2 a i $3n + 2$ ii $11n - 6$ iii $12 - 2n$
b i 7, 10, 13, 16, 19
ii $^-3$, 2, 7, 12, 17
iii 5, 8, 13, 20, 29
c You don't count the corners so an L × L square has 4 lots of (L – 2) crosses.
= (L – 2)

4 a i, ii and iii – they are all of the form $y = mx + c$
iv is a quadratic (x^2) graph
b False – (3, 5) is not on any of the lines as it doesn't fit any of the equations.

S3 Check in

1 **a** 120° **b** 95° **c** 280°

1 **a** Kite **b** regular pentagon

c regular hexagon

3 **a** **i** tetrahedron **ii** cuboid

b **i**

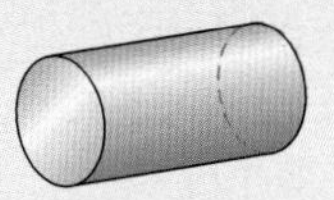

ii

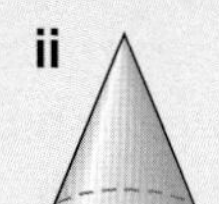

iii

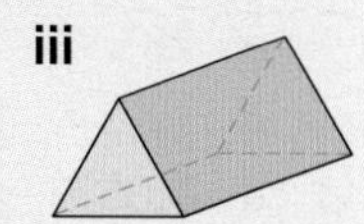

S3 Check out

1 **a** $a = 70°$, $b = 40°$

b $c = 40°$, $d = 90°$

2 **a** The bisector should cut the line at 2.85 cm

b

4 cm
40°
40°
4 cm

c

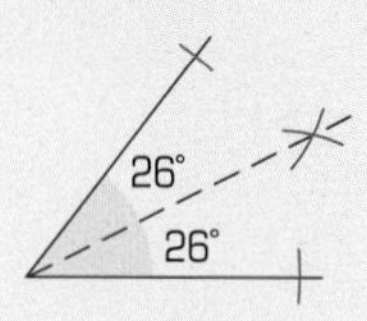

3 **a**

C
60°
70°
50°
A
B
10 cm

b

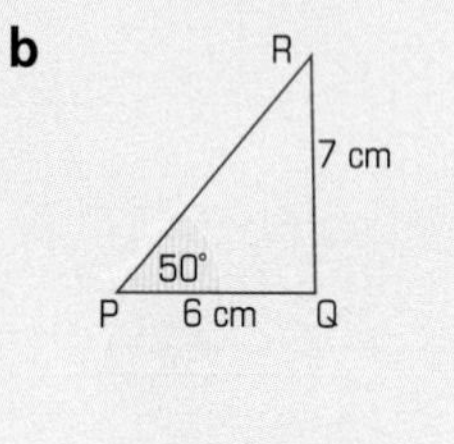

c

E
5 cm
3 cm
F
4 cm
G

4 **a** + **c** are unique as the lengths and angles are fixed.

There are 2 possible answers in **b** as shown in **3b** above.

5 The shape is a tetrahedron:

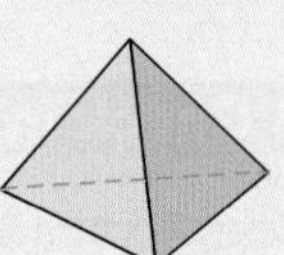

N4 Check in

1 **a** **i** 0.25, 25% **ii** 0.2, 20%

iii 0.3, 30%

b **i** $\frac{1}{20}$ **ii** $\frac{7}{40}$ **iii** $\frac{6}{5}$ or $1\frac{1}{5}$

2 **a** £6 **b** 96 m

N4 Check out

1 **a** 1.575 kg **b** 5.7 kg

2 **a** 13.2 cm **b** £36

3 24, 120, 16

4 **a** 832 **b** £17 742.86

c 9525 **d** £90

A4 Check in

1 **a** $6x + 7$
b $\frac{x - 4}{2}$
c $10(x^2 + 9)$

2 **a** $^-4$ **b** $^-6$ **c** 5
d $^-6$ **e** $^-25$ **f** 11

3 **a** **i** $5\frac{1}{7}$ **ii** $4\frac{3}{4}$ **iii** $^-6\frac{2}{3}$ **iv** $^-7\frac{2}{5}$
b $5\frac{7}{9}, \frac{43}{7}, 8\frac{2}{3}, \frac{71}{8}$

A4 Check out

1 **a** $a - 2b$
b $3ab - 3a + 2b$
c $3p - 5q$
d $3x^3y^2z$

2 **a** **i** $3x + 15$ **ii** $12x + 24$
iii $36x + 90$ **iv** $p^2 + 8p$
v $wz + z^2$ **vi** $12b^2 + 8ab$
b **i** $2x + 4$ **ii** $3x + 9$
iii $13x + 4$

3 **a** **i** 7 **ii** 10 **iii** 1
b **i** $6\frac{1}{5}$ **ii** $-1\frac{3}{14}$ **iii** $\frac{-8}{13}$
c **i** 10 **ii** 6 **iii** $5\frac{5}{7}$

4 **a** 20
b $61\frac{2}{3}°, 61\frac{2}{3}°, 56\frac{2}{3}°$

S4 Check in

1 **a** 4
b 0
c 3
d an infinite number

2 A parallelogram

S4 Check out

2 8

3 **b** (2.5, 2.5)

N5 Check in

1 a 3 b 3 c 20
2 a 0.125 b $\frac{7}{20}$ c $\frac{9}{50}$
3 a 9.6 km b £15.20

N5 Check out

1 a £23.40 b 28 c £982.25
2 a 8342 b 3875 c 29.308
d 139.612 e 27 f 29
g 26 h 16.5
3 a 2.3 b 16.1
c 1.61 d 0.161
e 230 f 2300
g 230 h 0.23
4 a 0.8
b 64%

D3 Check in

1 Colour of paper = categorical
weight = continuous
cost = discrete
2 a pie chart
b frequency diagram
c bar chart
3 mean = 5
median = 5
mode = 5
range = 5

D3 Check out

1 Boys: mean = 49, range = 5
Girls: mean = 7, range = 5

D4 Check in

1 Orange, Red or Green

2 a $\frac{3}{8}$ b $\frac{5}{8}$

3

Colour	O	R	G
Frequency	4	2	6

D4 Check out

1 SV, SC, SS, ST
LV, LC, LS, LT
OV, OC, OS, OT

2 a $\frac{2}{5}$ b $\frac{6}{10} = \frac{3}{5}$
The second estimate is better as it is based on a larger number of trials.

A5 Check in

1 a 12 b 3 c 5

2 a $3x + 16$ b $x^2 + 3x + 10$

3 a i $3n + 3$ ii $9n - 5$
b i 5, 12, 19, 26, 33
ii 7, 4, 1, $^{-}2$, $^{-}5$

A5 Check out

1 a $28\frac{1}{2}$ b 22

2 a 2700 cm^2 b 8.16 cm

3 a i $x + 20$ ii $4x - 80$
b $4y + 12$

4 (2, 3)

5 a $y = 3x + 4$
b $y = 7x + 1$
c $y = 5x + 2$

5 Someone eating a chocolate cake.

S5 Check in

1 a Regular pentagon
b Regular hexagon
c Rhombus
d Isosceles Trapezium

2 a One line of reflection symmetry
b Rotational symmetry of order 2.

S5 Check out

2 $x = 50°$, $y = 130°$,

3 a true
b true
c true

Index

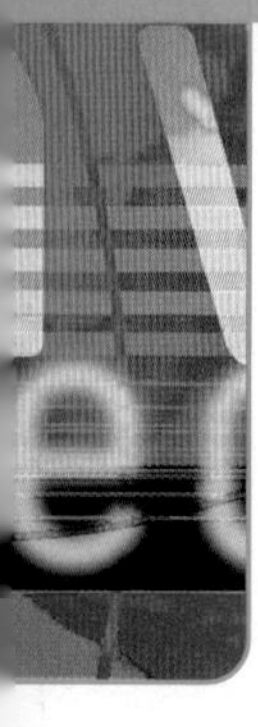